GENETICS

Second Edition

GENETICS
A Human Perspective

Linda R. Maxson
Penn State University

Charles H. Daugherty
Victoria University of Wellington

ᴡᴄᴃ
Wm. C. Brown Publishers
Dubuque, Iowa

Book Team

Editor *Kevin Kane*
Developmental Editor *Mary J. Porter*
Designer *David C. Lansdon*
Production Editor *Vickie Putman Caughron*
Photo Research Editor *Marge Manders*
Permissions Editor *Carrie Husemann*
Visuals Processor *Jodi Wagner*

web group

Chairman of the Board *Wm. C. Brown*
President and Chief Executive Officer *Mark C. Falb*

web

Wm. C. Brown Publishers, College Division

President *G. Franklin Lewis*
Vice President, Editor-in-Chief *George Wm. Bergquist*
Vice President, Director of Production *Beverly Kolz*
Vice President, National Sales Manager *Bob McLaughlin*
Director of Marketing *Thomas E. Doran*
Marketing Communications Manager *Edward Bartell*
Marketing Information Systems Manager *Craig S. Marty*
Executive Editor *Edward G. Jaffe*
Production Editorial Manager *Colleen A. Yonda*
Production Editorial Manager *Julie A. Kennedy*
Publishing Services Manager *Karen J. Slaght*
Manager of Visuals and Design *Faye M. Schilling*

Cover photograph and part openers © Peter Angelo Simon.

Chapter 4 © AP/Wide World Photos; **Chapter 6** © David Scharf/
Peter Arnold; **Chapter 8** © Biology Media/Photo Researchers;
Chapter 9 © George Lefevre and Academic Press; **Chapter 10**
Wolff, Sheldon and B. Rodin, *Science* 200 (5 May 1978) 543–545
© 1978 by the AAAS; **Chapter 12** © Neg. 333218 Courtesy
Department of Library Services, American Museum of Natural
History; **Chapter 13** © Dr. A. Jeffreys, *Nature* 314, 67–73.

Contents

Mendel, meiosis, and basic principles of inheritance
Chapter Three

Mendelian genetics of humans
Chapter Four

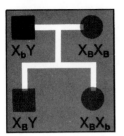
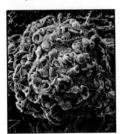

Reproduction: technologies and choices

Chapter Seven

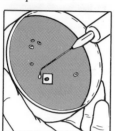

SECTION III—GENETIC INFORMATION and EXPRESSION

Informational macromolecules

Chapter Eight

Control of gene expression
Chapter Nine

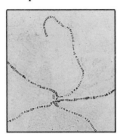

Misinformation
Chapter Ten

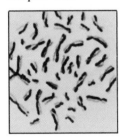

Immune genetics

Chapter Eleven

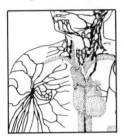

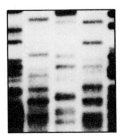

Boxes

Preface

The study of genetics provides "food" for both body and soul, with an immediacy unmatched by any other science. It is this immediacy that motivates us to write this book. Our goal remains unchanged from our first edition: To present the genetics that every person should know to live an informed life.

The science of genetics permeates every fundamental aspect of human life. Our health, reproduction, and sustenance rely to a large and increasing extent on our understanding of the hereditary mechanism of humans and other species. Furthermore, the universality of the hereditary mechanism underscores the evolutionary relatedness of all living organisms, including humans. All living species have a common biological origin, a crucial fact to consider when asking philosophical questions about what it means to be human. We are what we are partly because of our biological origins, and genetics provides evidence of those origins.

We are writing this book as a source of information for college and university students of all backgrounds and majors. The most important task is to describe in some detail the structure and functioning of the genetic apparatus. Because genetic knowledge has so many applications, in each chapter we relate directly the concepts and facts of genetics to important facets of everyday life. We also point to important issues of personal ethics or social responsibility raised by advancing genetic knowledge.

In our revision, the primary difficulty lies in what to exclude, not in what to include. Despite the rapid increase in genetic knowledge, this edition is considerably shorter than the first in order to sharpen our focus on the elegance and universality of the genetic mechanism. Information on advances in applications of genetics is treated in two new chapters (7 and 13) as well as throughout the text. Persons wishing more information on specific genetic topics should consult any textbook written for advanced genetics students.

Our choice of social and ethical issues related to genetics is highly selective. Issues raised by genetic knowledge and technologies are extraordinarily diverse and appear in the media daily. Instructors can choose topical issues to expand the coverage of any section of this book. We provide the genetic knowledge necessary for students to grasp the scientific basis of public controversies. We hope that such knowledge will also help the student develop a sense of the true excitement of modern genetics, as well as a deeper understanding of the scientific basis of contemporary life.

Aids to the student

Chapter outlines provide at a glance all the major headings and subheadings in the chapter. The outlines show students the structure and organization of a chapter and help them locate sections dealing with particular topics.

Key terms, indicated by boldfaced type in the text, identify concepts essential to an understanding of the subject. The terms help students locate and recall the basic elements that are the building blocks of a chapter.

Boxed readings present information that is of special interest or that provides more detail about specific topics in the text. This approach preserves the continuous flow of ideas in the text while providing related material that supplements the topic.

Chapter summaries encapsulate the chapter's main ideas in numbered lists that provide a convenient focus for both preview and review of a chapter. Students should scan the summary after completing the chapter or before taking a test in order to reinforce what they know and help identify the topics they need to look at again.

Lists of key words and phrases, at the end of each chapter, provide a concentrated look at the most important components of the chapter. Learning is reinforced when students go through the list systematically, defining the items they understand and making a note of the ones that are fuzzy. Students then know what sections they should reread.

Suggested readings are for students interested in detailed and interesting information on specific topics. Most of the suggested readings are taken from publications such as *Natural History, Scientific American,* and other science publications written for the educated layperson.

End of chapter questions review important concepts in the chapter. Students who can answer these questions should have a solid grasp of the material. A student having difficulty with a question will be able to identify sections that should be given additional attention.

Supplementary materials

A revised *Instructor's Manual* offers a chapter outline, a chapter summary, suggested answers to the text chapter questions, sources for further information on current genetics topics, and a test item file for each text chapter.

A separate set of helpful *acetate transparencies* is also available.

ueb TestPak rounds out the supplementary materials. It is a computerized system that enables you to make up customized exams quickly and easily. Test questions can be found in the Test Item File, which is printed in your instructor's manual or as a separate packet. For each exam you may select up to 250 questions from the file and either print the test yourself or have **ueb** print it.

Printing the exam yourself requires access to a personal computer—an IBM that uses 5.25- or 3.5-inch diskettes, an Apple IIe or IIc, or a Macintosh. TestPak requires two disk drives and will work with any printer. Diskettes are available through your local **ueb** sales representative or by phoning Educational Services at 319-588-1451. The package you receive will contain complete instructions for making up an exam.

If you don't have access to a suitable computer, you may use **ueb**'s call-in/mail-in service. First determine the chapter and question numbers and any specific heading you want on the exam. Then call Pat Powers at 800-351-7671 (in Iowa, 319-589-2953) or mail information to: Pat Powers, Wm. C. Brown Publishers, 2460 Kerper Blvd., Dubuque, IA 52001. Within two working days, **ueb** will send you via first-class mail a test master, student answer sheet, and an answer key.

Acknowledgments

We especially thank Richard D. Maxson, Kevin R. Maxson, Marie Daugherty, and Brendan Daugherty for patience, constant support, and generous help with all phases of the revision of this book, and for granting us the time that made it possible. Thanks are due to the University of Illinois Chancellor's Scholars in Linda R. Maxson's genetics class for their thoughtful critiques of the first edition. We thank Dr. Ann Bell, Dr. A. P. Blair, Dr. G. K. Chambers, and Dr. G. K. Rickards for helpful commentary, and Edith Warkentine, Dr. Joyce A. Resnick, and Dr. William J. Daugherty for expert advice. Special thanks are due Ruth and Al Resnick for their hospitality and attention to detail during the final revisions of this text. Portions of this revision were completed while LRM was on sabbatical leave in the Department of Ecology and Evolutionary Biology at the University of California, Irvine.

We'd also like to thank the following reviewers for their invaluable counsel on both editions of this book.

L. Herbert Bruneau
Oklahoma State University

Andrew G. Clark
Penn State University

Tommy L. Cole
Phillips County Community College

Jeffrey L. Doering
Northwestern University

Frank Einhellig
University of South Dakota

Miriam Golomb
University of Missouri

Thomas Gray
University of Kentucky

John C. Hartnett
St. Michael's College (Emeritus)

Caroll E. Henry
Chicago State University

George Hudock
Indiana University

Robert J. Huskey
University of Virginia

John D. Jackson
North Hennepin Community College

Genevieve D. Johnson
University of Iowa

George Labanick
University of South Carolina

Virginia G. Latta
Jefferson State Junior College

Tim Lyerla
Clark University

Judith Stone
St. Joseph's College

Gary Thorgaard
Washington State University

D. A. Whited
North Dakota State University

GENETICS

SECTION I

INTRODUCTION

Humans and alleles

Figure 1.1 (a) Gregor Mendel (1822–1884) discovered the principles of genetics through meticulous experiments on inheritance in peas. (b) Charles Darwin (1809–1882) proposed the theory of evolution by means of natural selection and spent the remainder of his life gathering and analyzing data to test his theory. (c) Francis H. C. Crick (b. 1916) and James D. Watson (b. 1928) discovered the double helical structure of DNA.
Source: (a) The Bettmann Archive; (b) National Library of Medicine, Bethesda, MD; (c) J. D. Watson, "The Double Helix," Atheneum, New York, 1968.

(a)

(b)

(c)

Genetics is the study of inheritance. Why are we each like our parents? Why do all organisms have an astonishing similarity to their parents? The first geneticists were persons who observed that "like begets like" and then used that basic observation to improve their lives. Domestication of wild animals—dogs, cats, cattle—occurred when humans learned how to control the matings of these species. Once animals were tamed, people could breed more of those that they found most useful. Docile cattle gave rise to more docile cattle, and control of the lineage remained with humans. Cultivation of crops required that seeds of one generation were saved to produce the next generation—and the seeds that were saved were from the largest, healthiest plants.

Genetics, however, involves a fundamental paradox: we are like our parents, but we are also different. Each person has many features that are similar to each parent, but also many features that resemble neither parent. Dark-haired parents usually have children with dark hair, but they often have blond children as well. Some seeds from the tallest plants give rise to short plants. Brown cows can produce white calves. How can this be?

While plant breeding and animal breeding have been practiced successfully for at least as long as humans have recorded their history, an understanding of the mechanisms of inheritance—genetic mechanisms—was almost nonexistent until about a century ago. In 1866, an Austrian monk named **Gregor Mendel** (figure 1.1) published the first paper accurately describing aspects of the hereditary mechanism ("Experiments with Plant Hybrids"). These studies were mostly ignored by the scientific community until 1900, when the importance of Mendel's work was independently recognized by three biologists, Hugo de Vries in Holland, Carl Correns in

Germany, and Erich von Tschermak in Austria. Since 1900, genetic knowledge has expanded with increasing rapidity. Genetics is a science that is truly a child of the twentieth century.

Thus, although humans have successfully practiced plant breeding and animal breeding for ten thousand years (and perhaps much longer), theoretical knowledge of principles of inheritance has existed for only 1% of that time: the past one hundred years. What is unknown about genetic mechanisms still dwarfs what we know. If you are now twenty years old, you can expect to live a half century longer. During the rest of your life genetic knowledge will increase immensely. The doubling time of scientific knowledge is now about ten years. What is known today will be only a small fraction of what is known fifty years from now. Furthermore, during the next fifty years genetic technology will be applied (and *is being applied*) to hundreds of problems that affect your daily life.

How rapidly is new genetic knowledge changing the lives of humans? Humans now use knowledge of genetic mechanisms for purposes unimaginable only a few decades ago: to cure disease, to improve crop yield, to create pest-resistant crops, to produce self-fertilizing crops, to produce organisms that can clean up oil spills, to conserve rare species, to insert genetic characteristics of one species into another species, and even to produce species that have never before existed on the face of the earth. The limits of the use of genetic knowledge, for good or bad, are the limits of the human imagination. Knowledge can indeed be power.

The hereditary mechanism

Long before Mendel's experiments, some scientists believed that tiny particles carrying hereditary information were somehow transmitted from parent to offspring. Mendel proved that such particles exist and, furthermore, that for some specific characteristics each individual receives one hereditary particle from each parent. These hereditary particles are now called **genes.** Mendel showed that flower color in garden peas, for example, was governed by two genes, one from each parent. Each organism possesses thousands of genes, half of which are contributed by each parent. Humans are estimated to have about 50,000–100,000 genes. **Genetics** *is the study of the structure and function of genes and of the transmission of genes between generations.*

One of the most important scientific discoveries of the twentieth century was the establishment of the physical nature of the gene. In a classic paper published in 1953, James Watson and Francis Crick (figure 1.1) proposed a structure for a complex molecule called **deoxyribonucleic acid**—usually simply called **DNA.** They suggested that DNA was a long double helix, like a twisted ladder (figure 1.2). Each of the uprights is a chain of alternating sugar molecules and phosphate molecules. The two uprights are linked together by steps of nitrogenous bases, much like the rungs of a ladder. Since the dramatic publication of the **Watson–Crick model,** further experimentation has confirmed the double helical nature of most DNA. DNA and RNA (a very similar molecule) constitute the genetic material of all living organisms.

Watson and Crick also proposed that the sequence of nitrogenous bases along the DNA molecule (the rungs of the ladder) carries the enormous amount of genetic information necessary for life. Their proposal turned out to be correct, and DNA is now often called an **informational macromolecule**—that is, DNA is a very long molecule that contains information encoded in its structure. The structure of DNA thus encodes the genetic message. A *gene* is a segment of DNA, and a DNA molecule may have a sequence of many genes along its length. The information encoded

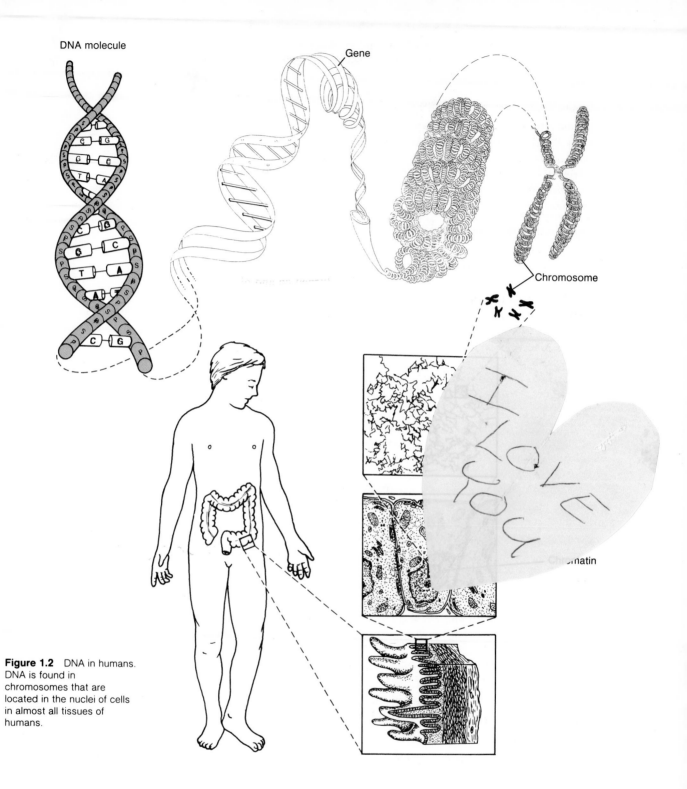

DNA molecule

Gene

Chromosome

Chromatin

Figure 1.2 DNA in humans. DNA is found in chromosomes that are located in the nuclei of cells in almost all tissues of humans.

in each gene is used to direct the production or synthesis of other types of molecules. The interaction of genes with the external environment and with the molecular products of other genes produces a living organism.

In addition to describing the mechanism by which DNA encodes the information of life, the Watson–Crick model also predicted other key features of the genetic mechanism: **replication,** a method for accurate self-reproduction of the genetic material; and **mutation,** the mechanism of genetic change. In conjunction with other available knowledge of reproduction, the Watson–Crick model also explained how genetic information is passed between generations, *transmission genetics*.

The enormous amount of information encoded in one hundred thousand or more genes makes apparent the precise control that the genes exert on the organism—its development, structure, and function. This recognition has in turn altered the science of genetics. The study of transmission of genes is no longer the only goal of geneticists. Many geneticists now study gene function and gene regulation: How is the information in genes used to influence and regulate various aspects of the life of the organism? How are the genes themselves turned on and off?

One aspect of transmission genetics has important implications for all of biology. Some genes are transmitted to succeeding generations far more frequently than other genes, which may be transmitted less efficiently. Genes that increase an individual's capacity to survive and reproduce will spread throughout a species, while other genes may disappear.

Differential survival and reproduction of individuals as a result of different genetic characteristics is called **natural selection.** In 1859, **Charles Darwin** (figure 1.1) proposed that natural selection was the mechanism responsible for evolution (*On the Origin of Species*). The discovery of the principle of natural selection by Darwin is one of the greatest achievements in the history of science. Darwin stressed that all species produce more young than survive, and that within each species is a vast reservoir of heritable variation. Moreover, conditions on earth are always changing. Through time, natural selection acts on the variation present in populations, changing populations, species, and their genes. Some species change into new species; some do not change fast enough and become extinct. Some species may appear not to change externally, but through time the genes of all species undergo some change. Darwin called this process "descent with modification." Today we call it evolution.

A genetic mechanism is fundamental to the evolutionary process. Traits favorable to survival and reproduction can increase in a population only if they are heritable. Furthermore, evolution can occur only when genetic variation exists. Although Darwin did not understand the mechanism of inheritance (Mendel's work was never seen by Darwin), he did understand that a mechanism of inheritance was required for the process of natural selection. Perhaps the most precise and simple definition of evolution is a genetic definition: Evolution equals change in genes; or Evolution equals "descent with modification" of genes.

The importance of genetics

Knowledge of the hereditary mechanism is essential information for any informed individual. To a large extent, the development of civilization has coincided with the capacity of humans to control their basic needs such as food and shelter. The enormous expansion of the human population in this century is directly related to genetic knowledge that allowed vast improvement in productivity of domesticated plant species used for food—rice, wheat, and corn, in particular. Genetic knowledge has also been a key component of the revolution in health and medical care in this century.

In the past decade, **bioengineering,** the ability to directly alter the genetic material of an organism, has begun to be applied to many aspects of human life. In the coming decades, bioengineering will play a large and increasing role in meeting fundamental human needs. Bioengineering will increase food, fiber, and wood production, and will be used to combat many diseases. These gains, however, will often have social costs, which will present society with complex questions. To deal with the exciting changes ahead, we will all need a basic understanding of genetics and the many ways in which the science of genetics affects our daily lives.

Health

About 3–5% of the world population—200 million people—are estimated to be afflicted by serious genetic disease (figure 1.3). Millions more have genetic anomalies that cause relatively few problems in their lives. Knowledge of the genetic basis of diseases such as phenylketonuria (PKU) has already allowed treatment. Other conditions such as Down's syndrome are clearly identified as genetic conditions though no treatment has yet been developed. However, for both PKU and Down's syndrome, as well as many other diseases, genetic testing and counseling can prevent the recurrence of certain genetic conditions.

Advances in medical genetics have also been accompanied by new problems. The wide use of antibiotics such as penicillin has led to the evolution of strains of disease-causing organisms resistant to these drugs. Thus, recent technologies have actually helped produce new organisms (chapter 13) that are less susceptible to treatment. Bioengineering not only offers the hope of creating more effective antibiotics, but also of producing useful chemicals, for example, human growth hormone, which is now being used to treat some forms of dwarfism. Eventually, it should be possible to directly alter defective human genes, the only true cure for a genetic disease.

Such technologies challenge human ethical codes. Many people fear that changing human genes is "playing God" and should thus be forbidden. Others argue that gene therapy is no different in principle from more conventional therapies and that withholding such treatment from the ill would be immoral. Even the ability to diagnose genetic disease presents difficult ethical choices. Should the results of screening for genetic diseases be made available to employers, insurance companies, or a future spouse?

Reproduction

Widely applied contemporary reproductive technologies such as artificial insemination and *in vitro* fertilization are not strictly genetic techniques. They do, however, raise numerous ethical questions and point to the widely held concern that identical copies of particular persons can be "mass produced," or **cloned.** Although cloning of humans may not be possible for some decades, it will certainly be subject to very close political and ethical scrutiny.

In the meantime, ethical aspects of present technologies present difficult choices. *In vitro* fertilization (*IVF*) is increasingly used to treat infertility, but it is expensive, inefficient, and appropriate only for a relatively uncommon type of infertility. Should the treatment be limited only to the affluent, who can afford it? Might the money spent on *IVF* be used more responsibly to prevent or treat more common types of infertility? *IVF* results in many excess frozen embryos. Should experimentation be allowed on these embryos?

(a)

(b)

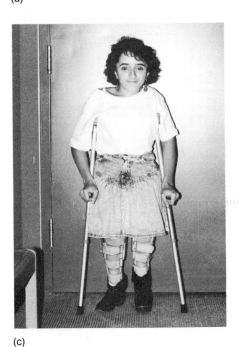

(c)

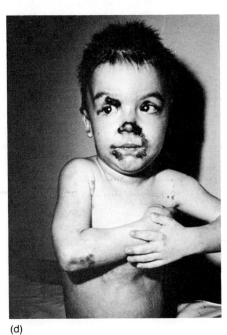

(d)

Figure 1.3 Examples of serious genetic diseases. (a) Folksinger Woody Guthrie (1912–1967) before onset of symptoms of Huntington's disease, a dominantly inherited degenerative disease, and (b) after effects of the disease became obvious. (c) Seventeen-year-old victim of osteogenesis imperfecta, characterized by bone fragility. (d) Young boy with rare, incurable skin disorder, epidermolysis bullosa.
Source: (a) The Bettmann Archive; (b) Woody Guthrie Publications, Inc.; (c) Dr. Joan Marini; (d) Lester V. Bergman & Associates, Inc.

Agriculture

Human life depends upon thousands of other species. Like most animals, humans use plants and animals for food. Additionally, we rely upon other species for countless, often highly imaginative, purposes: shelter, clothing, medicinal and recreational drugs, companionship, transportation, musical instruments, decoration, scientific experimentation, communication, and recreation.

An enormous variety of breeds of many species has been developed by the ingenuity of plant and animal breeders (figure 1.4). Since World War II, plant breeding has reached its culmination in programs of the Green Revolution, which have developed new, highly productive varieties of important grain crops for use in third world countries (figure 1.5). These programs have been scientifically successful—grain production has increased greatly. They have often had important social consequences, as well, which are widely debated. Ancient social systems have changed to meet new agricultural practices, and hunger and famine remain. Bioengineering will allow continued and probably dramatic increases in agricultural productivity. Ensuring that larger food supplies leads to a reduction in world hunger will involve all of us, not only geneticists and farmers.

Conservation

Genetic variation is a fundamental natural resource. Genetic differences exist both between different species and within the same species (between individuals, populations, and geographical races). Unique genetic types are often the basis for major advances in agriculturally important species. Genes for such important traits as disease resistance and increased productivity are identified from species and individuals in nature, and then introduced to domestic strains by hybridization. Natural genetic diversity is so important for supporting the development of new varieties of grain crops that gene banks containing hundreds of thousands of specimens of wild rice, wheat, and maize have been established in some tropical countries. Wild plant species are often the source of new drugs, as well. The active ingredients in about 25% of all prescription pharmaceuticals are derived from plants.

The potential value of the genetic information contained within the ten million or more living species has only begun to be exploited. Fewer than 1% of all living species has been examined for their benefit to humans; yet many species face *extinction* before they can be named by scientists. Their value to agriculture, for example, is irretrievably lost. Thus, **conservation** is increasingly recognized as a significant genetic issue. The fight to save species from extinction is of direct importance to agriculture, medicine, and business, and ultimately to human survival.

Genetics, evolution, and philosophy

The examples in the preceding sections give some idea of the extraordinary diversity of applications of genetic knowledge to practical aspects of human life. Genetics also plays a central role in larger issues of human experience.

Genetics was recognized by Darwin as the key to understanding evolution. Although Darwin's concepts of genetics are now known to be wrong, Darwin understood clearly that evolutionary change is the *inevitable consequence* of natural selection acting on an hereditary system. Genetics thus provides one of the foundations for the theory of evolution.

Evolution, in turn, is the central concept in biology. The famous biologist Theodosius Dobzhansky observed that "Nothing in biology makes sense except in the light of evolution." All aspects of all living organisms are influenced to some extent by their genes. And, because all organisms trace their origins back to a common ancestor several billion years ago, the studies of evolution and genetics merge with philosophy and help us understand ourselves and our place in the universe.

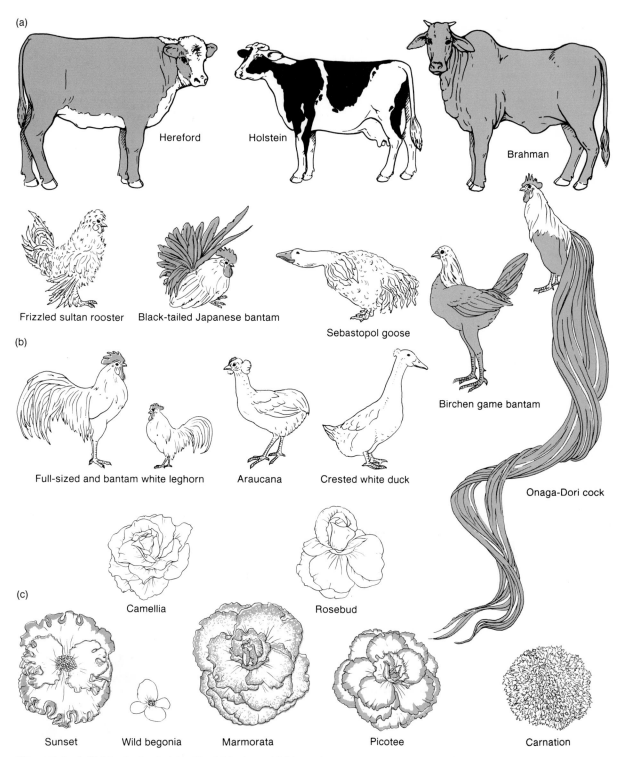

Figure 1.4 Artificial selection in (a) cattle, (b) fowls, and (c) begonias.
Source: (b and c) E. Peter Volpe, *Biology and Human Concerns,* 3d ed. Copyright © 1983 Wm. C. Brown Publishers, Dubuque, Iowa. All Rights Reserved. Reprinted by permission.

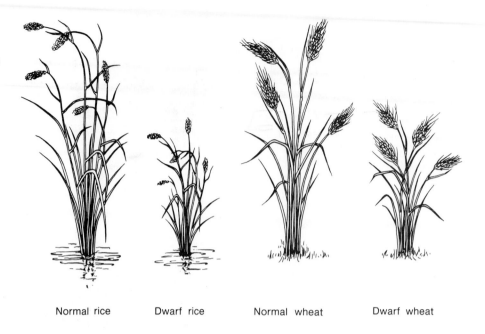

Figure 1.5 The development of genetic strains of food crops with high productivity has been a major goal of the Green Revolution.

Normal rice   Dwarf rice   Normal wheat   Dwarf wheat

Human evolution

Animals with backbones, **vertebrates,** have an exceptionally well-known evolutionary history because their bones can be preserved in the geological record as fossils. The first vertebrates were small fish, whose fossilized skeletons are found in rocks about 550 million years old. Descendants of these primitive fish diversified, and one lineage gave rise to an entirely new class of vertebrates, the *amphibians.* Further evolution produced the *reptiles,* the *birds,* and eventually the *mammals.*

The first *primate* appeared about fifty to sixty million years ago. Early primates were small-bodied, large-eyed, nocturnal tree dwellers whose evolutionary history has been confined primarily to the tropical regions of South America, Africa, and Asia. Most primates are small monkey-like creatures, but one lineage consists of a few species of relatively large-bodied forms—the **Great Apes,** which include gorillas, chimpanzees, orangutans, and gibbons. About five million years ago, a group diverged from the common ancestor of the Great Ape lineage and began to change rapidly in posture, size, and intelligence. This evolutionary lineage has thus existed for less than one-tenth of 1% of the total history of the earth, and only one species now survives: *Homo sapiens,* wise man.

In spite of the substantial differences in appearance between humans and chimpanzees, evidence of their recent common ancestry is revealed by their genes and gene products. The genes and proteins of humans and chimpanzees are over 99% identical in molecular sequence (see chapter 12). The genetic difference between humans and chimpanzees is less than that usually found between two species in the same genus, such as the coyote and the domestic dog.

Lucy

The extraordinary genetic similarity of humans and chimpanzees, described in the late 1960s, was the first evidence to indicate how very closely related humans are to the Great Apes. Such genetic similarity would be unlikely to exist unless the two species had diverged only recently. Previously, humans were believed to have originated as an independent evolutionary lineage at least fifteen million years ago.

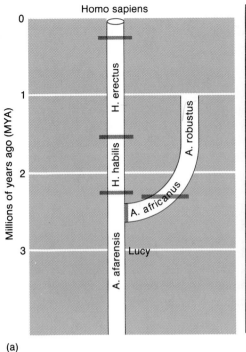

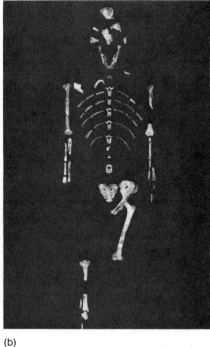

Figure 1.6 (a) Relationship of fossil "Lucy" to humans and other Australopithecine species. (b) Lucy, the most complete hominid fossil discovered.
Source: (b) Cleveland Museum of Natural History.

(a)

(b)

In 1974, one of the most famous fossils of all time provided further evidence of a recent origin. Working in the Afar triangle of northern Ethiopia, Donald Johansen and Tim White discovered the oldest hominid fossil yet found. Usually only a fragment of a jaw or a few teeth are found, but this fossil skeleton was 40% complete. Thus, Johansen and White could be certain of its size, posture, and sex. The extraordinary fossil was named **Lucy** after the song "Lucy in the Sky with Diamonds" (the discovery team celebrated by drinking beer and playing a Beatles' tape).

Lucy belongs to the species *Australopithecus afarensis,* and she is about 3.5 million years old (figure 1.6). The genus *Australopithecus* is the first hominid genus, a direct descendant of the Great Apes, and Johansen believes that Lucy's species is the common ancestor of other species of *Australopithecus* and of *Homo.* Thus, the genus *Homo* is probably less than three million years old. (This is the age for *Homo* suggested in 1967 by Vincent Sarich and Allan Wilson based on molecular studies of humans and Great Apes.) The genus *Australopithecus* is now extinct, perhaps due to the competition from *Homo habilis. Homo sapiens* is the only surviving species in the family Hominidae.

Modern human's common ancestor—"Eve"
Studies of the composition of human mitochondrial DNA (chapter 2) resulted in the announcement in 1987 that all humans living today (nearly five billion) may trace their ancestry back to the same ten thousandth-great-grandmother, dubbed "Eve." Most surprising, perhaps, is that this "mitochondrial mom" lived only some twenty thousand years ago, meaning that the differences in external appearance between humans developed quite recently. Although these findings are not without challenges, it appears clear that the human lineage appeared about five million years ago, with three million-year-old Lucy as our oldest hominid fossil. Modern populations of humans, including all the recognized races (see chapter 12), evolved in

response to pressures of natural selection only in the last twenty thousand years. Thus, all humans clearly share the same genetic heritage (see Tierney and Wright, 1988 for an excellent discussion).

Are humans still evolving?

All the genes that occur within one species are called the **gene pool** for that species. Many genes occur in alternate forms called **alleles**. All humans have the same basic genes (involved in the production of eye color, hair color, blood type, and so on). Different individuals, however, may have different forms or alleles of these genes. An individual may have alleles producing blue eyes or alleles that result in brown eyes, alleles producing blood type A or alleles producing blood type AB, etc. Thus, a gene pool contains many different alleles for each gene.

Evolution refers to changing proportions of alleles within the gene pool, as well as the introduction of new alleles and the loss of others. During the past four hundred years, the allelic content of the human gene pool has changed. Since 1600 the total population of the earth has increased dramatically, doubling many times. In 1850 the world population was about one billion; by 2050 it will be about ten billion. Different racial and ethnic groups, differing in allelic proportions, have contributed differentially to the total human gene pool.

Loss of alleles   The total genetic diversity of humans has decreased in recent history. Many racial and ethnic groups have simply ceased to exist. As Europeans explored the world and imposed their civilization on the native peoples of North America, South America, Africa, and Australia, systematic extermination of native tribes was common. When Captain Cook declared Australia to be the property of the King of England, about three million Aborigines lived in Australia. Today, the population of Australia is fifteen million, but only fifty thousand are Aborigines. In less than two hundred years, the proportion of alleles unique to Aborigines on the continent of Australia has decreased from 100% to less than 1%. A similar fate was suffered by Native Americans.

Such changes occurred wherever native peoples encountered more technologically advanced and expansionist societies. In some cases, native peoples were the object of genocide. In others, they simply slowly declined in numbers. Intermarriage with invaders quickly blended the surviving natives into the gene pool of the new culture. The result was almost always the disproportionate reduction of the genetic contribution of native people in succeeding generations.

In the twentieth century, native cultures have continued to disappear. Small tribes previously unknown to our civilization have been discovered in South America and, recently, in the Phillipines. Their discovery means that modern civilization has reached their homes, bringing changes that will probably mean their disappearance from the earth.

Increasing allele frequencies   As some genetic groups decline or disappear, others gain proportionately. The expansion of European cultures throughout the world increased the frequencies of alleles common to Caucasians. India, China, and Japan also maintained large and increasing populations. Since World War II, developed countries have become aware of the dangers of overpopulation, and birth control has increasingly been used to limit population growth. Now, the developed nations are stabilizing at low rates of growth, while many of the developing countries are experiencing extremely rapid rates of growth. In the coming decades, developing nations will comprise an increasingly greater proportion of the population of the world. The doubling time of populations in developed countries is 50–200 years, but only 20–35 years for other countries. Thus, the allelic composition of the human gene pool will continue to change significantly.

Within a heterogeneous population such as the United States, individual groups also change rapidly as a proportion of the total population. The proportion of blacks in the population of the United States has increased throughout the twentieth century and is expected to continue increasing. The proportion of alleles contributed by persons of Hispanic ancestry is also increasing rapidly. These racial, ethnic, and hence genetic changes now occurring reflect a continuation of historical trends in population composition of the United States. Successive periods in American history have been marked by pulses of immigration by nationalities and ethnic groups experiencing political and economic oppression and hardship.

Many persons consider the uncontrolled increase of the world's population to be the single largest problem facing humanity. For that reason, birth control has been widely advocated, and efforts have been made to distribute birth-control information and technology to those groups with the highest birth rates—minorities and people of the developing nations. While there is no question of the seriousness of the population problem, many persons are suspicious of birth-control programs, particularly those who have been the objects of oppression in the past. Many blacks, both in African countries and in the United States, view birth-control programs as racist. Many Jews remember the Holocaust and have similar fears. Most methods of birth control are opposed for religious reasons by the Roman Catholic Church and other religious groups. Resistance to population regulation favors continuing genetic change in the human gene pool.

Ethics

Ethics is the study of standards of morality and human conduct. Understanding our origins and place in the universe helps define standards of behavior. Racism, for example, is a prejudice mistakenly based on the belief that some humans are inherently—genetically—superior to others. The effects of racism in this century have included the mass murder of millions of Jews and other racial and national minorities. Today, racism persists in many forms in many countries, including advanced western countries. Genetics, however, provides a scientific framework for rejecting racism and all its pernicious influences.

Other forms of bias may affect behavior. Public policy toward victims of genetic diseases or of AIDS (acquired immune deficiency syndrome) are partly the result of society's understanding of the relationship of humans to nature. (For example, should victims of AIDS be reviled as sinners suffering a just "reward"? Or, should they be treated humanely as the victims of an especially deadly virus?) Thus, the influence of genetics lies far beyond mere scientific understanding of the hereditary mechanism and how it can be manipulated. Genetics provides one of the bases on which we decide many of the larger questions we face as a society. In combination with such fields as law, philosophy, politics, and theology, genetics can help us build the future we want.

Summary

1. Genetics is the study of inheritance, that is, the study of the structure, function, and transmission of genes. The modern study of genetics began in 1900 with the rediscovery of Mendel's work.
2. Genes are segments of deoxyribonucleic acid (DNA). DNA is an informational macromolecule that bears hereditary information encoded in its structure.
3. Evolution is defined most simply as "genetic change." Evolution is the inevitable consequence of natural selection acting on heritable variation.

4. The theory of evolution provides the foundation of modern biology. Genetics and evolution are interdependent sciences, and all other biological sciences draw upon them in understanding life on earth.

5. The rise of civilization has coincided with the development of human capacity to control the gene pool of domesticated plant and animal species. The most direct application of practical genetics has been to increase productivity of species used for food. Bioengineering will mean that productivity of agricultural systems should continue to increase significantly.

6. Genes that might be of value to humans are disappearing rapidly due to the extinction of many wild species, which have yet to be examined for their potential benefits. Conservation of genetic resources is now recognized as an important economic, as well as ethical, issue.

7. Serious genetic conditions affect 3–5% of all humans, and minor conditions probably affect most of us. Medical genetics has become an important health specialty in this century, and bioengineering is likely to enable treatment of many genetic conditions in the coming decades.

8. In conjunction with the humanities and other sciences, genetics can help us understand ourselves and our world.

9. Humans are a very recent group of large primates, most closely related to the Great Apes. Both fossil evidence and genetic evidence show the closeness of the relationship of humans to chimpanzees and the other Great Apes.

10. The human gene pool has changed greatly in recent history. Many human activities result in changing allele frequencies on a worldwide basis.

Key Words and Phrases

allele

bioengineering

Charles Darwin

cloning

conservation

deoxyribonucleic acid (DNA)

ethics

evolution

extinction

gene

gene pool

genetics

Great Apes

Gregor Mendel

informational macromolecule

Lucy

mutation

natural selection

replication

vertebrates

Watson–Crick double helix model

Questions

1. Consider our statement that "for some specific characteristics each individual receives one hereditary particle from each parent." What particular types of characteristics might be governed directly by genes? Might some characteristics be governed only indirectly? not at all? Are some features, such as an arm or a leg, likely to be controlled by one or by many genes?

2. The significance of the chemical DNA in defining life on earth has been recognized only since the 1950s. While genetic factors unquestionably play an important role in human behavior, why is it incorrect to attribute either good or evil actions to the genes one possesses?

3. Look around you where you now sit. Identify products of biological origin. Name the species from which these products were derived that might have been subjected to artificial selection of genes to improve their useful characteristics. Name the species that you ate at your last meal that have been genetically modified.
4. Why might medical and genetic techniques that improve human health or lengthen life be considered controversial? Give examples.
5. Why do you suppose there was controversy over the evidence of a relatively recent common ancestry of humans and other primates?
6. Speculate about other simple characteristics (such as height) that might be or have been the object of natural selection in human evolution.

For More Information

Attenborough, D. 1979. *Life on Earth*. William Collins Sons & Co. Ltd., London.

"Evolution." 1978. A special issue of *Scientific American,* 239.

Gonick, L., and M. Wheelis. 1983. *The Cartoon Guide to Genetics*. Barnes and Noble Books, New York.

Gould, S. J. 1977. *Ever Since Darwin*. W. W. Norton & Co., New York.

Howard, J. 1982. *Darwin*. Oxford University Press, New York.

Johanson, D., and M. Edey. 1982. *Lucy: The Beginnings of Mankind*. Warner Books, Inc., New York.

Miller, J., and B. Van Loon. 1982. *Darwin for Beginners*. Pantheon Books, New York.

Orel, V. 1984. *Mendel*. Oxford University Press, New York.

Sarich, V. M., and A. C. Wilson. 1967. "Immunological Time Scale for Hominid Evolution." *Science,* 158:1200–1203.

Sayre, A. 1975. *Rosalind Franklin and DNA*. W. W. Norton & Co. Inc., New York. (An alternative to Watson's view of events.)

Tierney, J., and L. Wright. 1988. "The Search for Adam and Eve." *Newsweek,* Jan. 11:46–52.

Volpe, E. P. 1985. *Understanding Evolution,* fifth edition. Wm. C. Brown Publishers, Dubuque, Iowa.

Watson, J. D. 1968. *The Double Helix*. Atheneum, New York.

Watson, J. D., and F. H. C. Crick. 1953. "Genetical Implications of the Structure of Deoxyribonucleic Acid." *Nature,* 171:964–67.

SECTION II

REPRODUCTION AND INHERITANCE

Cellular basis of structure and growth

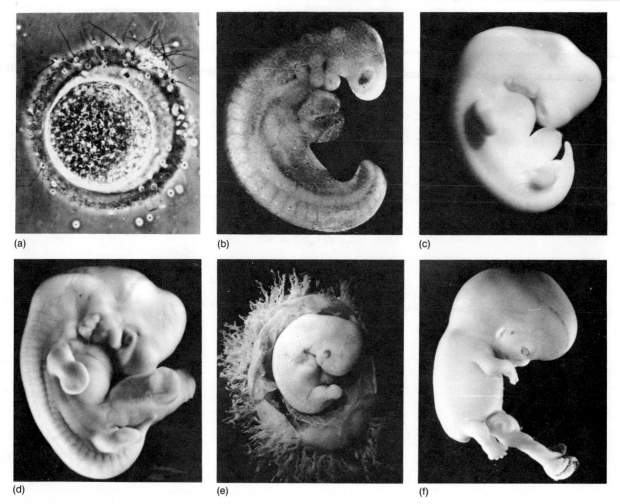

(a) (b) (c)

(d) (e) (f)

Figure 2.1 (a) Human egg and sperm (magnification 400×); stages in the development of a human embryo—approximate age after ovulation (b) twenty-eight days, (c) thirty-two days, (d) thirty-eight days, (e) forty-two days, (f) forty-nine days.
Source: (a–f) Dr. Landrum Shettles.

All living things are composed of cells. Each of us arose from a single cell through a complicated process of development (figure 2.1). This special cell, the **zygote**, results from the fertilization of an egg from the mother by a single sperm from the father. The cells of the developing **embryo** divide many times to produce, at eight weeks of age, a *fetus*. A fetus has essentially all the structures that the adult human will possess. Rudiments of the external sex organs are the last to develop, not appearing clearly until about sixteen weeks. At birth, thirty-eight weeks after conception, the newborn infant consists of approximately one trillion cells (10^{12}). Remarkably, the nuclear genetic content of almost all cells in the body is identical.

Organization of the body

The physical structure of a human being, like that of any organism, is complex and highly organized. The structure consists of a series of levels of organization and is thus *hierarchical* (figure 2.2).

The basic unit of all matter, both living and non-living, is the *atom*. One hundred five types of atoms (or elements) have been discovered, but living organisms require only about one-fourth of those. The elemental requirements vary between species, but humans, for example, need twenty-four elements. Even so, most of these twenty-four elements are not needed in large quantities. Four fundamental elements—carbon, hydrogen, oxygen, and nitrogen—comprise 99% of the atoms in the body.

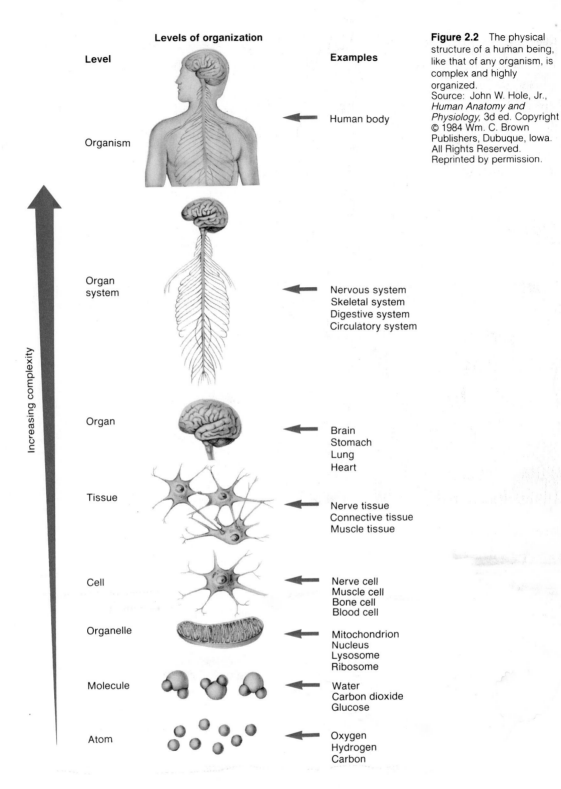

Levels of organization

Level

Examples

Figure 2.2 The physical structure of a human being, like that of any organism, is complex and highly organized.
Source: John W. Hole, Jr., *Human Anatomy and Physiology,* 3d ed. Copyright © 1984 Wm. C. Brown Publishers, Dubuque, Iowa. All Rights Reserved. Reprinted by permission.

Organism → Human body

Organ system → Nervous system
Skeletal system
Digestive system
Circulatory system

Organ → Brain
Stomach
Lung
Heart

Tissue → Nerve tissue
Connective tissue
Muscle tissue

Cell → Nerve cell
Muscle cell
Bone cell
Blood cell

Organelle → Mitochondrion
Nucleus
Lysosome
Ribosome

Molecule → Water
Carbon dioxide
Glucose

Atom → Oxygen
Hydrogen
Carbon

Increasing complexity

Atoms of the same or different types combine to form *molecules*. Molecules occur in an astonishing variety of sizes and shapes (figure 2.3), but again, living organisms require specific amounts of only a few basic types. Over 60% of the molecules in the human body are water—a molecule composed of two hydrogen atoms and one oxygen atom.

Figure 2.3 Examples of atoms and molecules showing the components of sugars, starches, and proteins. (See colorplate 1.)

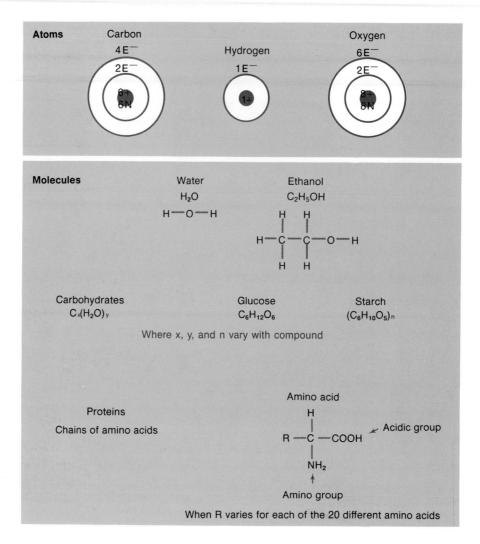

The majority of the remaining molecules are *organic molecules,* whose structure and chemistry depend primarily on the carbon atom. Carbon has the capacity to bind with as many as four other atoms at the same time (figure 2.3), and the chemistry of carbon is often considered to be the chemistry of life. There are four basic types of organic molecules—carbohydrates, lipids, proteins, and nucleic acids. Proteins and nucleic acids are macromolecules crucial to genetic function, which we examine in detail in chapter 8.

Molecules do not combine randomly to form living organisms. Instead, they occur in highly organized structures called **cells,** the fundamental unit of structure and function for life on earth. Two basic types of cells can be distinguished (figure 2.4).

Prokaryotic cells (*pro* meaning before; *karyon* meaning cell or nucleus) are small and simple in appearance, with few distinguishable features under the light microscope. With an electron microscope, DNA of prokaryotes appears as a long

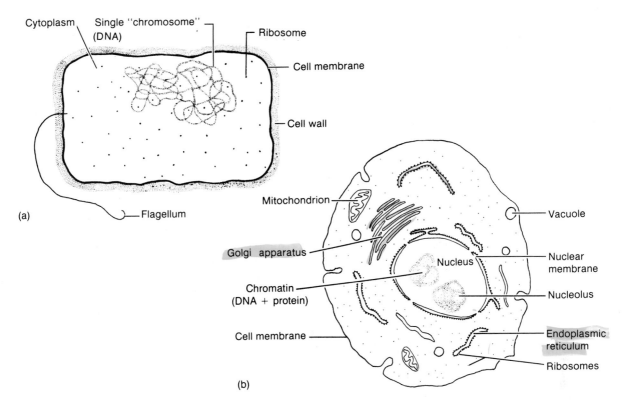

Cytoplasm   Single "chromosome"
            (DNA)
                              Ribosome

                              Cell membrane

                              Cell wall

(a)
            Flagellum

Mitochondrion   Vacuole

Golgi apparatus   Nuclear
                              Nucleus   membrane

Chromatin   Nucleolus
(DNA + protein)

Cell membrane   Endoplasmic
                                        reticulum

                                        Ribosomes

(b)

Figure 2.4 Comparison of typical prokaryotic cell and typical eukaryotic animal cell: (a) generalized diagram of a bacterial cell; (b) generalized representation of an animal cell. Note the increased complexity of the animal cell with its numerous organelles and internal membranes. (Size not to scale.) Source: James L. Sumich, *An Introduction to the Biology of Marine Life,* 4th ed. Copyright © 1988 Wm. C. Brown Publishers, Dubuque, Iowa. All Rights Reserved. Reprinted by permission.

ring. No proteins have been found associated with prokaryotic DNA. For support, prokaryotic cells are surrounded by a polysaccharide (complex sugar) cell wall. They reproduce primarily by fission, or equal division, and not by sexual reproduction. On an average, prokaryotic cells are about one micrometer in diameter (1,000 micrometers = 1 millimeter). Bacteria and blue-green algae are examples of prokaryotes.

Eukaryotic cells (*eu* meaning true; *karyon* meaning cell or nucleus) usually have a volume one thousand times greater than prokaryotic cells. They average about ten micrometers in diameter and are highly variable in shape (figure 2.5). Unlike prokaryotes, they have a very well-defined nucleus, and within the nucleus, DNA occurs in association with proteins. During cellular reproduction, the DNA-protein complex appears as many dark rods instead of as a single ring. The rods of DNA and protein are known as **chromosomes.** Many other structures are visible within the cell.

In multicellular organisms, cells must meet their own needs and those of the entire organism as well. Often, cells of the same type grow together or combine to form *tissues.* Tissues usually are identified by the same name as the cells that form them. For example, muscle tissue is composed of muscle cells. Tissues of several types are organized into larger units called *organs,* such as kidneys, lungs, or the brain. Organs are in turn grouped into *organ systems.* The nervous system consists of the brain, spinal cord, peripheral nerves, and all the sense organs, such as the eyes and ears.

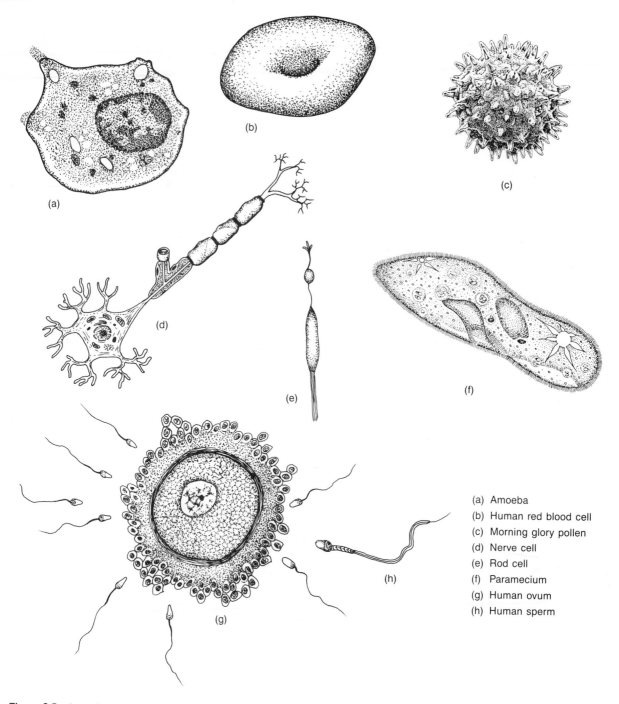

(a) Amoeba
(b) Human red blood cell
(c) Morning glory pollen
(d) Nerve cell
(e) Rod cell
(f) Paramecium
(g) Human ovum
(h) Human sperm

Figure 2.5   A small sample of the diversity of eukaryotic cell types: (a) an amoeba, (b) a red blood cell (when mature, these cells are unusual in totally lacking a nucleus), (c) morning glory pollen, (d) a nerve cell (with long projections—nerve cells are among the longest known cells), (e) a rod cell from the retina of the eye, (f) a paramecium, (g) a human ovum, (h) a human sperm.

Finally, of course, organ systems are combined to form the *organism.* A primary feature of a living organism is its high level of structural and functional organization. The science of genetics used to be considered primarily the study of *inheritance,* the transmission of characters from one generation to the next. Now, genetics is equally viewed as the mechanism of *control* of the extraordinary organization of the individual organism as well.

The eukaryotic cell

Cells come in many shapes and sizes (figure 2.5). Within each cell are structural and functional units called *organelles* ("little organs"), which carry out basic metabolic functions for the cells. Approximately two hundred types of cells have been described in humans, differing not only in size and shape, but also in the relative proportion of various types of organelles. This variety is not surprising because cells fulfill many different functions. Thus, cells may be spherical, flattened, or tubular. They may have a relatively smooth surface or possess numerous projections. They are usually too small to be seen without a microscope, but some nerve cells are more than a meter long, and an ostrich egg (which is only one cell) weighs several kilograms.

All cells conduct the same basic metabolic and reproductive processes common to all life; thus, all share certain common features. Biologists have learned much about cells by examining them under microscopes. Often, special stains are applied to the cells, which make particular organelles highly visible. Figure 2.6 shows the structure of a generalized eukaryotic cell.

The *cell membrane* encloses and supports the cell and its organelles. The cell membrane regulates the passage of molecules into and out of the cell. Prokaryotic, fungal, and plant cells possess an additional external supportive structure, the *cell wall.*

The nuclear membrane surrounds the largest organelle within the cell, the **nucleus.** The nucleus is usually visible through a microscope as a central darkened area, except when the cell is undergoing division. External to the nucleus is the **cytoplasm,** a viscous fluid containing those organelles responsible for the metabolic functions of the cell. The cytoplasm consists of water and many complex molecules, including sugars, lipids, and amino acids.

Several organelles within the cytoplasm are crucial to genetic function. A complex series of folds of the cell membrane within the cytoplasm is called the *endoplasmic reticulum.* Attached to some sections of the endoplasmic reticulum are sequences of small, spherical bodies called **ribosomes.** Ribosomes are the sites of the synthesis of proteins.

In many (including human) eukaryotic cells, a small spherical *centrosome* lies close to the nucleus. It contains two cylindrical structures oriented at right angles to each other, the *centrioles.* The centrioles play an important role in organizing the process of cell division.

Throughout the cytoplasm are numerous **mitochondria.** Mitochondria release the energy from carbohydrate molecules, thus supplying the cell with energy. They also contain their own genetic information, independent of the DNA within the nucleus. They are inherited only from the female parent. Plant (but not animal) cells

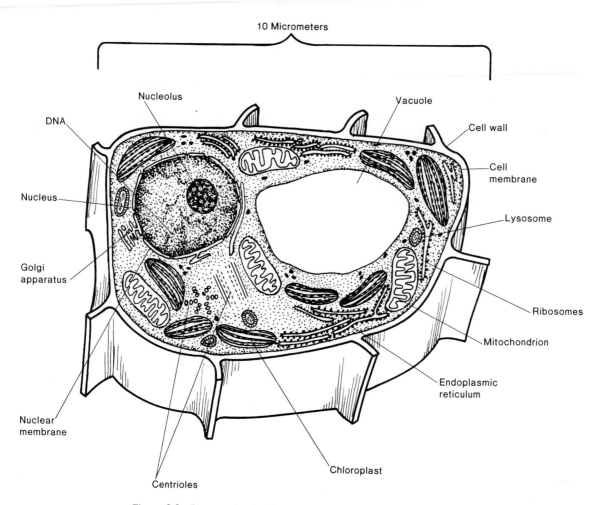

10 Micrometers

Nucleolus

DNA

Nucleus

Golgi
apparatus

Nuclear
membrane

Centrioles

Vacuole

Cell wall

Cell
membrane

Lysosome

Ribosomes

Mitochondrion

Endoplasmic
reticulum

Chloroplast

Figure 2.6 Diagram showing the structure of a typical eukaryotic cell.

often contain **chloroplasts.** Chloroplasts are the site of photosynthesis, where sunlight is used to produce carbohydrate molecules from carbon dioxide and water. (All human life is ultimately dependent upon plants' ability to capture energy and make food.) Like mitochondria, chloroplasts also contain their own DNA.

The nucleus

Although some cytoplasmic organelles contain DNA, the nucleus is the primary site of genetic information within the cell. Both mitochondria and chloroplasts require the use of nuclear genetic information, as well as their own. Two structures are usually visible within the nucleus: a very dark **nucleolus** ("little nucleus") and filamentous **chromatin.** Chromatin that stains lightly is called **euchromatin;** darkly staining chromatin is called **heterochromatin.**

Chromatin is distributed throughout the nucleus and consists of two types of molecules: proteins and nucleic acids. **Histones** are the major type of nuclear protein. There are five types of histones, and each shows almost no variability in structure among the eukaryotes. Prokaryotes have no histones. Histones are usually bound tightly to the surface of the long chains of nucleic acids. Approximately twenty other types of proteins are found in the nucleus, also bound to the nucleic acids. All nuclear proteins appear to play a role in controlling the function of nucleic acids.

The nucleic acids in chromatin are of a particular type, deoxyribonucleic acid, or DNA. The genetic information that organizes and governs every aspect of the life of a cell is encoded in its DNA. DNA is the fundamental genetic material and provides the physical basis of inheritance.

For most of the life of the cell, DNA and its associated proteins occur as extremely long, diffuse threads only barely visible microscopically. At this time, the DNA-protein complex is called chromatin. During cell division, however, the chromatin becomes highly condensed and clearly visible for a brief time. These highly condensed lengths of chromatin are then called *chromosomes*.

Chromosomes

The appearance of chromosomes is governed by the coiling of the chromatin. When the chromatin is most tightly coiled, or condensed, the chromosomes are the shortest and most visible. This normally occurs during cell division, when the chromosomes appear as dense, compact rods (figures 2.7 and 2.8).

Chromosomes can be seen in living cells, but they are best studied in cells that have been killed (carefully, so that they are not damaged) and then treated with special stains (figure 2.8). Stains are simply chemicals that adhere to specific molecules and make them more visible. In fact, chromosome means "colored body," referring to the appearance of chromosomes when stained. Along the length of a chromosome, certain segments stain darkly because they are tightly coiled, giving the chromosome a characteristic black- and white-banded appearance.

Most chromosomes have a unique pattern of bands that allows identification of specific chromosomes. Several types of stains are used to visualize different bands. Giemsa is a common stain that seems to adhere to the tightly coiled chromosome segments and produces G bands. C-banding uses a modification of the Giemsa technique to darken the centromeric region of the chromosome. Most (but not all) chromosomes produce either G or C bands when stained appropriately.

The primary constriction that occurs somewhere along the length of the chromosome is the **centromere.** The centromere plays a crucial role in directing chromosome movement during cell division. Chromosomes are classified according to the location of the centromere (figure 2.7): **metacentric chromosomes** have a centromere in the middle of the chromosome; **submetacentric chromosomes** have a centromere slightly off-center; **acrocentric chromosomes** have the centromere almost at one end; and **telocentric chromosomes** have a terminal centromere. Some chromosomes have a second constriction (usually near one end) that pinches off a small segment known as a **chromosome satellite.**

The term **karyotype** refers to the total chromosome complement of an individual. A karyotype is also a formal arrangement of photographed chromosomes of a single cell from one individual, in order of decreasing size (figure 2.9). Among eukaryotes, each species possesses a unique karyotype. Members of a species usually have the same number of chromosomes. Exceptions are found in species such as bees, in which females have twice the number of chromosomes as males, and many

Figure 2.7 (a) Human chromosomes as seen at mitotic metaphase. Note each chromosome consists of two sister chromatids held together by a single centromere. (b) Classification of chromosomes based on gross morphology. For each type, the chromosome illustrated in (a) is nondividing; that in (b) has been arrested in metaphase of mitosis.
Source: (a) Kleberg Cytogenetics Laboratory, Institute for Molecular Genetics, Baylor College of Medicine, Houston, Texas.

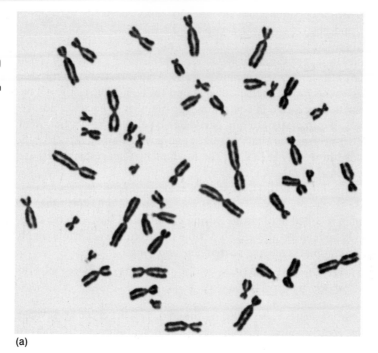

(a)

Metacentric	Submetacentric	Acrocentric	Telocentric	Chromosome with satellites
(a) (b)	(a) (b)	(a) (b)	(a) (b)	Satellite (a) (b)

(b)

species with varying numbers of tiny *supernumerary chromosomes*. Individual chromosomes are identified on the basis of length, location of the centromere, and the banding pattern. Each pair of chromosomes is given a number. Within the karyotype, groups of chromosomes are often recognized by shared characteristics, for example, large metacentrics, short acrocentrics, etc.

In many species, the karyotype of some individuals contains one pair of chromosomes that are *not* identical in appearance. These are the *sex chromosomes* (figure 2.9). In species that have sex chromosomes, members of that chromosome pair will appear identical in one sex, but different in the other sex. The sex chromosomes are always given the last (highest) number in the karyotypic designation. All chromosomes that are not sex chromosomes are called **autosomes.**

The number of chromosomes is not related to the apparent complexity or evolutionary position of a species. Humans have forty-six chromosomes—which, if not condensed, would stretch two meters long. Most species possess between ten and fifty chromosomes. One species of roundworm has only two chromosomes, and some ferns have over twelve hundred chromosomes.

For many species the total number of chromosomes in an adult is known as the **diploid** or **2N number** for that species. In diploid species the chromosomes occur in pairs. One member of each pair originally came from the female parent and is referred to as the maternal chromosome; the other chromosome from the male parent

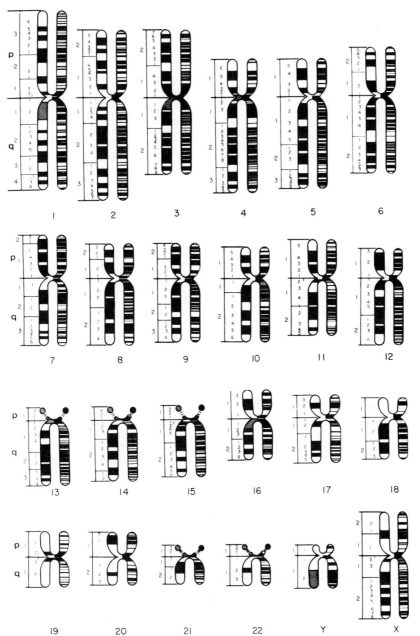

Figure 2.8 Haploid karyotype of human male, as shown by diagrammatic representation according to Paris Conference nomenclature. The left chromatid represents the G-banding pattern seen in mid-metaphase, and the right chromatid shows the pattern in late prophase.
Source: J. J. Yunis, "High Resolution of Human Chromosomes," *Science*, Vol. 191, Fig. 2, pp. 1268–1270, March 26, 1976. Copyright 1976 by the AAAS.

is known as the paternal chromosome. Each member of a chromosome pair is said to be **homologous** to the other. This means that the chromosomes are similar in appearance and structure and possess the same sequence of genes along their length (figure 2.10).

Recall that a gene occupies a particular position, or **locus,** on a chromosome. A gene is a segment of DNA, which can take one of several different forms called **alleles.** Because most chromosomes occur in pairs, an individual possesses two alleles for most genes, one on the maternal chromosome and one on the paternal chromosome (exceptions will be discussed in later chapters). An individual with two copies of the same allele at a locus is **homozygous** at that locus, with two different alleles, **heterozygous.** This pair of alleles is the individual's **genotype.**

Figure 2.9 Karyotypes of
(a) human male and
(b) human female.
Source: (a and b) The
Kleberg Cytogenetics
Laboratory, Baylor College of
Medicine, Houston, Texas.

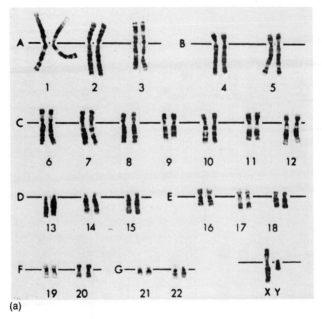

(a)

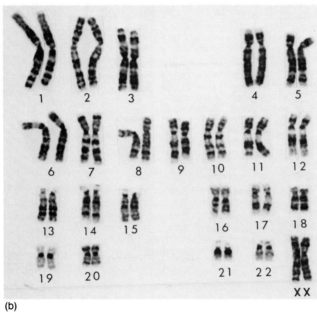

(b)

Figure 2.10 A homologous
pair of human chromosomes,
chromosome number 9,
showing the heterozygous
condition at two loci and the
homozygous condition at
one locus. The person with
these chromosomes would
be a carrier of the recessive
condition galactosemia,
would have blood type AB,
and would not have the
dominant condition of nail-
patella syndrome.

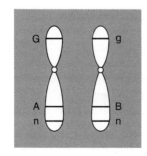

The number of pairs of chromosomes in a species is called the **haploid or N number.** The haploid number is one-half the diploid number. Each new individual receives a complete haploid *set* (one member of each pair of chromosomes) from each parent. The egg carries the maternal set of chromosomes, and the sperm carries the paternal set of chromosomes. At fertilization, the two haploid sets unite to form the diploid number that characterizes each individual.

On occasion genetic mistakes occur in the process of cell division. An individual who receives too many or too few chromosomes is **aneuploid** ("without the true number"). Such karyotypes are designated $2N + 1$, $2N - 1$, $2N + 2$, and so on. Aneuploidy usually has serious medical consequences for humans.

Instead of having an excess or deficit of single chromosomes, some individuals possess entire extra sets of chromosomes. They are called **polyploid.** Polyploidy is fatal in humans. Specific names identify how many haploid sets a polyploid individual has: $3N = $ *triploid,* $4N = $ *tetraploid,* $5N = $ *pentaploid,* and so on. Sometimes the polyploid condition is not harmful. In fact, among plants, some polyploids are exceptionally healthy and vigorous. Agricultural geneticists have used this fact to select many of our common food crops. Cereal grains, such as wheat and rye, and many other commercial crops are polyploids.

Evolution of chromosomes

Closely related species, such as members of the same genus, usually have similar but not necessarily identical numbers of chromosomes. For example, dogs, wolves, foxes, and coyotes (all members of the genus *Canis*) have seventy-eight chromosomes; humans (*Homo*) have forty-six chromosomes and the closely related chimpanzees (*Pan*) and gorillas (*Gorilla*) have forty-eight. The process of speciation is often accompanied by changes in chromosome number as a result of *fusion,* the union of two or more pairs of chromosomes, which reduces the diploid number. Human chromosome number 2 is homologous to two chromosomes in chimpanzees and gorillas, accounting for our lower chromosome number (see figure 12.10). *Fission* is also common, in which a pair divides into two or more pairs. Thus, closely related species possess many of the same genes, but those genes may be organized into chromosomes very differently.

The genetic similarity of closely related species means that they are sometimes capable of **hybridizing,** or interbreeding. The mule is a hybrid between a male donkey (*Equus asinus,* $2N = 62$) and a female horse (*Equus caballus,* $2N = 64$), but like many hybrids, it is usually not fertile. New mules are made by continued hybridization, not by breeding mule to mule.

An overview of reproduction

An organism may use *sexual* or *asexual* processes to reproduce itself, and some species use both. Sexual reproduction involves the fusion of genetic material from two parents to produce an offspring, while in asexual reproduction a single parent produces offspring that are all genetically identical.

During sexual reproduction in most eukaryotes, a new organism is formed by the union of two reproductive cells called **gametes,** each containing one haploid set of chromosomes from one parent. The offspring thus begins life as a single cell with a complete diploid complement of chromosomes.

Gametes are formed within special reproductive organs, the *gonads*. The female gonad, the *ovary,* produces the female gamete or *ovum* (egg). The male gonad, the *testis,* produces *sperm.* Formation of haploid gametes by diploid parent cells within the gonad requires a type of cell division that is unique to reproductive or *germinal* tissues: **meiosis.** The distinguishing feature of meiosis is the reduction of the karyotype from a diploid to a haploid condition, and it occurs nowhere in the body except the gonads. In fact, meiosis is derived from a Greek word meaning "diminution." Meiosis is discussed in detail in the next chapter.

Genetic recombination also occurs in prokaryotes. It involves the transfer of genetic material between two bacteria, who may or may not undergo fission to produce new individuals at the time. Thus, the key element of sexual reproduction is the union of DNA from two individuals to produce new genetic combinations, rather than actual reproduction of the individual itself. Transfers and exchanges of DNA may also occur among viruses that infect a bacterial cell, and two, three, or more individual viruses may participate.

All nonreproductive cells of the body are called *somatic* cells (*soma* meaning body). Somatic cells and some gonadal cells of some species reproduce by **mitosis** (*mitos* meaning thread—a reference to the appearance of the chromosomes). During mitosis, a single diploid cell divides equally to produce two diploid daughter cells. Mitosis thus generates new somatic cells for multicellular organisms, and it is also the mechanism of *asexual reproduction.* One-celled organisms and some multicellular organisms reproduce simply by mitotic fission or budding of cells. In contrast to sexual reproduction, asexual reproduction involves a single diploid parent giving rise to a diploid offspring. Asexual reproduction thus produces offspring that have nuclei that are genetically identical to—a **clone** of—their parents. A clear understanding of mitosis and the life cycle of the cell is essential to an understanding of genetics.

The mitotic cell cycle

The complete life history of a single cell, from its origin as one of two daughter cells to its own mitotic division, is called the *mitotic cell cycle* (figure 2.11). There are two major segments of the cell cycle: interphase and mitosis.

Interphase

For almost all its life, the cell nucleus is clearly visible, and the chromosomes appear as diffuse chromatin. This period of cell growth, called **interphase,** is a time of high metabolic activity. The cell is in a period of vegetative growth. Some cells, such as nerve cells and adult muscle cells, never divide and spend their entire life in interphase. For most cells, however, interphase is a time of preparation for mitosis. The key genetic event of interphase is the precise replication of the chromosomes.

Three distinct phases occur during interphase. Immediately following mitosis, the cell enters the first gap (G_1) phase, the longest single period of the cell cycle. Then it enters the synthesis (S) phase, during which its DNA is replicated. Then a second gap (G_2) phase follows. Remember that during the gap phases no chromosome or DNA replication occurs, but other aspects of cell life continue normally. After the second gap, the cell enters the shortest period of its life cycle, mitosis.

Mitosis

All cells require genetic information to control their activities. Mitosis insures that each cell receives a complete set of chromosomes identical to the chromosomes of its parent. An extraordinary feature of multicellular organisms is that every cell in

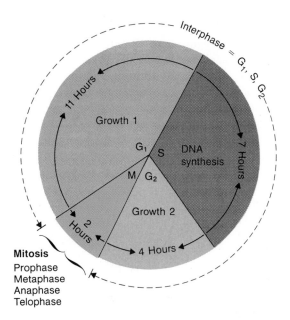

Figure 2.11 The cell cycle of the human white blood cell. Note that interphase is by far the longest part of the cycle and consists of three segments. The times of cell cycle phases vary with cell type and species. (See colorplate 2.)

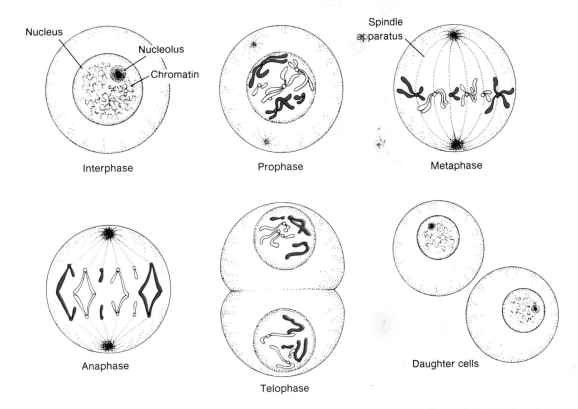

Figure 2.12 Mitotic cell cycle for a species with diploid number of 6 (2N = 6). (See colorplate 3.)

the body possesses all the genetic information required to control the activities of every other cell in the body. In an adult, about fifty million cells die every second and are immediately replaced by mitotic division of other cells.

Mitosis is a continuous sequence of events. For purposes of analysis, however, four stages have been defined on the basis of nuclear and, particularly, chromosomal behavior (figure 2.12).

Prophase

As mitosis begins, the chromatin begins to coil and condense. **Prophase** begins when the condensing chromosomes become microscopically visible. At this stage, each chromosome appears to be double stranded. Each of the dense strands of a chromosome is called a *chromatid,* and each chromatid is attached to its **sister chromatid** at the centromere. Although each chromosome consists of two sister chromatids, the structure is still considered to be just one chromosome as long as only one functional centromere exists. In other words, the number of chromosomes equals the number of centromeres. The existence of two chromatids at this point reflects the replication of each chromosome that has occurred during interphase. Except for the lack of its own centromere, each chromatid is in fact a complete chromosome and identical to its sister.

Also during prophase, the nucleoli and nuclear membrane disappear. In the cytoplasm a new structure forms—the **spindle apparatus.** The spindle, shaped like its name suggests, appears to consist of a series of lines or spindle fibers stretched between two poles. In animal cells, the centrioles separate to form the poles of the spindle. In plants, centrioles do not exist, but a spindle still forms. The spindle serves as a structure that organizes the movement of chromosomes during later phases of mitosis.

Metaphase

The pace now quickens. As **metaphase** begins, spindle formation is completed. The chromosomes appear to be attached to the spindle fibers by the centromere of each pair of sister chromatids. Then, the chromosomes move to the *equatorial* or *metaphase plate,* a plane located midway between the centrioles (or spindle poles) and at right angles to the line connecting them. At the conclusion of metaphase, all chromosomes are lined up along the metaphase plate. Chromosomes are most condensed and most visible during metaphase. Therefore, karyotypes are normally studied from preparations made at metaphase of mitosis.

Anaphase

At the beginning of **anaphase,** each centromere divides. Each sister chromatid becomes a full-fledged chromosome with its own centromere. The spindle fibers contract, and the two daughter chromosomes (former sister chromatids) are drawn toward opposite poles of the cell. Because it appears to be pulled by its centromere, each chromosome now has a V, J, or I shape, depending upon whether it is metacentric, submetacentric, or telocentric. Imagine the complexity of twenty-five hundred chromosomes of one fern cell going their separate ways.

Anaphase is the shortest period of mitosis. The result of anaphase is the exactly equal division of chromosomes. The dividing cell now possesses two daughter groups of chromosomes, each with a complete diploid complement equivalent to that of the parent cell.

Telophase

The last stage of mitosis, **telophase,** begins when the nuclear membranes reappear and enclose each set of daughter chromosomes. Nucleoli also reappear. In animal cells, a furrow appears in the cell membrane that deepens until finally the dividing cell is pinched in two. The developing furrow, of course, also divides the cytoplasmic contents of the parent cell, but the division of cytoplasm does not appear to occur with the precision of nuclear and chromosomal division. In plants, a cell plate forms dividing the cytoplasm among the daughter cells. Mitochondria and, in plants, chloroplasts are also divided among the daughter cells. If the daughter cell does not receive any mitochondria or chloroplasts, it will die.

Development

Plants, animals, and fungi all begin life as one-celled organisms, but then enter a multicellular stage for most of their lives. The process by which a single, unspecialized cell becomes a large, complex, and highly organized aggregation of specialized cells of many types is called *development*. Development consists of two processes—*growth* and *differentiation*.

Growth

Individual cells exist only within a small range of sizes. Organisms attain large size through mitotic division and the subsequent aggregation of many thousands of daughter cells. Such enlargement via mitosis is the primary mechanism of *growth*.

Mitotic division is a continuous occurrence throughout the life of multicellular organisms, even after full size has been attained. Surface cells are continually sloughed off so the cells of our skin or digestive tract, as well as blood cells and hair cells, grow throughout life. Some cells, however, do not: nerve cells and muscle cells stop dividing, once fully formed.

As with single cells, there are physical limits to the size of multicelled organisms. The largest terrestrial animals, now extinct, were not much larger than elephants. The largest animal ever, the blue whale, lives in the ocean, where water helps support the huge mass. The largest known organisms are the giant sequoias of California. Almost all the tissues of these trees are dead supportive tissue; only a small amount of the total mass of a tree is living, growing tissue.

Differentiation

Multicellular organisms do not consist only of one type of cell. *Differentiation* occurs when cells become specialized, both structurally and functionally, to meet particular needs of the entire organism. As development progresses, the organism not only becomes larger as a result of mitotic increase in the number of cells, but differentiation produces a series of highly specialized cells, then tissues, organs, and systems.

We noted earlier that all cells possess all the genetic information necessary to produce the entire organism. Each cell is thus potentially **totipotent**—"all powerful" genetically. A truly totipotent cell has the capacity to recreate the entire organism on the basis of the genetic information it contains. However, most cells use only a fraction of their genetic information.

A key aspect of development is precise control of specialization of cells. Each bit of genetic information guiding development is used only in a specific sequence, for a specific duration of time, and is then shut down or "turned off." At conception a new human consists of only one relatively nondescript cell; at birth a human consists of approximately one trillion cells, and an adult consists of perhaps seventy-five trillion cells. What distinguishes humans or any species from all others is the pattern of organization of specialized cells that occurs during development. Thus, *control* and *organization* of development have become fundamental aspects of the study of genetics.

One aspect of the control or limit to cell growth is **contact inhibition.** When normal cells are grown in a cell culture, they grow only in a single layer or monolayer. When the enlarging monolayer encounters a barrier such as the wall of the dish or another cell, it will grow around but not over that barrier. When the surface of the culture dish is covered with cells, growth stops. Thus, contact inhibition is an expression of orderly and limited cell growth.

Growth and healing in a live organism are equally orderly. For example, if a portion of the liver of a rat is removed surgically, the remaining liver cells grow to replace the removed section. Growth stops when the liver attains its normal size. This process can be repeated many times with the same result. Thus, normal liver cells are able to respond to environmental changes. The genetic information in liver cells directs the orderly replacement of missing tissue until normal size is reached, at which point growth stops.

One characteristic of cancerous cells is unlimited growth, and indeed cancer cells do not exhibit contact inhibition. At present the changes in the normal mechanism of cellular control that cause loss of contact inhibition are unknown. One possibility is that molecules on the cell surfaces play a major role in contact inhibition. Changes in these molecules coincide with changes in the behavior of cells that contact one another.

Summary

1. Eukaryotic cells are larger and more complex than prokaryotic cells. They possess many organelles and have a discrete, well-organized nucleus.
2. Within the nucleus are one or more nucleoli and many chromosomes. Eukaryotic chromosomes consist of segments of DNA bound to nuclear proteins, such as histones.
3. Chromosomes are highly organized. A centromere is always present, and G bands and C bands are often apparent with appropriate stains. A satellite may be present at the end of a chromosome.
4. In diploid species, chromosomes occur in homologous pairs. The two homologues are similar in size, location of the centromere, and banding patterns, and they possess the same sequence of genes on each (except for sex chromosomes).
5. A specific form of a gene is known as an allele. An allele is one of many possible molecular sequences that govern a particular character.
6. An individual almost always has two homologous genes governing a specific characteristic. A diploid individual can have two identical alleles or two different alleles for each gene. Within a species there may be dozens of different alleles for a given gene.
7. Chromosome number and structure is usually similar in closely related species. Often, speciation involves fusion or fission of pairs of chromosomes. Closely related species can often hybridize, but such hybrids are usually sterile.
8. Species may reproduce either sexually or asexually. Prokaryotes most commonly use asexual reproduction; eukaryotes, sexual.
9. The growth and division of one cell is called a mitotic cell cycle. Interphase is a prolonged period of vegetative growth and the replication of genetic material. Mitosis is a brief period of actual cell division in the cell cycle.
10. Development has two basic aspects—growth and differentiation. Growth is achieved primarily by division of existing cells to produce more cells, while differentiation involves the structural and functional specialization of individual cells.

Key Words and Phrases

acrocentric chromosomes

allele

anaphase

aneuploid

autosomes

cell

centromere

chloroplast

chromatin

chromosome

chromosome satellite

clone

contact inhibition

cytoplasm

diploid (2N) number

embryo

euchromatin

eukaryotic cell

gametes

genotype

haploid (N) number

heterochromatin

heterozygous

histones

homologous chromosomes

homozygous

hybridization

interphase

karyotype

locus

meiosis

metacentric chromosomes

metaphase

mitochondria

mitosis

nucleolus

nucleus

polyploid

prokaryotic cell

prophase

ribosome

sister chromatids

spindle apparatus

submetacentric chromosomes

telocentric chromosomes

telophase

totipotent

zygote

Questions

1. The diploid number of donkeys is sixty-two and that of horses is sixty-four. What is the diploid number of a mule? Why are mules usually sterile?
2. Diagram mitosis in a species with 2N = 8, where two pairs of chromosomes are metacentric and two pairs are telocentric. Color paternal chromosomes solid.
3. How many chromosomes are present in prophase in a species with 2N = 30? How many in anaphase?
4. Diagram a pair of metacentric chromosomes at mitotic metaphase. Label two genetic loci on this chromosome pair by A and B. Place alleles A and a on genetic locus A and alleles B and b on genetic locus B.
5. Most treefrogs have a diploid number of twenty-four. How many chromosomes would be found in a triploid treefrog? How many in a tetraploid treefrog?

For More Information

Brachet, J. 1961. "The Living Cell." *Scientific American,* 205:50–62.

"The Cell." 1961. A special edition of *Scientific American,* 205.

Mazia, D. 1974. "The Cell Cycle." *Scientific American,* 230:54–69.

CHAPTER THREE

Mendel, meiosis, and basic principles of inheritance

Gregor Mendel

Gregor Mendel was born on a farm in northern Moravia, part of modern Czechoslovakia near the Polish border. As a youth, he performed well in school, especially in natural history (biology) and physics. After a period of hardship as a teenager, he entered the Augustinian monastery at Brno at age twenty-one in order to (in his own words) "free [me] from the bitter struggle for existence." The monastery was administered by Abbot Napp, who for many years had pursued an interest in plant breeding and hybridization. Abbot Napp organized a course in hybridization, which was taught for the first time during the year that Mendel entered the monastery.

When he was twenty-nine Mendel was sent by the Abbot to the University of Vienna to study math and physics. Biologists had long sought unsuccessfully to understand mechanisms of heredity, and Abbot Napp was convinced that progress would result from the techniques of disciplines other than biology. Mendel studied at the University for two years before returning to the monastery. He then spent the next seven years growing and hybridizing common garden peas. Peas had been the subject of breeding studies since the 1820s. They offered many advantages for experimental study and were already under cultivation in the monastery.

Mendel reported the results of his breeding studies to the local natural history society in 1865, and his paper ("Experiments with Plant Hybrids") was published in their journal the following year. In 1868 Abbot Napp died and Mendel was elected to succeed him as Abbot, thus ending his scientific research. He died in 1884, his work still unknown to the scientific community.

In 1865 Gregor Mendel (box 3.1) presented his now classic paper describing the mechanisms of inheritance. In it, he demonstrated the particulate basis of heredity and laid the foundation of modern genetics. Mendel showed that physical traits of an organism corresponded to invisible *elementen* in the cells. More important, however, he showed that the genes (*elementen*) controlling these characters exist in pairs, and that only one member of each pair is incorporated into each germ cell during gametogenesis. As we will see, these concepts allowed Mendel to suggest a mechanism for explaining the genetic structure of hybrids, as well as to develop the laws of segregation and independent assortment.

During the subsequent "Golden Age of Cytology" (1880–1900), the discovery of the behavior of the chromosomes during meiosis provided a physical explanation for the segregation and independent assortment of genes described by Mendel in 1865. It is important to remember that Mendel's work appeared at a time when there was little understanding of cell structure and function; and nothing was known about chromosomes, mitosis, meiosis, or fertilization. Thus, Mendel's work was in a sense premature—the scientific community did not possess a framework within which his principles could be understood. Furthermore, Mendel used mathematical language and concepts to explain his findings, which probably helped to make his work relatively inaccessible to the biologists of his day.

Mendel's principles

Mendel could not study genes directly. Instead he studied their physical expression, what we now call the **phenotype** (box 3.2). Mendel wanted to know how phenotypic characters such as color were transmitted between generations. Were individual characteristics transmitted unchanged from parent to offspring, or were they somehow altered or recreated between generations? Or, said another way, do hereditary particles exist, and, if so, what is their nature?

BOX 3.2

Genotype and phenotype

Geneticists distinguish between genes and their expressions. The actual genes of an individual are referred to as its **genotype;** the expressions of those genes are its **phenotype.** An individual has alleles that govern hair color; those specific alleles constitute the individual's genotype for hair color. The actual hair color that the individual exhibits is its phenotype (blond, brown, red, etc.).

Why is such a distinction necessary? Few things are ever as simple as they seem. The phenotype is determined not only by the genotype, but also by the interaction of the genotype with the **environment.** While the influence of the environment may be difficult to assess, *all phenotypic characteristics are determined by the joint effects of genotype and the environment.*

In the case of hair color, for example, how might an individual whose alleles encode information for brown hair actually possess a phenotype other than brown? In most environments, the genotype will be expressed as the phenotype brown hair, but if the environment includes long periods of intense sunlight (such as a spring trip to a Florida beach), the hair may become bleached and not at all brown.

A deficiency of protein may result in the inability to produce hair at all, in which case baldness rather than brown hair is the phenotype. Furthermore, part of the environment for one gene is all other genes. That is, there is a genetic environment, as well as an external environment, in which all genes function. Thus, with increasing age, the effects of alleles encoding brown hair will be superceded by the effects of alleles governing the aging process, and gray hair results.

Phenotype should be used in the broadest possible sense. Phenotype commonly refers to visible characteristics, such as size, shape, and color of the parts of an organism—height, weight, eye color, length of limb, and so on. But other, less obvious characteristics also comprise the phenotype: behavioral traits (twitches, yawns, posture, smiles per hour, psychosis), physiology (heart rate, blood pressure, level of basal metabolism), and biochemical characteristics (cholesterol level, presence or absence of particular enzymes, blood type). In fact, any characteristic of an organism that can be described and measured is a phenotypic character.

To answer such questions, Mendel conducted a series of experiments that are a model of scientific simplicity. His experiments succeeded because of their careful design, and because Mendel collected his data very systematically. Previous investigators had attempted to examine the big picture, collecting much data on inheritance of many characters over many generations. But Mendel categorized his data, focusing on the transmission of a few carefully chosen characters. Most importantly, Mendel counted. He simply tallied the number of offspring from different types of matings. His systematic work resulted in the first demonstration of principles of inheritance.

Mendel chose for his experiments the simple garden pea, *Pisum sativum* (figure 3.1). It was easy to cultivate domestically, and it produced one generation with many offspring per year. Many different varieties of peas had been developed and were available for study. Usually, peas are self-fertilizing (pollen from the anthers of one flower fertilize the stigma of that same flower by simply falling on the stigma) (figure 3.1); but when it was necessary to cross-fertilize, the stamens (or male, pollen-bearing organs) could be removed from one plant and used to pollinate another. Often, such manipulation was used to produce **reciprocal crosses,** so that females of one phenotype were mated with males of another, and vice versa. Reciprocal experiments can help detect the influence of sex in governing genetic traits.

Geneticists still select experimental organisms that can be easily cultured (or grown and bred under strict control), have short generation times, and produce many offspring. Laboratory rats and mice, *Drosophila* (fruit flies), *Neurospora* (bread

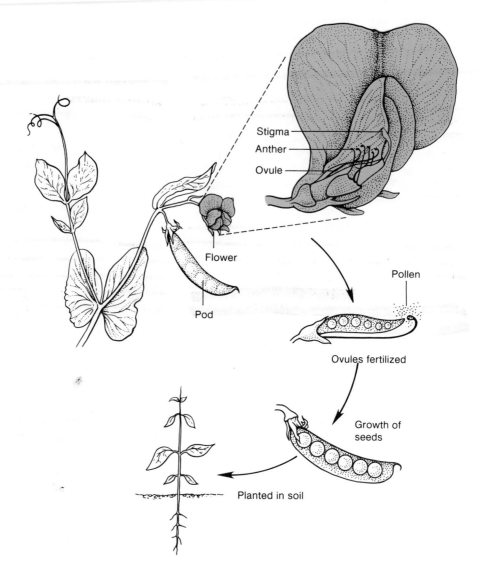

Figure 3.1 Anatomy of a pea plant showing the reproductive features that make artificial pollination possible.

Stigma
Anther
Ovule

Flower

Pod

Pollen

Ovules fertilized

Growth of seeds

Planted in soil

mold), yeast, and the common gut bacterium, *Escherichia coli,* all meet those criteria, and they are the organisms that are best known genetically (figure 3.2).

Mendel studied seven different traits of pea plants (figure 3.3). Each plant showed one of two possible states or phenotypes for each trait. For example, stems were either long (about six to seven feet) or short (about one foot). Most importantly, the type of difference shown by each of the seven characters was **qualitative.** The two phenotypes occurred in discrete categories, and an individual plant could be easily and accurately assigned to one of the categories. Stems were long or short, never intermediate. Thus, Mendel could count the number of offspring from each cross possessing each phenotype.

The alternate type of variation is **quantitative** or continuous variation. Quantitative characters are those that occur on a continuous scale, such as height and weight. Within a large population, almost all possible values of a character will be shown if the character is a quantitative variable. (Consider, for example, the range of weights shown by all the adult males in New York City.) In most instances, stem length of peas would be a quantitative variable. Although limited variation occurred within the short phenotype and within the long phenotype, the differences found by Mendel were so great that there was no overlap *between* the two phenotypes. Plants clearly fell into either the long category or the short category.

Figure 3.2 Organisms commonly used in genetic research: (a) *Neurospora* (bread mold), (b) *E. coli* infected with virus, (c) mouse, (d) fruit flies, and (e) *Cryptococcus* (yeast). Source: (a) Omikron/Photo Researchers; (b) Lee D. Simon/Photo Researchers; (e) Martin M. Rotker/Taurus Photos.

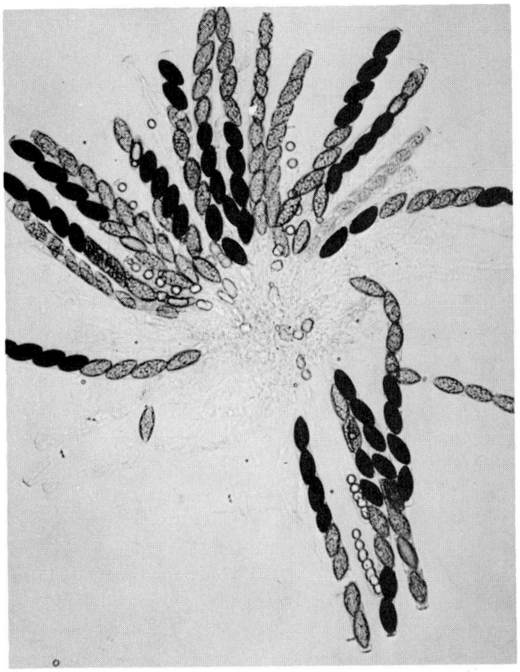

(a)

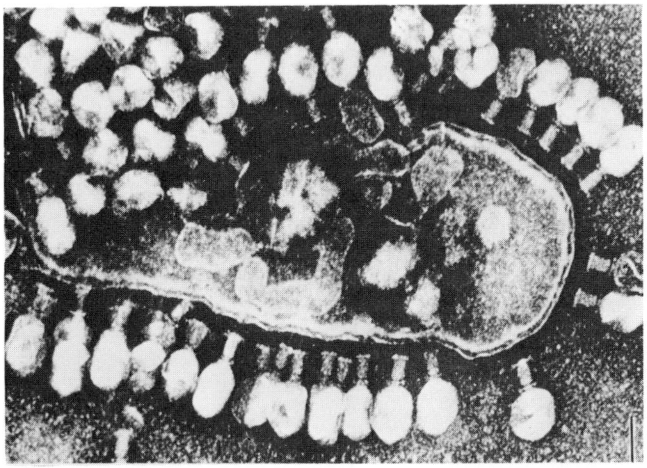

(b)

(c)

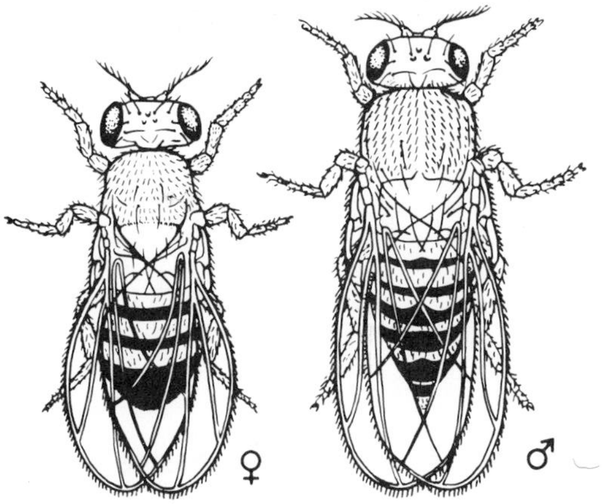

(d)

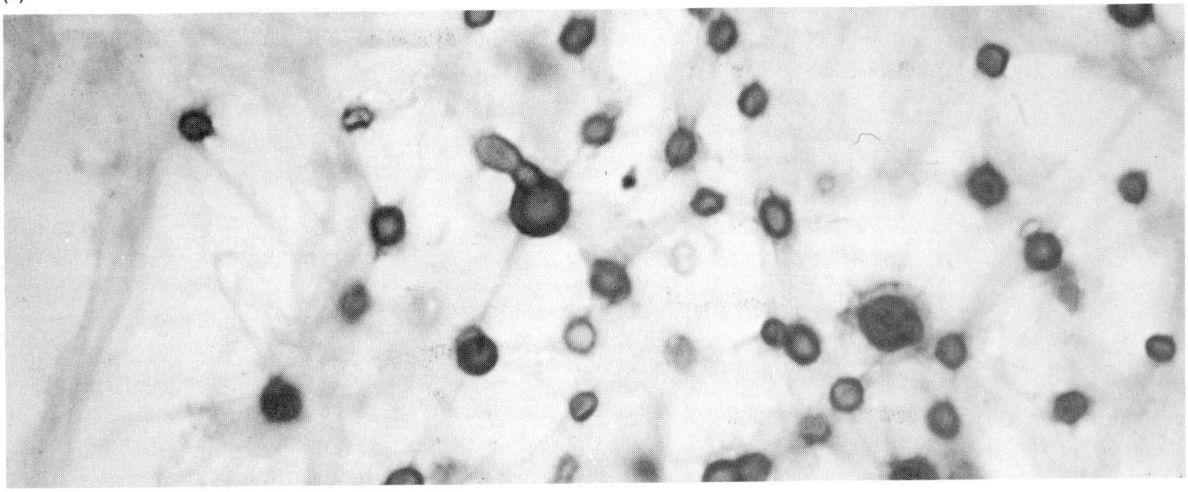

(e)

Figure 3.3 Traits of the peas that Mendel grew and studied in the monastery garden.
Source: E. Peter Volpe, *Biology and Human Concerns*, 3d ed. Copyright © 1983 Wm. C. Brown Publishers, Dubuque, Iowa. All Rights Reserved. Reprinted by permission.

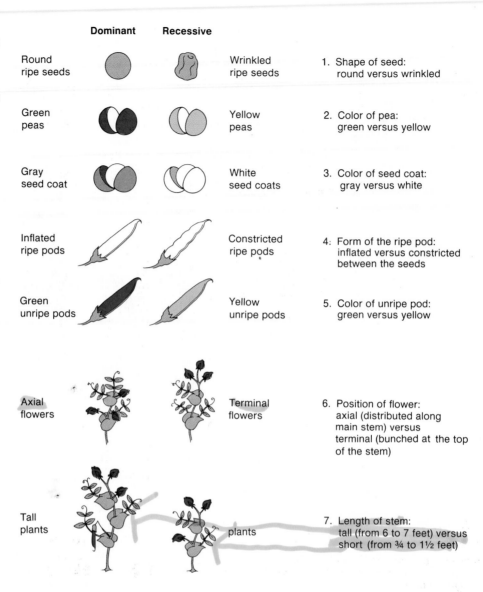

	Dominant	Recessive	
Round ripe seeds		Wrinkled ripe seeds	1. Shape of seed: round versus wrinkled
Green peas		Yellow peas	2. Color of pea: green versus yellow
Gray seed coat		White seed coats	3. Color of seed coat: gray versus white
Inflated ripe pods		Constricted ripe pods	4. Form of the ripe pod: inflated versus constricted between the seeds
Green unripe pods		Yellow unripe pods	5. Color of unripe pod: green versus yellow
Axial flowers		Terminal flowers	6. Position of flower: axial (distributed along main stem) versus terminal (bunched at the top of the stem)
Tall plants		plants	7. Length of stem: tall (from 6 to 7 feet) versus short (from ¾ to 1½ feet)

Mendel began his experiments with a set of **pure-breeding** pea plants. For each trait, he obtained seeds that always gave rise to plants with phenotypes identical to the parents. In other words, pure-breeding round seeds always grew into plants that produced more round seeds, and so on. Mendel then conducted a series of **monohybrid crosses.** For each of the seven characters he studied, he crossed pure-breeding parents of opposite phenotypes—the parental or P generation (figure 3.4). The experiments involved one trait at a time, hence were monohybrid. The descendants of the P generation are called the F1 or first filial (meaning child) generation. Descendants of the F1 generation are called the F2 generation, and so on.

For each of the seven monohybrid crosses (figure 3.3), similar results were obtained. All F1s in each cross resembled one parental type, and none resembled the other parental type. For example, all the offspring from the "round seed × wrinkled seed" cross possessed round seeds. In other words, one phenotype was not expressed in each of the F1 generations. The trait that was not expressed is thus **recessive** to the trait that was expressed, the **dominant** trait. This phenomenon is known as **Mendel's Law of Dominance.**

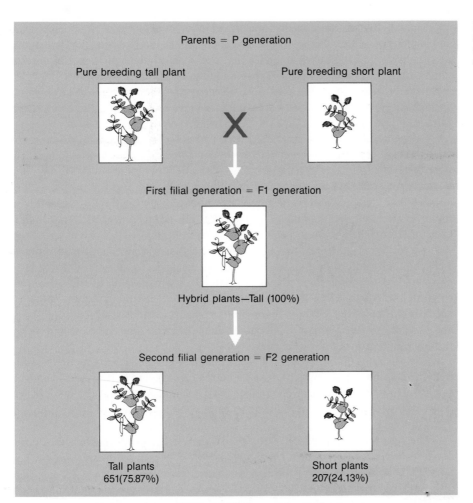

Figure 3.4 Results of one of Mendel's monohybrid crosses. Tall plants are dominant to short plants.

Parents = P generation

Pure breeding tall plant   Pure breeding short plant

X

First filial generation = F1 generation

Hybrid plants—Tall (100%)

Second filial generation = F2 generation

Tall plants
651(75.87%)

Short plants
207(24.13%)

Further, the sex of the parent had no influence over which phenotype was absent and which expressed, because reciprocal crosses produced the same results. A round-seed female × a wrinkled-seed male produced a round-seed F1, as did a wrinkled-seed female × a round-seed male. This result is called **Mendel's Law of Parental Equivalence.**

When the F1s were crossed with each other, however, an extraordinary thing happened: the phenotype absent in the F1 generation reappeared in the F2. Furthermore, it always occurred as about one-fourth of the total progeny. Thus, round-seed F1 plants, when bred with one another or self-fertilized, always gave rise to $\frac{3}{4}$ round and $\frac{1}{4}$ wrinkled F2s. It was impossible to predict with certainty the phenotype of a particular individual, but collectively the F2s always occurred in 3 round: 1 wrinkled ratio. Furthermore, the 3:1 ratio held true for all seven traits.

Mendel's results and modern concepts

Mendel explained his findings by postulating the existence of hereditary particles or factors (*elementen*) that are transmitted unchanged between generations. These factors control particular traits, but each factor is not necessarily expressed in each generation. A factor—what we call an *allele*—that is always expressed when present is called *dominant*, while one that can be suppressed or masked is called *recessive*.

Mendel's hypothesis is easily interpreted in terms of our understanding of DNA and homologous chromosomes introduced in chapter 2. Each phenotypic trait studied by Mendel is governed by two alleles that occur on a homologous pair of chromosomes. If the same allele is present on each homologous chromosome, the individual is said to be **homozygous** for that trait. Homozygous individuals—called *homozygotes*—are pure breeding for that trait, if they are self-fertilizing like pea plants. They can only transmit one type of allele to their offspring.

On the other hand, an individual with two different alleles of a gene is called **heterozygous.** Heterozygous individuals—called *heterozygotes*—transmit one of two different types of alleles to their offspring (one allele per gamete). All of Mendel's F1 progeny were heterozygous for a particular trait; and in every case, one allele was dominant to the other and was the only one expressed (figure 3.4). Nonetheless, the recessive allele retained its integrity. It was not changed in the F1 and could be transmitted by the F1 to succeeding generations.

We will use a shorthand to label alleles. The alleles at a gene *locus* (position on a chromosome) will be designated by a letter or a set of a few letters. The first letter of abbreviations of dominant alleles is capitalized, while recessive alleles are indicated in lower case. Thus, the allele for round seeds is represented by R, while r is used for wrinkled seeds. Where there are more than two alleles per locus, subscripts and superscripts are often used. Genetic nomenclature can be complex, and we will explain other variations as necessary.

The existence of two alleles at a locus means that three genotypes are possible: RR (the dominant homozygote), Rr (the heterozygote), and rr (the recessive homozygote). However, because of dominance both RR and Rr exhibit the same phenotype (producing round seeds). These two genotypes cannot be told apart simply by appearance.

The results of Mendel's F1 crosses are most clearly represented by a visual device called a **Punnett square**—named after the first person to use it (figure 3.5). Each parent possesses both R and r alleles, and each offspring receives one allele from each parent. Which allele is received from each parent is a **random** event. In this context, random means that the chance that the progeny will receive one allele is identical to the chance it will receive the alternate allele. Transmission of an allele is analogous to the toss of a coin—heads for R, tails for r (Appendix B).

Genotypes can be formed in four ways. An offspring receiving one R from each parent has the homozygous dominant genotype. If it receives one r from each, it is homozygous recessive. Finally, it could receive one R from the mother and one r from the father (Rr), or vice versa (rR), but in both cases it is heterozygous and always represented as Rr. Each of the four possibilities is equally likely, that is, each has a 25% chance of occurrence, because the chance of receiving R from each parent is the same as that of receiving r. However, since three of the four genotypic possibilities produce a round-seed phenotype, the offspring of a mating of two heterozygotes occur in a 3:1 phenotypic ratio—exactly Mendel's result.

Mendel's data were so important and so compelling precisely because he expressed his results in terms of ratios. His 3:1 ratio expresses a general phenomenon of the genetic mechanism, and this 3:1 ratio has been duplicated in many experiments by other scientists and by students.

Furthermore, the fact that the recessive phenotype reappears in the F2 after being absent in the F1 argues strongly for the integrity of hereditary particles. Alleles are transmitted between generations without being changed, even though they may not always be expressed. Thus, Mendel's results provided the first strong evidence for the existence of discrete hereditary particles, transmitted unchanged from generation to generation.

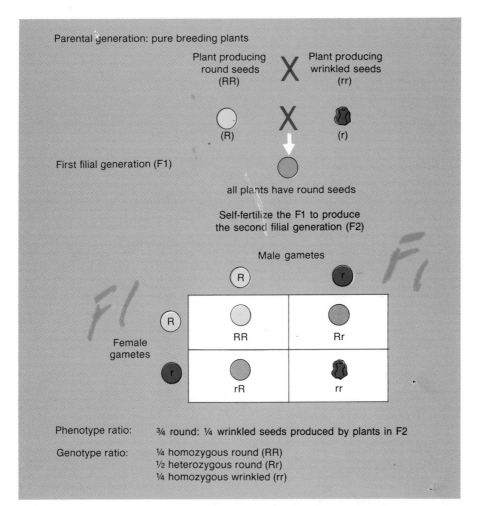

Parental generation: pure breeding plants

Plant producing round seeds (RR) X Plant producing wrinkled seeds (rr)

(R) X (r)

First filial generation (F1)

all plants have round seeds

Self-fertilize the F1 to produce the second filial generation (F2)

Male gametes

R   r

Female gametes

	R	r
R	RR	Rr
r	rR	rr

Phenotype ratio: ¾ round: ¼ wrinkled seeds produced by plants in F2

Genotype ratio: ¼ homozygous round (RR)
½ heterozygous round (Rr)
¼ homozygous wrinkled (rr)

Figure 3.5 A Punnett square, used to determine the types and proportions of offspring of a cross between two F1 hybrid plants. Each of the F1 parents produces round seeds, but are heterozygous for the recessive wrinkled trait. The two types of gametes produced by each parent combine at random, giving rise to three genotypes and two phenotypes in the F2 generation.

Mendel's hypothesis to explain his data fulfills a key requirement of a good scientific hypothesis: it makes predictions that can be tested by further experiments, thus helping to validate the hypothesis. For example, in the F2 generation, Mendel postulated the existence of three different genotypes, RR, Rr, and rr. Individuals with the wrinkled phenotype should be rr, and if permitted to self-fertilize they should produce only more wrinkled individuals, rr, in the F3 generation. In other words, recessive homozygotes should breed true.

On the other hand, round individuals should be either RR or Rr, occurring in a ratio of 1:2 (because heterozygotes may be either Rr or rR.) Thus, if permitted to self-fertilize, one-third of round-seed pea plants should breed true and produce only more round peas. The remaining two-thirds should produce both round and wrinkled progeny in a ratio of, again, 3:1 (figure 3.6). In Mendel's experiments using self-fertilizing F2s, he obtained just such results for all seven traits. Mendel's genetic model not only explained results of past experiments, but also predicted results of future experiments.

An additional experiment was performed by Mendel that verified the genotype of his F1 generation. Mendel predicted that all F1s were heterozygous. Therefore, the mating of the F1 hybrid with the recessive homozygous parental type should produce two types of progeny, Rr and rr (figure 3.7). No RR will occur, because the recessive homozygous parent can transmit only r alleles to the next generation.

Figure 3.6 The results of Mendel's Law of Segregation over many generations. Offspring of the F1 generation are either round or wrinkled. If self-fertilized, the homozygous wrinkled F2 plants will breed pure, producing only more wrinkled seeds in all succeeding generations. However, because the round-seeded F2 plants represent two differing genotypes, two differing results occur if these plants are self-fertilized. Roughly $\frac{1}{3}$ of the F2 plants with round seeds are homozygous and will breed pure, producing only more plants with round seeds. The other $\frac{2}{3}$ of F2 plants with round seeds are heterozygous and thus continue to produce progeny in a ratio of $\frac{3}{4}$ round and $\frac{1}{4}$ wrinkled, just like their F1 parents.

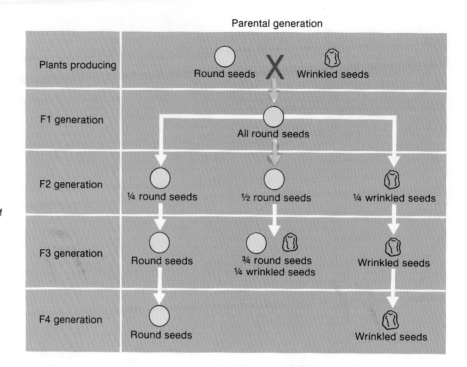

Figure 3.7 A *testcross*, a mating of a heterozygote and a recessive homozygote. A testcross produces heterozygous and recessive homozygous offspring in approximately equal proportions. No dominant homozygotes are produced.

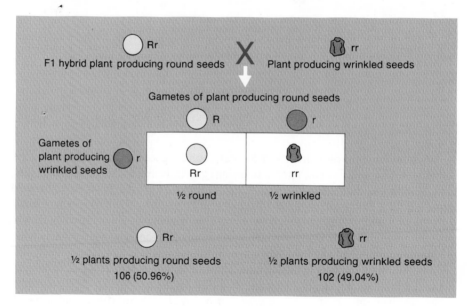

Furthermore, because the probability of a progeny receiving the R allele is equal to the probability of receiving the r allele, the two types of progeny should occur in approximately equal numbers.

This type of cross—mating a heterozygote to the recessive homozygote—is called a **testcross.** A testcross is a diagnostic test for the heterozygous genotype in the F1 generation. Mendel performed a series of testcrosses using the seven traits in peas. In every case, his data show the expected 1:1 ratio of the two phenotypes.

Mendel's Law of Segregation

The reappearance of the recessive phenotype in the F2 generation led to **Mendel's Law of Segregation.** This principle of inheritance states that each member of a pair of alleles governing a character maintains its own integrity, and that at reproduction only one allele of a pair is transmitted to each gamete. Normally, the allele does not change between generations. Both types of alleles are incorporated into gametes, but only one per gamete. Which allele occurs in a particular gamete is a random event (i.e., impossible to predict), although in general each of the two types occurs in 50% of the gametes. As we shall see, meiosis is the process governing segregation of alleles into gametes.

Mendel's Law of Independent Assortment

Each of the seven traits Mendel examined demonstrated a 3:1 phenotypic ratio in the F2 generation. As a logical development of these experiments, Mendel conducted a series of **dihybrid crosses;** that is, crosses between strains breeding true for two characters. As an example, consider the cross between seeds differing for both seed shape and seed color. In addition to seeds showing round or wrinkled phenotypes, Mendel also used seeds that were either yellow (allele Y) or green (allele y) in color, where Y is dominant to y.

In the P generation, the double dominant homozygote, RRYY, was crossed with double recessive homozygote, rryy. Because each F1 received one dominant and one recessive gene for each character, all F1s expressed a round/yellow phenotype (genotype RrYy). Mendel then permitted the doubly heterozygous F1s to self-fertilize and produce the F2 generation.

Examination of these data for one trait at a time shows that the expected monohybrid ratios still occur (figure 3.8). For seed shape, the data show 423 smooth: 133 wrinkled, and for color, 416 yellow: 140 green, both very close to 3:1. Thus, the Law of Segregation applies independently to each trait in the dihybrid cross. Next, if the same 556 seeds are classified by both shape *and* color simultaneously, four distinct phenotypes are possible: round/yellow, round/green, wrinkled/yellow, and wrinkled/green. Mendel observed 315 seeds of round/yellow phenotype, 108 round/green, 101 wrinkled/yellow, and 32 wrinkled/green. If the number in each category is divided by the number in the smallest category (32), the result approximates the expected 9:3:3:1 phenotypic ratio of a *dihybrid cross* (figure 3.8).

The simplest possible explanation of a dihybrid cross is that the two characters, seed shape and seed color, do not interact in the formation of gametes. Thus, each F1 produces equal proportions of four types of gametes for the pollen and eggs, RY, Ry, rY, and ry. A Punnett square (figure 3.8) shows that nine different genetic combinations may occur within the sixteen (4×4) possible ways of randomly combining the gametes. However, because of dominance, only four phenotypic classes are possible. These do not occur in equal proportions, but in a ratio of 9:3:3:1, almost exactly what Mendel observed.

This 9:3:3:1 dihybrid ratio can be obtained by using the laws of probability (Appendix B). For example, $\frac{3}{4}$ of all seeds are round, and $\frac{3}{4}$ of those are also yellow. Therefore, the proportion of seeds that are *both* round *and* yellow is $\frac{3}{4} \times \frac{3}{4}$, or $\frac{9}{16}$. The proportion that are round and green is $\frac{3}{4} \times \frac{1}{4}$, or $\frac{3}{16}$; wrinkled and yellow, $\frac{1}{4} \times \frac{3}{4}$, or $\frac{3}{16}$; and wrinkled and green, $\frac{1}{4} \times \frac{1}{4}$, or $\frac{1}{16}$—9:3:3:1. Thus, not only does the Law of Segregation apply to dihybrid crosses, but traits for characters controlled by different genes also assort during reproduction independently of each other. This phenomenon is now referred to as **Mendel's Law of Independent Assortment.**

Figure 3.8 Mendel's Law of Independent Assortment, shown by a dihybrid cross of genes controlling seed shape and seed color. The Punnett square shows the nine different genotypes and their proportions. Because of dominance, only four phenotypes occur. The table at the bottom shows Mendel's actual data.

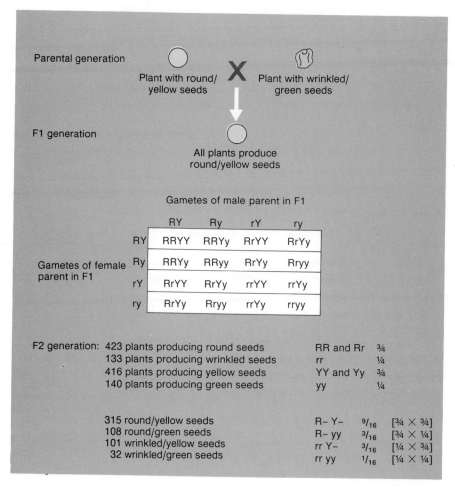

Parental generation

Plant with round/
yellow seeds

✕

Plant with wrinkled/
green seeds

F1 generation

All plants produce
round/yellow seeds

Gametes of male parent in F1

	RY	Ry	rY	ry
RY	RRYY	RRYy	RrYY	RrYy
Ry	RRYy	RRyy	RrYy	Rryy
rY	RrYY	RrYy	rrYY	rrYy
ry	RrYy	Rryy	rrYy	rryy

Gametes of female parent in F1

F2 generation:

423 plants producing round seeds	RR and Rr	¾	
133 plants producing wrinkled seeds	rr	¼	
416 plants producing yellow seeds	YY and Yy	¾	
140 plants producing green seeds	yy	¼	

315 round/yellow seeds	R– Y–	⁹/₁₆	[¾ × ¾]
108 round/green seeds	R– yy	³/₁₆	[¾ × ¼]
101 wrinkled/yellow seeds	rr Y–	³/₁₆	[¼ × ¾]
32 wrinkled/green seeds	rr yy	¹/₁₆	[¼ × ¼]

Mendel demonstrated the generality of the Law of Independent Assortment by conducting experiments involving *trihybrid crosses*—hybridizing strains breeding true for three separate traits. In this case, the triply heterozygous F1s produced eight different gametic types, twenty-seven different genotypes, and eight phenotypes. The eight phenotypes occurred in a 27:9:9:9:3:3:3:1 ratio, just as Mendel's laws predict.

The importance of Mendel's laws

Be sure that you understand the distinction between Mendel's Law of Segregation and Mendel's Law of Independent Assortment. These principles describe the ways in which genes and alleles occur in the gametes of an individual.

The Law of Segregation refers to the distribution of alleles of one pair of homologous genes among gametes. Because a single diploid individual has two alleles for one gene, a heterozygote produces two different types of gametes with respect to that gene, and those two types of gametes are produced in equal numbers. A homozygote produces only one type of gamete.

The Law of Independent Assortment refers to the distribution of non-homologous chromosomes to gametes (genes on the same chromosome may be an exception; see the following discussion of linkage). Genes on non-homologous chromosomes occur in gametes independently of each other. In the case of Mendel's experiment, knowledge that a gamete contains allele R gives no information as to whether that gamete contains a Y allele or a y allele. In other words, alleles on separate chromosomes occur in gametes (assort) randomly with respect to one another.

In combination, these two laws demonstrate how immense amounts of genetic variation are produced in every generation, simply by the random mixing of alleles of a single gene and the mixing of genes over the entire genome. Remember that humans have about one hundred thousand genes. The number of possible genetic combinations that may occur in a gamete from one individual is immense: if a person is heterozygous at only one gene on each pair of chromosomes, he or she will be capable of producing 2^{23} or over eight million different types of gametes. A new individual is produced, of course, by the union of two gametes from parents who probably possess a far greater amount of genetic variation than in this example, essentially guaranteeing that every human fertilization produces a unique combination of alleles. Thus, these laws explain much of the variation that is observed among individuals of a species.

Mendel's laws also show that a large component of randomness exists in the genetic mechanism. It is impossible to predict which combinations of alleles will occur in a particular gamete and, if fertilized, whether the combination of the two parental alleles will be viable. Clearly, some of the billions of possible combinations must be extremely vigorous and fertile, while some must be lethal.

Further considerations of Mendel's laws

Mendel's laws apply to all sexually reproducing diploid organisms. Many traits of many organisms have been examined for agreement with the predictions of these laws. In many cases, data have not agreed completely with the laws as Mendel stated them. Such cases have led to modification or elaboration of the principles as first stated by Mendel. Here, we mention some of those modifications, which will be discussed further in later chapters.

Multiple alleles and codominance

Mendel experimented with seven genes, each having two different alleles, one dominant to the other. Nature, of course, is not designed for the convenience of scientific experimentation, and it is hardly surprising that many variations on this pattern occur.

Although a diploid individual can have only two alleles of one particular gene, within a population, more than two alleles commonly occur at many loci. A particularly well-known example of **multiple allelism** is the ABO blood group locus of humans, which has four common alleles and several other extremely rare ones. The ABO locus will be discussed in detail in chapter 4. Many other genes in humans and other species are known to have multiple alleles (table 3.1), and some genes are known to have several dozen alleles.

With so many alleles commonly occurring, dominance relationships are also often variable. In many cases, both alleles are expressed in a heterozygote, and neither suppresses the other. In such a case, the alleles are said to be **codominant** with respect to each other. Codominance is a common type of expression of alleles at loci governing enzyme production. Some alleles at the ABO blood group locus exhibit codominance in their relationships to other alleles, while other alleles exhibit dominance or recessiveness.

Table 3.1
Table 3.1
Examples of allelic diversity in a few organisms

Organism	Trait	Locus	Number of Alleles
Human	ABO blood group	ABO	4 (common)
Human	HLA antigens*	HLA-A	23 (at least)
		HLA-B	47 (at least)
Fruit fly	Eye color	White	12
Cat	Coat color	C-full color	5
Rabbit	Coat color	Color	4
Tobacco	Self-sterility	Sterility	15
Snapdragon	Flower color	Solid color	9

*See chapter 4.

Quantitative inheritance

Prior to the discovery of the particulate nature of inheritance, one genetic theory was that of **blending inheritance.** This theory claimed that hereditary particles from both parents combine in the progeny. The particles then lose their integrity and are permanently changed. Evidence for blending came from quantitatively varying characters, such as height and weight, where offspring are often intermediate with respect to their parents. Such events may in fact be explained by variations on Mendelian principles.

One of the discoverers of Mendel's work, Carl Correns, conducted experiments similar to Mendel's, but examining flower color in four-o'clocks (figure 3.9). Upon hybridizing true-breeding red and white flowering parents, he discovered that all F1s had pink flowers. The F2 plants resulting from self-fertilizing the F1s produced red, pink, and white flowers in the ratio 1:2:1. The pink flowered plants were all heterozygotes. Thus, each genotype was represented by a unique phenotype, explaining the F2 ratio.

The pink phenotype resulted because the heterozygote did not express red pigment as strongly as did the homozygote. When the expression of a character is reduced in the heterozygote as compared to the homozygote, the alleles exhibit **partial** or **incomplete dominance.** Incomplete dominance thus produces F1 hybrids that appear intermediate to the parental phenotypes. However, the segregation of alleles in the F2 generation demonstrates clearly that the genes themselves remain unchanged between generations, retaining their integrity in accordance with Mendelian principles, but contrary to ideas of blending inheritance.

With incomplete dominance, phenotypes still may occur in discrete categories—red, pink, and white, for example. However, if the gene controlling flower color in four-o'clocks had many alleles, instead of just two as in Correns' experiment, and if each allele gave rise to differing shades of pink, then it is possible to imagine a population of different individuals exhibiting the entire spectrum of pink, grading imperceptibly from red to white.

Yet another genetic phenomenon also may explain continuous variation: many characters are controlled not by one but by many genes. This type of inheritance is called **polygenic inheritance.** If many different genes contribute to a character such as height, the value of that character will depend upon the way in which those genes interact. Their expressions may be cumulative, or they may interact in more complex ways. Most phenotypic characters are the result of the interactions of many, perhaps dozens to hundreds of different genes. The result of these interactions is the enormous amount of variation shown by most phenotypic traits (figure 3.10).

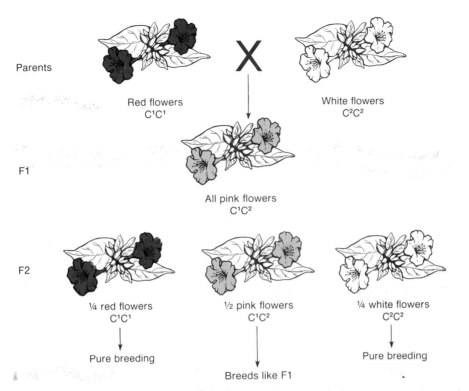

Parents

Red flowers
C^1C^1

White flowers
C^2C^2

F1

All pink flowers
C^1C^2

F2

¼ red flowers
C^1C^1

½ pink flowers
C^1C^2

¼ white flowers
C^2C^2

Pure breeding

Breeds like F1

Pure breeding

Figure 3.9 Incomplete or partial dominance among the alleles for color in four-o'clocks. Because each genotype is represented by a unique phenotype, Mendel's Law of Segregation produces a 1:2:1 ratio of phenotypes.

Number of individuals in each rank	1	0	0	1	5	7	7	22	25	26	27	17	11	17	4	4	1
Height in feet and inches	4:10	4:11	5:0	5:1	5:2	5:3	5:4	5:5	5:6	5:7	5:8	5:9	5:10	5:11	6:0	6:1	6:2

Figure 3.10 Frequency distribution of heights of male students at Connecticut Agricultural College. Source: Library of Congress.

Linkage: a modification of the Law of Independent Assortment

Genes do not exist independently in the nucleus. Genes occur sequentially on chromosomes, and most chromosomes have hundreds or thousands of genes. Genes located on the same chromosome are *linked* and are transmitted as a group to gametes. Linkage groups may be broken during meiosis, but the existence of **linkage** means that the Law of Independent Assortment does not always apply to transmission of linked genes.

Independent assortment occurs when genes are located on non-homologous chromosomes, or when genes on the same chromosome are relatively distant from each other. Genes close to one another on the same chromosome do *not* show the 9:3:3:1 ratio of the F2 generation of a dihybrid cross, because certain allelic combinations occur only rarely. Linkage always alters the phenotypic ratios from those expected if genes assort independently. Mendel was either exceptionally lucky or highly selective in the seven traits he reported examining in peas. Although several of the characters he reported were linked, they were far enough apart that recombination during meiosis caused them to assort independently. Therefore, Mendel obtained the appropriate ratios in all dihybrid and trihybrid crosses.

Mutations

Mendel's laws demonstrate the constancy of genetic factors between generations. Genes are transmitted from parent to progeny unaltered—most of the time. In fact, **mutations** or changes in the genetic material occur regularly. The probability of a particular gene changing in one cell cycle (meiosis or mitosis) is less than one in a million in most cases. However, because there are so many genes, and because cell division occurs continuously, mutation is a common occurrence. Many birth defects, for example, result from gene mutations.

Mutation was not a major concern for Mendel because it would have affected only a tiny fraction of the individuals and traits he examined. Thus mutation would not be expected to alter his ratios significantly. However, the phenomenon of mutation is of great importance to modern genetics, medicine, and agriculture. We discuss the causes of mutation in chapter 10.

Meiosis: the cellular basis of Mendel's laws

Shortly after the discovery and acceptance of Mendel's work in the early 1900s, it was noticed that the behavior of chromosomes during sexual reproduction corresponded to the behavior of the hereditary factors postulated by Mendel. In chapter 2 we showed how somatic cells divide mitotically to produce daughter cells that have a chromosome complement identical to the parent cell. The germinal cells residing in the gonads, however, divide to give a very different result.

Meiosis is a sequence of two divisions of a diploid germinal cell to produce haploid cells. The continued development and differentiation of the haploid cell to produce gametes is called **gametogenesis.**

Meiosis is the cellular basis for the process of sexual reproduction, and it has three crucial results. First, the number of chromosomes is reduced by one-half during meiosis from the diploid to the haploid state. This serves to maintain a constant number of chromosomes in all diploid members of a species. Because sexual reproduction involves the union of cells from two individuals, the absence of a reduction in chromosome number during the formation of gametes would produce individuals with immense numbers of chromosomes in only a few generations.

Second, halving of the chromosome number is not a random process. The reduction to the haploid state involves the orderly assortment of one member of each homologous pair into daughter cells. This insures that each gamete contains one of each type of chromosome and, hence, one copy of each gene. Without such balance, most gametes would not produce viable progeny. However, the separation of members of any homologous pair of chromosomes occurs randomly with respect to all other pairs. Thus, some chromosomes of maternal and some of paternal origin occur together in individual gametes. This phenomenon is the basis for independent assortment observed by Mendel and explains much of the genetic variation that occurs between individuals.

Third, during meiosis, chromosomes routinely exchange homologous segments by a process called **crossing-over,** causing genetic **recombination.** Crossing-over brings together entirely new combinations of alleles on a chromosome. The chromosomes that an individual passes on to its offspring via its gametes are almost never identical to the chromosomes received from either of its parents. The genes of each parent are recombined on chromosomes during meiosis to form a new combination transmitted to the next generation.

Meiosis occurs only during gametogenesis of sexually reproducing animals. In many respects it resembles mitosis, but in key aspects it is very different. It consists of two separate divisions, called meiosis I and meiosis II. Each division has the same four phases as occur in mitosis, but each is labelled with a Roman numeral to distinguish the phases of the two divisions—for example, prophase I, metaphase I, and so on. We present here an outline of meiosis as it occurs in a human male. The process we describe is a general model for meiosis in diploid organisms, although there are many variations on this pattern. Meiosis in the female is a slightly modified version of male meiosis, and we will discuss meiosis in females in chapter 6.

Meiosis I

Interphase prior to the first meiotic division is a time of vegetative growth and replication of the chromosomes, just as in mitosis. When the chromosomes first become visible at the beginning of meiosis, each consists of a pair of sister chromatids, held together by a common centromere. The first meiotic division separates members of each homologous pair into two daughter cells, but each chromosome remains in the duplicated condition. Recall that the number of chromosomes equals the number of centromeres. Because meiosis I reduces the number of centromeres in each of the daughter cells to the haploid number, it is considered a **reduction division.**

Prophase I (figures 3.11a and 3.12a)
The first phase of meiosis is the most complex of the entire cycle. Prophase I is a period of high chromosome activity, with several distinct segments. It begins with the appearance of the chromosomes as long, single filaments. Their duplicated condition is not yet obvious under the microscope.

Next, the members of a homologous pair line up tightly side by side, a process called **synapsis** (meaning juncture). Synapsis is a key feature that distinguishes meiosis from mitosis. The chromosomes progressively condense throughout prophase I, and as they do, a number of attachments between homologues become visible. Because of the X-shaped appearance of these attachments, they are called **chiasmata.** Formation of chiasmata is a crucial event, because they are the mechanism of crossing-over. At a chiasma, chromosomes break and then usually re-attach to another broken end. If a chromosome heals to its own broken end or that of its sister chromatid, it remains identical to its original condition. However, if it heals to the end of an homologous but *non-sister* chromatid, the result can be *genetic recombination* due to the *equal exchange of segments among homologous chromosomes.* Such recombination is called *crossing-over.* Usually, at least two and often more chiasmata form between homologues, indicating the exchange of many segments.

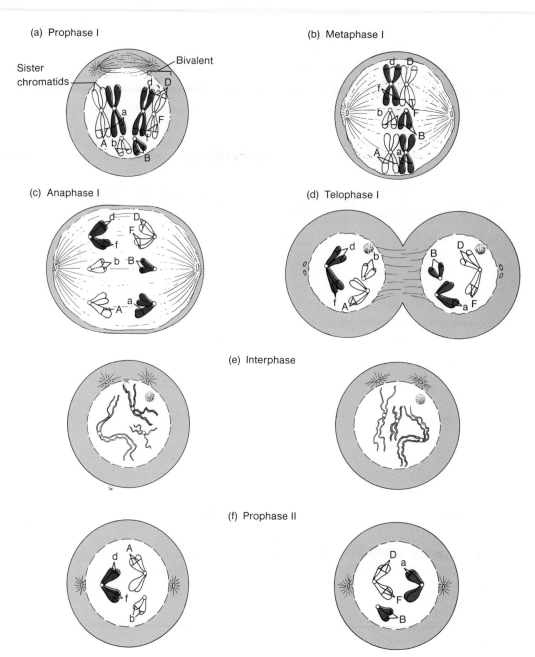

Figure 3.11 Meiosis in a male with 2N = 6: no crossover. (See colorplate 4.)

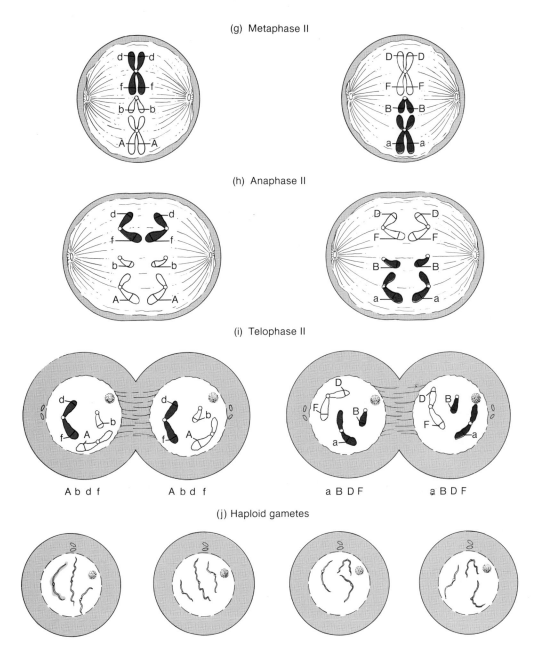

(g) Metaphase II

(h) Anaphase II

(i) Telophase II

A b d f   A b d f   a B D F   a B D F

(j) Haploid gametes

Figure 3.11 *(continued)* (See colorplate 4 continued.)

Figure 3.12 Meiosis in a male with 2N = 6: one crossover.

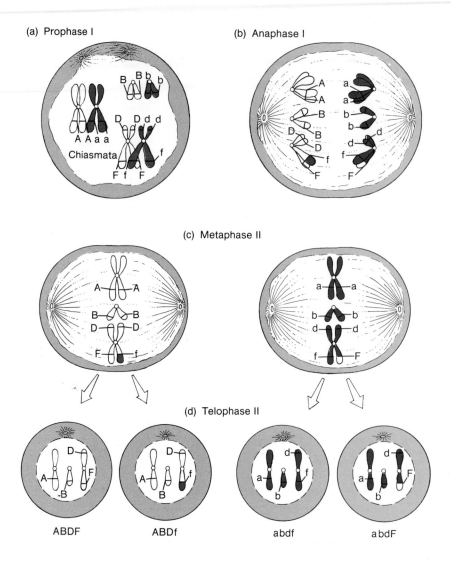

(a) Prophase I

(b) Anaphase I

(c) Metaphase II

(d) Telophase II

ABDF   ABDf   abdf   abdF

Because crossing-over changes the results of meiosis so significantly, we have diagrammed the process of meiosis both without (figures 3.11a-j) and with (figures 3.12a-d) crossovers. Through the remainder of the discussion of meiosis, compare the results of the two figures to appreciate fully the importance of crossing-over in generating genetic variation.

At about the time of chiasma formation and crossover, the existence of sister chromatids becomes apparent. Each pair of homologues is seen to possess four chromatids, and together the duplicated pair is called a **bivalent.** Prophase I ends when the nuclear membrane disappears, a spindle begins to form, and the centromere of each bivalent attaches to the spindle.

Metaphase I (figure 3.11b)
At this stage, the chromosomes are highly condensed. Spindle formation is complete, and the bivalents line up along the center of the spindle at the metaphase plate. Homologues appear to repel each other and the chiasmata are now at terminal positions for each bivalent. For example, this would leave two chiasmata on a metacentric chromosome bivalent, one at each end. An acrocentric chromosome bivalent would have only a single chiasma remaining. Homologues are thus held together only by their chiasmata.

Anaphase I (figures 3.11c and 3.12b)

Pulled by their centromeres, members of each pair of homologues separate and move to opposite poles of the spindle. In contrast to mitotic anaphase, centromeres do not divide in meiotic anaphase I. Thus, the two daughter groups now consist of complete haploid sets of duplicated chromosomes. Because the number of centromeres is halved, the first meiotic division is a reduction division. A further contrast to mitosis is apparent when considering the condition of the two sister chromatids in each duplicated chromosome. Because of crossing-over, these two sister chromatids may no longer be identical (figure 3.12b).

Telophase I (figure 3.11d)

As in mitosis, the cytoplasm and its constituents are also divided between the new cells. In many species, a nuclear membrane forms within each daughter cell.

Interphase (figure 3.11e)

Usually, interphase is only a brief pause between the two meiotic divisions. In some species, both telophase I and interphase may be skipped entirely, and meiosis II ensues directly from anaphase I. There is never any replication of chromosomes during this interval.

Meiosis II

Because the genetic material was duplicated once prior to initiation of meiosis, there must be two divisions to achieve haploid gametes whose chromosomes are not duplicated. The second division involves division of the centromeres and the separation of the two sister chromatids. Bear in mind, as pointed out above, that the sister chromatids may not be identical at this point, if crossing-over has occurred. Meiosis II is less complex than meiosis I and similar in many respects to mitosis.

Prophase II (figure 3.11f)

The nuclear membrane (if present) disappears. The chromosomes, which had become relatively diffuse late in meiosis I, once again begin condensing. Unlike prophase I, there is no pairing of chromosomes because the duplicated chromosomes have no homologous partners with which to pair. A spindle begins to form.

Metaphase II (figures 3.11g and 3.12c)

The still duplicated chromosomes move to the equator of the spindle, each attached by its single centromere.

Anaphase II (figure 3.11h)

The centromere divides, and now two new and single chromosomes exist. Each new set of haploid daughter chromosomes moves toward opposite poles of the spindle.

Telophase II (figures 3.11i and 3.12d)

A nuclear membrane forms around each daughter set, and the chromosomes once again become diffuse. The contents of the cytoplasm are divided, and meiosis is complete (figure 3.11i).

The results of meiosis

Mitosis involves one division that produces two daughter cells, each chromosomally identical to the parent. In meiosis, each parent cell undergoes two divisions to produce four daughter cells in males (one functional gamete in females), each containing a potentially unique haploid set. Because of independent assortment of chromosomes in anaphase I, the number of potential combinations of chromosomes in a gamete depends upon the number of pairs of chromosomes, the haploid number, n. The number of different possible combinations is 2^n. For humans, that number is 2^{23}, or 8,388,608.

Even that amount of variation is small when compared to that due to crossing-over. While independent assortment produces different combinations of chromosomes, crossing-over changes the allelic combinations that occur within each chromosome. These two factors thus assure an immense amount of variation within the gametes of each individual and between individuals. With the single exception of monozygotic twins (discussed in chapter 4), the meiotic mechanism virtually guarantees that no two humans will ever have the same genetic constitution. A geneticist views each human as a unique result of the interaction of a unique set of environmental influences and a unique combination of alleles and chromosomes. Each of us is a genetic experiment that will be tried only once.

Meiosis and Mendel's laws

Chromosome behavior during meiosis is the cellular basis for the Law of Segregation and the Law of Independent Assortment. Mendel's hereditary factors, genes, are located on chromosomes. Each gamete randomly receives one member of each homologous pair of chromosomes, ensuring segregation of alleles and also providing for independent assortment of unlinked genes. Crossing-over provides a measure of independent assortment even for linked genes, although *how much* independence is related to the proximity of the linked genes on a chromosome (chapter 9).

In retrospect, it is extraordinary indeed that Mendel discovered the Law of Segregation and the Law of Independent Assortment with no knowledge of the cellular mechanism underlying them. The combination of Mendel's inheritance experiments and cytological examination of meiosis provided the evidence that quickly led to acceptance of these laws as the foundation of modern genetic theory. Thus, modern genetics dates from the first years of the twentieth century, when Mendel's work was "discovered" and linked to the newly understood behavior of chromosomes.

Summary

1. In 1865 Mendel described the basic laws of particulate inheritance in diploids. His laws combined with the cytology of meiosis are the foundation of modern genetics.
2. The phenotype is the expression of the combined effects of the genotype and environmental influences.
3. All phenotypic characters exhibit either quantitative or qualitative variations. Mendel studied qualitatively variable characters in the garden pea.
4. Mendel's Law of Dominance describes the interaction of alleles at a locus. Dominant alleles are always expressed, while recessive alleles are expressed only when homozygous.
5. Mendel's Law of Parental Equivalence states that the results of a cross do not depend upon which parent has the genetic characters or alleles being studied.
6. Mendel's Law of Segregation states that genes governing a phenotypic character occur in pairs and that one member of each pair is transmitted to each progeny. The two alleles that each individual has are transmitted to the next generation in approximately equal proportions.
7. Mendel's Law of Independent Assortment states that unlinked genes are distributed among gametes independently with respect to each other. This law does not apply to genes that are closely linked.

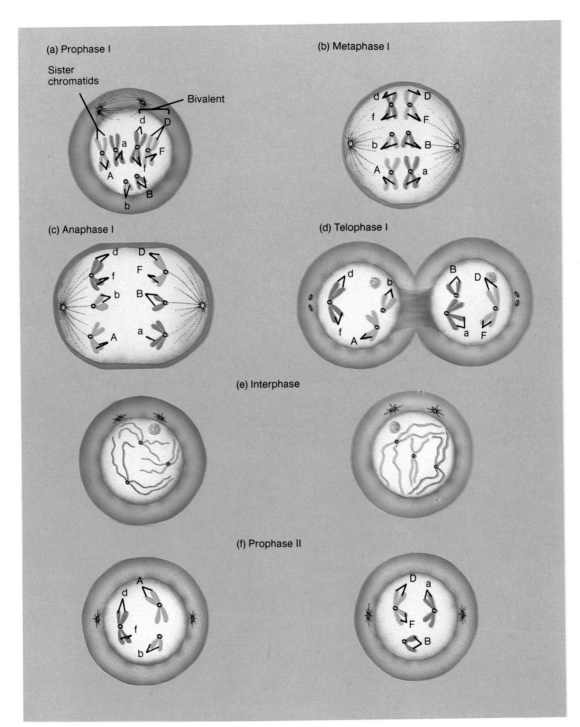

Plate 4 Meiosis in a male with 2N = 6: no crossover.

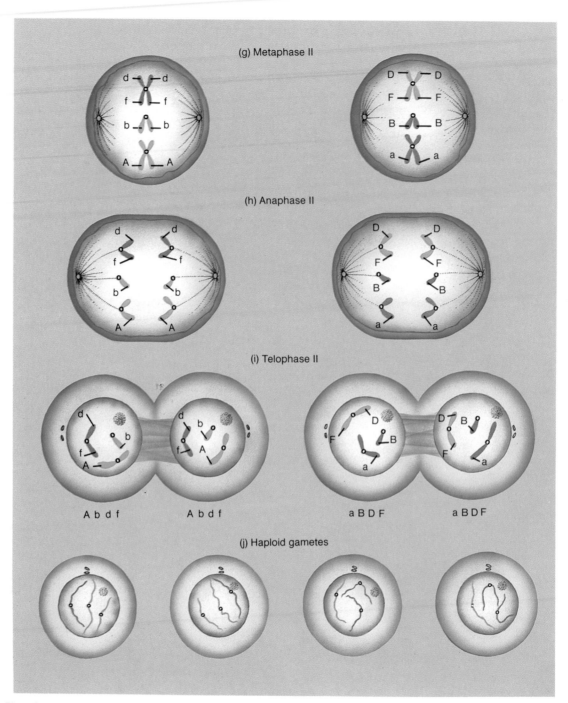

(g) Metaphase II

(h) Anaphase II

(i) Telophase II

A b d f   A b d f   a B D F   a B D F

(j) Haploid gametes

Plate 4 *(continued)*

8. The primary results of meiosis are the production of haploid daughter cells that will become gametes and the generation of huge amounts of genetic variation.
9. Partial dominance and polygenic inheritance are mechanisms that can account for quantitative variation.
10. Linkage alters the gametic ratios obtained in crosses involving more than one character.
11. Meiosis is the cytological process that provides the physical basis of Mendel's laws. Meiosis consists of one replication of the chromosomes, followed by two divisions.
12. Independent assortment of chromosomes during meiosis I distributes random mixtures of parental chromosomes, but exactly one of each homologous pair, to daughter cells. Crossing-over produces new combinations of alleles within one chromosome.

Key Words and Phrases

bivalent	linkage	polygenic inheritance
blending inheritance	meiosis	Punnett square
chiasma	Mendel's Law of Dominance	pure breeding
codominance		qualitative variation
crossing-over	Mendel's Law of Independent Assortment	quantitative variation
dihybrid cross	Mendel's Law of Parental Equivalence	random
dominant		recessive
environment	Mendel's Law of Segregation	reciprocal cross
gametogenesis		recombination
genotype	monohybrid cross	reduction division
heterozygous	multiple alleles	synapsis
homozygous	mutation	testcross
incomplete dominance	partial dominance	
independent assortment	phenotype	

Questions

1. Assume that in Asian yaks a gene on chromosome one controls appearance. At that locus, allele C for cuteness is dominant to c for ugliness. On a second chromosome is the locus controlling hair color: allele P for pink hair is dominant to allele p for blond hair. With respect to these two characters answer the following:
   a. If an individual possesses the genotype CcPP, how many different types of gametes can it produce? What are they?
   b. What are the possible genotypes of offspring of a mating between this individual (genotype CcPP) and an individual of genotype CcPp? ccpp?
   c. What phenotypes will result from these different matings?
2. About 70% of Americans get a bitter taste from the drug phenylthiocarbamide (PTC). The ability to taste this compound is dominant to the inability to taste it, and this is an autosomal trait. Albinism is an autosomal recessive trait. A normally pigmented woman who is a nontaster has a father who is an albino taster. She marries an albino taster whose

mother is an albino nontaster. Assuming that the genetic factors for albinism and the ability to taste PTC assort independently, what types of offspring can this couple produce, and what is the probability of each type?

3. An individual possesses the genotype AaBbCcDdEE. All the genes are unlinked except DdEE, which are so closely linked that no crossover occurs. How many types of different gametes can this individual produce? What if there was no linkage at all?

4. How many ways are discussed in this chapter that can cause genetic variation? What are they?

5. Name some phenotypic characters of humans that are obviously caused by the effects of both genes and the environment. What are those environmental factors?

6. In addition to the ABO blood group, many others have been identified in humans. One such group is the MN group, controlled by two codominant alleles, M and N, at one locus. What genotypes and phenotypes would be produced in crosses involving the following phenotypes: M $\times$ N; M $\times$ MN; N $\times$ N; and MN $\times$ MN?

7. Could a child of type N result from the mating of M $\times$ MN?

8. In a cell with 2N = 14, how many chromatids are present at meiotic prophase I? At meiotic prophase II? At mitotic metaphase?

9. In a cell with 2N = 14, how many chromosomes are present during meiotic anaphase I? During meiotic metaphase II? During mitotic anaphase?

10. Re-examine figure 3.6. If all offspring in the F3 generation are only allowed to self-fertilize for ten more generations, will the proportion of heterozygotes in the experimental population increase, decrease, or remain the same in succeeding generations?

For More Information

Dunn, L. C. 1965. *A Short History of Genetics.* McGraw-Hill, New York.

"Gregor Mendel's Autobiography." Reprinted in English in *The Journal of Heredity,* 1954. Volume 45, pp. 231–34.

Klug, W. S., and M. R. Cummings. 1986. *Concepts of Genetics,* second edition. Scott, Foresman & Company, Glenview, Illinois.

Novitski, E. 1982. "The Formation of the Germ Cells." *Human Genetics,* second edition (chapter 3). Macmillan Publishing Co., New York.

Mendelian genetics of humans

Figure 4.1 Some Mendelian traits in humans: (a) polydactyly (extra digits on hands and feet); (b) hereditary ptosis (drooping eyelid, due to muscle degeneration); (c) progeria (premature aging and greatly reduced life expectancy); (d) Ellis van Creveld syndrome (a type of dwarfism common in the Pennsylvania Amish, resulting in shortened limbs, malformed hearts, and six fingers on each hand). Source: (a) Lester V. Bergman and Associates, Inc.; (b) Ira A. Abrahamson, Jr., M. D., "Opthamologic Manifestations of Medical and Neurologic Diseases," 1972, by Medcom, Inc. Reprinted by permission of Medcom, Inc.; (c and d) Victor McKusick.

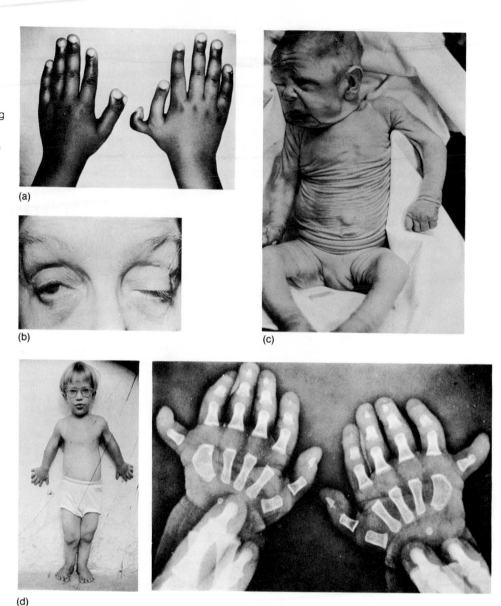

(a)

(b)

(c)

(d)

Mendel analyzed the proportions of offspring of various types of matings to deduce fundamental laws of genetics. He chose a species that produced many offspring, that had a short generation time, and whose matings he could carefully control. Most genetic experimentation is still conducted on organisms meeting these criteria.

Pedigree analysis

Many examples of Mendelian inheritance have now been discovered in humans (figure 4.1), but not by using the techniques of Mendel. In fact, it is hard to imagine a species more difficult to examine experimentally than humans. Humans produce a relatively small number of offspring, they have one of the longest generation times of any species, and religious and ethical considerations prohibit controlled matings. Therefore, discovery of traits exhibiting Mendelian patterns of inheritance has depended upon a methodology not used by Mendel—*pedigree analysis*.

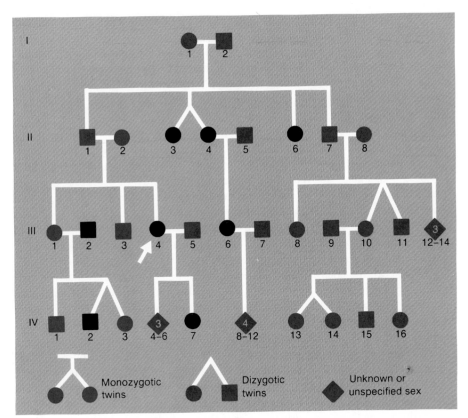

Figure 4.2 An example of the more common symbols used in human pedigree analysis. Hypothetical case of a trait caused by an autosomal recessive gene. The index case, III-4, is indicated by an arrow. Horizontal lines connect reproductive partners and vertical lines connect offspring of a mating. Generations are indicated by Roman numerals and individuals are given Arabic numbers. Circles represent females, squares represent males, and diamonds represent unknown or unspecified sex. When persons do not need to be indicated individually, numbers are inserted into the symbol, as in generation III, numbers 12–14.

A **pedigree** is a diagrammatic representation of a genealogy, a set of familial relationships. The word *pedigree* is derived from the Latin words *pes* and *grus*, meaning literally "foot of the crane," a reference to the lines in the pedigree specifying particular relationships. Figure 4.2 shows a hypothetical pedigree that illustrates the shorthand symbolism used by geneticists.

Pedigrees are usually constructed with reference to a particular phenotypic (potentially genetic) trait. Squares represent males, circles represent females, and diamonds represent offspring of unknown or unspecified sex. The symbols are solid for affected individuals, open for individuals who are normal (for a particular trait). The affected individual who first called attention to the pedigree is known as the *proband* or **index case** and is indicated by an arrow. Mated pairs are directly connected by a horizontal line. A vertical line descending from the marriage line connects the progeny (brothers and sisters, *siblings,* or members of a *sibship*) to the reproductive couple. Offspring are listed in order of birth from left to right. Each individual within a pedigree is identified with a unique number: a Roman numeral indicates the generation (row) and an Arabic numeral indicates the position in the row. In figure 4.2, for example, the lineage begins with female I-1, who is married to male I-2. Their offspring are II-1, II-3, II-4, II-6, and II-7.

Relatedness

Individuals who share genes derived from a common ancestor are said to be related, or **consanguineous** (meaning of the same blood). In an evolutionary sense, all humans are descended from a few common ancestors that gave rise to our species. In a genetic sense, however, consanguinity, or relatedness, usually refers to the presence in two or more individuals of particular alleles that are *identical by descent*, that is, identical copies of a gene or genes from a recent common ancestor. Geneticists usually trace genetic identity no further than four or five generations.

The proportion of alleles that are identical by descent is a measure of relatedness between individuals, called the **coefficient of relationship,** r. Recall that as a result of meiosis, each parent transmits $\frac{1}{2}$ of their genes (and $\frac{1}{2}$ of their chromosomes) to each offspring. The proportion of shared alleles between parent and offspring, r, is thus $\frac{1}{2}$ (figure 4.3a). (Unless there is reason to believe otherwise, all matings with which a pedigree begins are believed to be random, thus having an initial relatedness of 0.)

Full siblings, sibs with the same parents, normally share some alleles derived from the mother, and some derived from the father. On average, each should share $\frac{1}{2}$ of the genes received from each parent. Thus, a pair of siblings share $\frac{1}{2}$ of the $\frac{1}{2}$—or $\frac{1}{4}$—of alleles derived from the mother, and similarly $\frac{1}{4}$ of the alleles derived from the father. Their total relatedness is the sum of the probability that they share alleles from their mother ($\frac{1}{4}$) and the probability that they share alleles from their father ($\frac{1}{4}$) or $\frac{1}{4} + \frac{1}{4} = \frac{1}{2}$ (figure 4.3b). While the relatedness of parent to offspring is the same as the relatedness of sibling to sibling, note that a different set of alleles is shared in each case (figures 4.3a and 4.3b).

In general, the coefficient of relationship can be determined by analysis of the pathway of relationships shown in a pedigree. Because the degree of relatedness decreases by $\frac{1}{2}$ with each step (usually a generation) in a pedigree, the probability of a particular allele being shared between two individuals is $\frac{1}{2}^n$, where n is the number of steps in the pathway of relationships connecting two individuals.

Where the relationship in question involves only one ancestor and a direct descendant—for example, parent-child (figure 4.3a)—only one pathway of relatedness exists for sharing alleles, and the degree of relationship is simply $\frac{1}{2}^1 = \frac{1}{2}$ for that single pathway. The relationship of half-siblings (figure 4.3c) also involves only one pathway, in this case with two steps; the relatedness of half-siblings is thus $\frac{1}{2}^2 = \frac{1}{4}$. For many other relationships, genes are shared through two or more ancestors, as in the case of full siblings (note there are two pathways of two steps each in figure 4.3b). As a result of *inbreeding*—the mating of related individuals—it is possible for relatives to share alleles through three or more pathways of relationship. The calculation of relatedness thus involves the sum of relationships along all possible pathways of shared descent.

As indicated, the relatedness of an individual to almost all others is 0. Genetically identical individuals have a relatedness of 1. Among humans, only **monozygotic** (or *identical*) **twins** are genetically identical. They arise as the result of a single zygote splitting very early during development, giving rise to two individuals of identical genotype. The phenotypic similarities of monozygotic twins are extraordinary, but it should be remembered that they are usually raised in similar environments (including the same womb for the first nine months of life), so a part of the phenotypic similarity is due to environmental factors. Monozygotic twins are always of the same sex.

Nine-banded armadillos are extraordinary because they always give birth to identical quadruplets, the result of two sequential fissions of one zygote. The genetic identity of the offspring have made this species a valuable laboratory organism for genetic and medical studies. Among invertebrates and plants, many species reproduce asexually, often producing thousands or millions of genetically identical copies—**clones**—of themselves.

Twins may also arise as a result of the fertilization of two separate eggs by two separate sperm. These are **dizygotic** (or *fraternal*) **twins.** Dizygotic twins are no more closely related genetically than are normal full siblings—$r = \frac{1}{2}$.

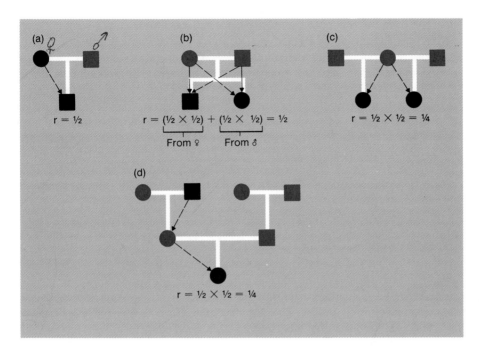

Figure 4.3 Determination of coefficients of relationship (r) for (a) parent-child, (b) full siblings, (c) half siblings, and (d) grandparent-grandchild. Note that the coefficient decreases by 1/2 for each step in the chain of relationship, and that, as in (b), two pathways of relatedness may connect one individual. (See colorplate 5.)

(a) $r = \frac{1}{2}$

(b) $r = (\frac{1}{2} \times \frac{1}{2}) + (\frac{1}{2} \times \frac{1}{2}) = \frac{1}{2}$ From ♀   From ♂

(c) $r = \frac{1}{2} \times \frac{1}{2} = \frac{1}{4}$

(d) $r = \frac{1}{2} \times \frac{1}{2} = \frac{1}{4}$

Consanguineous matings

Most *deleterious* (harmful) recessive genes occur at a low frequency among humans. Because the expression of recessive alleles requires that they occur in a homozygous condition, relatively few individuals expressing the trait actually occur. For example, consider a locus with a normal allele D and a recessive allele d (which is lethal when homozygous). If one in one hundred alleles in the population is d (an allele frequency of 1%), then one in every fifty persons is a heterozygous **carrier** of this allele (because each person in the population has two alleles of all autosomal genes). The probability of two carriers randomly mating is $\frac{1}{50} \times \frac{1}{50} = \frac{1}{2,500}$. The probability of one of their offspring being homozygous recessive is $\frac{1}{4}$, and therefore the probability of an affected offspring from a random mating is $\frac{1}{2,500} \times \frac{1}{4} = \frac{1}{10,000}$.

Another way of calculating this probability is to consider a random mating and the probability of a zygote receiving two d alleles, one from each parent. The probability of any parent contributing the allele to a gamete is the frequency of the allele in the population, or $\frac{1}{100}$. Thus, the probability of a random zygote receiving two d alleles is $\frac{1}{100} \times \frac{1}{100} = \frac{1}{10,000}$. Recessive alleles in low frequency are expressed only rarely. They are maintained in the population almost exclusively in the heterozygous condition.

The key concept in the example above is *random mating*. If mating is random, and if the allele occurs in low frequency, the probability of one carrier mating with another carrier is about twice the frequency of the deleterious allele in the population. However, under some types of non-random mating, especially **consanguineous matings,** the probability of one carrier encountering another is much higher.

Descendants of an individual carrying a particular allele have a much higher probability of carrying that same allele than the population at large. Matings between close relatives greatly increase the chance of both parents having identical alleles, thereby increasing the probability of progeny carrying recessive alleles in the homozygous condition, and in turn expressing deleterious traits.

The harmful effects of consanguineous matings (inbreeding) are well known and are called **inbreeding depression.** The effect of inbreeding depression increases with the relatedness of the mated individuals. On the average, full sibling matings are far more likely to produce negatively affected progeny than random matings.

It is believed that, on average, each of us carries four or five deleterious alleles in the heterozygous condition. Inbreeding greatly increases the probability that one of these will be expressed in our descendants. In a few societies, brother-sister and father-daughter matings have been practiced, but in most societies an **incest taboo** is observed (tacit recognition of the harmful effects of inbreeding). In the United States, most states restrict the degree of consanguinity permitted in marriages. These prohibitions are generally based on religious custom, not adherence to genetic information.

Autosomal recessive inheritance

At least 610 traits in humans have been positively identified as the result of homozygous recessive alleles on the autosomes (table 4.1). Another 810 traits are strongly suspected to be due to homozygous recessive alleles. Most, but not all, of these cause serious phenotypic conditions. They are now understood as Mendelian genetic variants because they differ qualitatively from the normal condition, just as did the traits of peas that Mendel examined. Certainly, many more or less normal conditions are also governed by Mendelian variants, but they are less clearly phenotypically defined, and therefore less easily studied. Many other traits (such as height and weight) are governed by the interaction of many alleles and the environment and thus are extremely difficult, if not impossible, to subject to normal Mendelian analysis.

Enzymes, one of the many types of proteins produced by genes, are important as catalysts in the thousands of biochemical reactions continually occurring in the body. These reactions proceed in highly controlled sequences called *metabolic pathways,* whose functions are the production of crucial chemicals such as pigments, the digestion of our foods, and the breakdown of various body waste products. In the homozygous state, many recessive alleles result in a defective or absent enzyme in a metabolic pathway, often with serious physiological consequences. The inability to produce an important enzyme, thus interrupting a metabolic pathway, is called an **inborn error of metabolism.** Many of the best known genetic diseases are inborn errors of metabolism. Alkaptonuria was the first such disease to be described in the early 1900s (table 4.1).

Pedigree examination can reveal the mode of inheritance of a particular trait. Begin by checking the parents. If both are normal, yet they have an affected child, the trait is probably recessive. Similarly, if one parent is affected, but all offspring are normal, the trait is again recessive (see figures 4.2, 4.4, and 4.5). Dominant traits, on the other hand, are expressed in all individuals who inherit the allele, and dominant traits do *not* skip generations.

Albinism

Albinism is an hereditary disorder involving an absence or reduction of melanin, the brown pigment responsible for the coloration of our skin, hair, and eyes. Albinism is one of the most common inborn errors of metabolism, having been observed in plants, insects, and all major groups of vertebrates. One of the first genetic traits identified as an autosomal recessive (box 4.1), it occurs in all major races of humans. The word *albino* is derived from the Latin word *alba,* meaning white. Due to the absence of pigment, albinos are a ghostly white in color. In some species, the eyes of albinos may appear pink from the color of *hemoglobin* (a pigment in red blood cells) visible through the unpigmented eyes.

Table 4.1
Some autosomal recessive genetic diseases identified in humans

albinism Absence of the pigment melanin (figure 4.5).

alkaptonuria A metabolic defect characterized by black urine due to the absence of a key enzyme in a metabolic pathway.

amyotrophic lateral sclerosis Spasticity and muscle atrophy; Lou Gehrig's disease.

atrichia Nearly complete absence of hair; individuals are born with hair that falls out and is not replaced.

Bloom's syndrome Dwarfism with skin changes; characteristic rash from exposure to sun; increased susceptibility to death from cancer.

Chediak-Higashi syndrome Partial albinism associated with death by age seven.

total color blindness Day blindness; cones in the retina of the eye are missing. Persons affected can see better at night than in the day.

congenital heart block Lethal heart defect leading to neonatal death.

cystic fibrosis Exocrine gland malfunction.

deafmutism Deaf from birth.

Ellis-Van Creveld syndrome Dwarfism associated with polydactyly, partial harelip, and cardiac malfunction (figure 4.1d).

fructose intolerance Inability to metabolize fruit sugar, fructose.

galactosemia Inability to metabolize galactose, with mental retardation and blindness resulting from buildup of galactose.

hypotrichosis Failure to replace intrauterine hair that is shed shortly before birth or after birth. No pubic or axillary hair develops at puberty.

lactose intolerance Individuals unable to metabolize milk sugar, lactose.

mast syndrome Presenile dementia with onset occurring in the late teens or early twenties.

Niemann-Pick disease Accumulation of lipids, leading to retarded physical and mental growth. Death usually occurs by age three.

phenylketonuria Defect in phenylalanine metabolism leading to mental retardation.

progeria Precocious aging and senility. Death often results from coronary artery disease by age ten (figures 4.1c and 10.14).

Tay-Sachs disease Onset of developmental retardation in infancy, followed by paralysis, dementia, blindness, and death by age three.

Wilson's disease Defect in copper metabolism leading to cirrhosis of the liver.

xeroderma pigmentosum Sun sensitivity leading to skin tumors and early death. Patients are unable to repair ultraviolet light damage to DNA (figure 10.13).

From V. A. McKusick, *Mendelian Inheritance in Man,* 7th ed. Copyright 1986 Johns Hopkins University Press. Reprinted by permission.

The production of melanin requires the enzyme tyrosinase. At least three different genes are known that can interfere with the production of tyrosinase. The most common form of albinism, caused by a recessive allele at one genetic locus, affects one in thirty-three thousand whites in the United States and one in twenty-eight thousand blacks. Different populations and racial groups are often characterized by differing frequencies of many alleles.

Melanin serves as a protective screen against the harmful effects of sunlight in normally pigmented individuals. Lacking pigmentation, albinos are highly susceptible to sunburn and skin cancer. Albinism leaves the retina of the eye unprotected by pigment, so visual problems and blindness are also common. About 9% of the visually handicapped students in the United States are albinos.

In addition to classic (tyrosinase negative) albinism, there are several types of albinism where small amounts of tyrosinase are present. These tyrosinase positive albinisms are also inherited as autosomal recessive traits. Tyrosinase positive albinos

BOX 4.1

Discovery of albinism as an autosomal recessive trait in humans

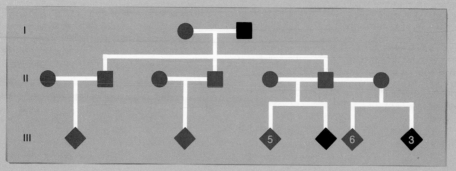

Figure 4.4 Pedigree of albinism in a black family in Mississippi.

By 1903 breeding tests had established that albinism was an autosomal recessive trait in mice, rabbits, and guinea pigs. In that year, W. E. Castle, a geneticist at Harvard, analyzed the pedigree of a black family in Mississippi that demonstrated that albinism was almost certainly an autosomal recessive trait in humans, also (figure 4.4). Castle subsequently hypothesized that all mammals have a similar inheritance pattern for albinism.

The grandfather in this family was an albino who married a normal woman. They had three sons, all normal—recessive conditions often "disappear" for one or more generations. The first two sons had only normal children. The third son married twice, having one albino and five normal children by the first wife, and three albino and six

normal children by the second. How do these data fit a Mendelian pattern?

All the children of the albino grandfather were heterozygous carriers for the albino allele. The wives of the first two sons were probably homozygous normal, resulting in all normal children for them. However, both wives of the third son were carriers. (Castle's paper notes that albinos were not uncommon in the area.) The Mendelian prediction for the mating of two heterozygotes is that $\frac{1}{4}$ of the offspring will be homozygous recessive. Because $\frac{4}{15}$ (as close to $\frac{4}{16}$ as possible) of the offspring of the third son were albino, Castle reasonably concluded that albinism is inherited as an autosomal recessive trait in humans.

have white hair in infancy, but with age the hair turns yellow, light brown, or red. This is the most frequent type of albinism in American blacks (one in fifteen thousand), and in some Native American tribes it occurs at truly extraordinary frequencies (box 4.2). Tyrosinase positive albinism in whites occurs at about the same frequency as tyrosinase negative albinism. The two forms are not allelic—they are controlled by separate genes.

Phenylketonuria

Phenylalanine is an amino acid necessary for human nutrition. The proper metabolism of phenylalanine requires the enzyme phenylalanine hydroxylase, whose production is genetically controlled. In the United States about one in fifteen thousand whites is born unable to produce this enzyme due to a homozygous recessive allele, causing **phenylketonuria** (PKU). In Ireland and Scotland, PKU is more common, occurring once in every five thousand to six thousand births. Among blacks, PKU is extremely rare, affecting one in every three hundred thousand births.

BOX 4.2

Natural selection sometimes favors
the allele for albinism

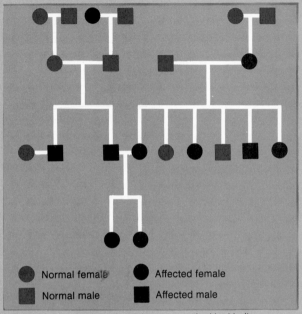

Normal female   Affected female

Normal male   Affected male

Figure 4.5 An albino Hopi girl and a pedigree of a family with a high frequency of albinism among the Hopi Indians. Source: Field Museum of Natural History.

The frequencies of many alleles differ greatly among human populations. Some diseases or genetic traits are restricted to certain populations or are far more characteristic of some populations than others. Throughout this book, we will note the frequency of occurrence of various genetic conditions in different human populations.

The genetic disease albinism illustrates the selective differences that may exist for disease-causing alleles in different human societies. Among most populations in North America, the common form of albinism is due to a relatively infrequent allele found in about 1 in 30,000 individuals. This is not surprising given the many harmful effects of the condition and the shortened life expectancy of albinos. However, among the Hopi Indians of Arizona, the frequency of tyrosinase positive albinism is about 1 in 200 individuals (figure 4.5), and in the Cuña Indians of San Blas, Panama, the frequency is 1 in 140. How can such a harmful allele become so abundant?

In many societies, unusual individuals are ostracized or subject to some forms of discrimination. Both the Hopi and Cuña have strong taboos against marriage between albinos. Thus, the high frequency of the disease results from matings of albinos with carriers, of carriers with other carriers, or of illicit liasons between albinos. The Cuña and Hopi are tribes of farmers, and in general the

men cultivate the fields, while the women work at home. Because albinos cannot tolerate sunlight, albino men are also permitted to remain at home during the day. These circumstances apparently offer albino men increased opportunities for reproduction. Thus, cultural practices have resulted in a high frequency of a normally deleterious allele. An additional factor may be small population size, which increases the probability of consanguineous matings.

This example offers two lessons about the way in which natural selection occurs. First, alleles can be favored that may not cause increased survival, but that result in increased reproduction of the carriers. This phenomenon is called *sexual selection*. Albinos still do not survive as well as normal individuals, but in some societies they breed more frequently. Second, the selective value of an allele depends upon the environment in which it occurs. Many alleles are extremely valuable in some circumstances, but harmful in others. The way in which an allele functions depends upon the circumstances in which it occurs. In most places and societies, albinos have a reproductive disadvantage, but in a few, albinos have a reproductive advantage. The allele for albinism, however, is neither good nor bad—only different from others. Genetic diversity, the existence of many different alleles, is a sign of a genetically healthy species.

One effect of PKU is the accumulation of phenylalanine in the blood and phenylpyruvic acid, a related chemical, in the urine. PKU was discovered in 1934 by a Norwegian who noticed that the urine of her two mentally retarded children had a distinctive odor which, upon analysis, turned out to be phenylpyruvic acid. The abnormal accumulation of particular chemicals in the urine and blood is characteristic of many inborn errors of metabolism.

The most serious effect of PKU is mental retardation, caused by the high levels of phenylalanine in the brain and spinal fluid in the first few weeks after birth. About 1% of all Caucasians in mental institutions suffer from mental retardation due to PKU. Infants with PKU can now be identified by a simple blood test immediately following birth. In the United States, more than forty states now require PKU screening of all newborns.

Victims of PKU, if identified at birth, can be treated with a diet carefully regulating the amount of phenylalanine. A small amount is necessary for adequate nutrition, but too much phenylalanine causes brain damage. Brain damage can be prevented if dietary regulation begins immediately after birth and is continued until about age seven. Growth of the brain is extremely rapid early in life, at which time irreversible damage can occur. However, recent studies indicate that it may be advisable for all phenylketonurics to continue a low-phenylalanine diet throughout their lives. New evidence suggests that personality and intellectual capacity may be adversely affected by the sustained high levels of serum phenylalanine typical of phenylketonurics not on special diets.

Dietary therapy for phenylketonurics is extremely effective (although any parent knows the difficulties of controlling a child's eating habits). PKU cannot be cured— an individual's genes cannot be changed—but the phenotypic consequences can be significantly ameliorated. In the late 1970s the United States government estimated that an efficient national system of screening and treatment of phenylketonurics would cost about $3.3 million annually. The result of *not* screening would be an average of 270 cases per year of preventable mental retardation, costing $189 million annually to treat. In 1987, recombinant DNA technologies (see chapter 13) made it possible to recognize more than half of all carriers of PKU. It is anticipated that soon it will be possible to identify more than 90% of all (PKU) carriers.

The effectiveness of PKU therapy has meant that many female phenylketonurics who might otherwise have been institutionalized have developed normally and begun to reproduce. However, PKU mothers who have been on diet therapy even from birth almost always produce children who are mentally retarded. The children are usually heterozygous for the PKU allele, so their condition is not due to their genotype, but to the abnormal biochemistry of their mother. The concentrations of phenylalanine in the affected mother's blood are so great (unless she is on a special diet during pregnancy) that the brain of a fetus she carries cannot develop normally. Thus, a genetically normal fetus is irreversibly damaged by its environment—the womb of its affected mother—and it exhibits the phenotype of a genetically affected child. The phenomenon, in which an environmentally caused condition mimics a genetically caused condition, is called a **phenocopy.** Women who are phenylketonurics are now advised to remain on a low phenylalanine diet, starting at least three months prior to conception and throughout the entire pregnancy.

Cystic fibrosis

The most common serious genetic disease among Caucasians in the United States is **cystic fibrosis (CF)**, caused by an autosomally inherited recessive allele. Approximately one in every two thousand white births is affected by CF, an exceptionally high frequency (compared to PKU or albinism, for example), considering that affected individuals usually die before reproducing. This means that all cystic fibrosis births occur as the result of the mating of two heterozygotes; essentially no homozygotes reproduce.

The high frequency of carriers in the population (perhaps as high as one in twenty) may result from *overcompensation;* parents of CF children continue to reproduce until they have the number of normal children they desire. The result is that the average number of offspring in families with a CF child is about 25% higher than in normal families. Of the normal children in CF families, two-thirds are carriers and continue to transmit the harmful allele.

The precise genetic defect causing CF has not yet been established. The disease itself is characterized by a malfunction of the mucous producing glands, producing a *syndrome* (a large set of symptoms associated with one pathological condition). The products of many different glands are altered, affecting particularly the respiratory and digestive systems. The mucous produced by CF victims is extremely viscous, much like cotton candy, clogging the lungs and digestive tract. The accumulation of mucous in the lungs causes a persistent cough and difficulty in breathing. The residual mucous is a good medium for bacterial growth, making pneumonia and other infections a common and sometimes fatal complication.

The mucous also clogs the ducts of the pancreas, blocking the pathway for important digestive enzymes into the gut. Thus, even if CF children eat a balanced diet, they cannot digest their food properly and consequently suffer from malnutrition. Malnutrition can be alleviated by massive doses of vitamins and other nutrients, and such therapy has caused some increase in life span. The average life span is now about fourteen years, and as of 1978 about one hundred CF children had reached age twenty.

Another effect of CF is an unusually high concentration of salt in the sweat of affected children—about two to five times normal concentrations. The disease is sometimes first detected when parents kissing their children taste the salty perspiration. The disease can be diagnosed by a *sweat test,* which measures the concentration of salt in a sample taken from a warmed patch on the child's back or forearm. Parents of CF children are heterozygotes, and their levels of salt are intermediate to levels found in affected persons and in normal persons. The different phenotypic aspects of this disease thus have different types of expression. For the ultimately fatal set of symptoms that are called CF, the allele is expressed as a recessive. For the single biochemical trait of salt concentrations in sweat, the CF allele is expressed as a codominant or incompletely dominant allele.

Although CF, like PKU, can be identified at birth, there is no cure. Modern treatment has somewhat ameliorated the effects of this disease, but few CF victims live long enough to reproduce. CF is the single greatest cause of death among white children in the United States. In 1986, the CF gene was localized to a small region of chromosome 7, opening the way for the development of a prenatal test for CF in families who already have an affected child. Siblings of CF patients can be tested to determine if they are carriers of the CF allele, although the test is definitive only in 50–80% of the siblings. With improvements in DNA technologies (see chapter 13), tests should become 100% definitive.

Sickle-cell anemia

The most common genetic disease among blacks in the United States is **sickle-cell anemia (SCA).** Victims of SCA possess an abnormal form of the protein *hemoglobin.* Hemoglobin occurs in the *red blood cells* (RBCs), imparts color to them, and carries oxygen to all tissues of the body. It is composed of four chains of amino acids

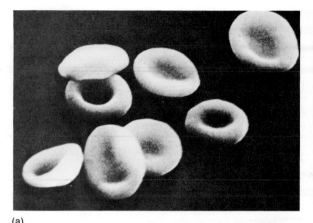

(a)

(b)

Figure 4.6 (a) Normal red blood cells; (b) sickled red blood cells; (c) diagrammatic representation of molecule of hemoglobin, showing alpha and beta globin chains and heme groups.
Source: (a and b) AP/Wide World Photos; (c) Stuart Ira Fox, *Human Physiology*, 2d ed. Copyright © 1987 Wm. C. Brown Publishers, Dubuque, Iowa. All Rights Reserved. Reprinted by permission.

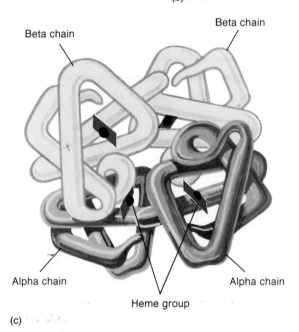

Beta chain

Beta chain

Alpha chain

Alpha chain

Heme group

(c)

(two alpha chains and two beta chains) and four heme groups, an iron containing portion that functions to bind and release oxygen to the tissues. Each RBC contains about 280 million molecules of hemoglobin, and each molecule contains 574 amino acids. The hemoglobin found in sickle-cell anemics differs from normal hemoglobin by one amino acid difference in each beta chain—two differences that have a profound physiological effect. They cause the hemoglobin to crystallize, which in turn causes malformation of the RBCs. The RBCs assume a crescent or sickle shape, hence the name of the disease (figure 4.6).

SCA has two primary effects: (1) the deformed RBCs become fragile and are easily destroyed; (2) sickle-shaped cells tend to become entangled with each other, increasing blood viscosity and thus reducing blood flow. The impact of these effects is felt throughout the system (figure 4.7). The high rate of cell destruction induces *anemia* (low RBC count), reflected in a decreased oxygen supply to the tissues, which causes the victim to tire easily. Anemia stresses the heart, which must work harder to pump adequate amounts of blood to the tissues, eventually leading to heart failure. The spleen is taxed trying to break down defective RBCs; eventually it begins to deteriorate, losing its capacity to remove bacteria from the blood. SCA victims are continually subject to infection due to failure of the spleen.

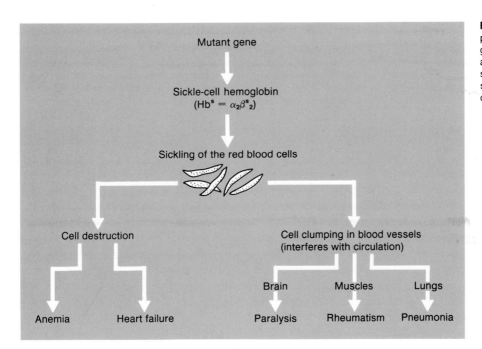

Figure 4.7 Some of the pleiotropic effects of the gene causing sickle-cell anemia. Collectively, these symptoms constitute the syndrome characterizing the disease.

Increased blood viscosity interferes with circulation of the blood, causing further stress on the heart and localized damage to many organs. Circulatory impairment in the brain may cause strokes and paralysis. Fluids accumulate in the lungs, leading to pneumonia. Decreased circulation in the muscles leads to rheumatism. A gene that has many phenotypic effects is called **pleiotropic;** there is no better example than the allele for SCA.

SCA is usually lethal. Most victims die by age ten, but medical treatment is permitting an increasing number to survive into adulthood. As with cystic fibrosis, the incidence is extraordinarily high, considering the lethality of the allele when homozygous. Among blacks in the United States, about one birth in four hundred is a child with SCA, and one in every ten blacks is a carrier of the disease.

Although the disease is normally considered to be a recessive condition, heterozygotes show some phenotypic manifestations and are said to have **sickle-cell trait.** About 35% of the RBCs of individuals with sickle-cell trait have defective hemoglobin. However, the remaining normal cells permit the carrier to lead a relatively normal life, subject to occasional episodes of anemia and complications during times of stress. As with cystic fibrosis, this allele shows a recessive pattern of expression in terms of the disease itself, but some physiological traits show incomplete dominance.

One trait of heterozygotes may explain the high frequency of the allele among blacks. Carriers of the sickle-cell allele are relatively immune to a type of malaria common in Africa. Historically, malaria has been the largest single cause of death for humans in the world, and it is still one of the largest causes of mortality in Africa. In parts of Africa where malaria was common, heterozygotes for the sickle-cell allele survived longer and produced more offspring than either homozygous genotype, thus perpetuating the sickle-cell allele in spite of its lethal effects in homozygotes.

Malaria is caused by a parasitic protozoan that lives in RBCs. The defective hemoglobin in individuals with sickle-cell trait appears to make the RBCs an unsuitable environment for the protozoan, thus conferring relative immunity on carriers. In parts of East Africa, as much as 45% of the black population may possess sickle-cell trait. In North America, immunity to malaria no longer enhances survival and the sickle-cell allele has therefore decreased in frequency among blacks.

There is no cure for SCA. New methods of DNA analysis now provide a safe, quick, and reliable **prenatal screening** technique, which allows parents to know early in the pregnancy whether their fetus carries zero, one, or two copies of the SCA allele. The test can also screen adults who wish to know if they are carriers of the disease.

Genetic counseling

The ability to diagnose certain diseases as heritable in predictable ways has direct medical implications. Most people have a limited knowledge of genetics, however. In addition, genetic knowledge is expanding so rapidly that techniques of diagnosis and treatment of genetic diseases are constantly improving. In response to such rapid change, professional **genetic counseling** has arisen to make the latest information on medical genetics available to all persons.

Most commonly, a family is referred to a genetic counselor after the birth of a child afflicted with a genetic disease or when the family has a history of genetic disease. A genetic counselor helps the individual or family understand the genetic basis of specific disorders, the probable course of the disease, and available management or treatments. The counselor also advises clients of the risk of recurrence in specified relatives and options for dealing with the risk of recurrence. Finally, counselors help families adjust to genetic disorders in family members, suggesting courses of action appropriate to the family goals and backgrounds.

Traditionally, a counselor educates potential parents but does not make choices for them. The counselor explains medical risks and results of diagnostic procedures, as well as the probabilities for genetic affliction in future children. In all instances, final decisions rest with the couple. In many genetic counseling centers, psychologists and social workers help the geneticist and physician in the counseling process.

While advances in medical genetics offer new methods of treatment and prevention of certain diseases, they also present affected families with difficult and often highly emotion-laden choices. Perhaps the most extreme case is the prenatal diagnosis of a serious genetic defect. If the pregnancy is allowed to continue, the family must accept a severe emotional and financial burden in caring for or institutionalizing the defective child. The decision to continue the pregnancy may also impose costs on society, which must often assume the responsibility of caring for the child.

On the other hand, parents have the option of terminating the pregnancy by **therapeutic abortion.** If the condition is serious and the child will have a short, painful life, a decision to abort may seem relatively straightforward. However, many serious conditions are being treated with increasing success; children with cystic fibrosis, for example, now may live two decades or more. Thus, the decision to abort is not simple, even in very serious cases. Furthermore, because of ethical and religious reasons, some families find abortion unacceptable under any circumstance.

In cases where abortion is not a consideration, the use of genetic information may still present difficult choices. The ability to diagnose the gene for sickle-cell anemia, for example, has not been greeted with uniform enthusiasm. Some employers have suggested that SCA screening be a part of the hiring procedure, but because such tests would only be conducted on blacks, these practices have been called discriminatory. Similarly, some insurance companies have required tests for SCA before issuing health and life insurance policies.

The treatment of carriers, such as individuals with sickle-cell trait, is particularly contentious because many of these individuals show few symptoms. Should they receive different treatment simply because they carry a gene that probably will cause them little harm? And, if employers and insurance companies are given access to genetic information, what are the limits? Should a possible future spouse be allowed such information?

Table 4.2
Some human traits inherited in an autosomal dominant fashion

absence of fingerprints   No dermal ridges in fingers.

achondroplasia   Dwarfism due to abnormality of ossification centers of the long bones (figure 4.8).

amyloidosis   Congestive heart failure.

brachydactyly   Short fingers and toes.

cleft chin   Bony peculiarity underlying a Y-shaped fissure in the chin.

congenital night blindness   No night vision, but normal vision in daylight.

deafness   Progressive high-tone deafness.

Ehlers-Danlos syndrome   Loose jointedness and fragile, easily bruised skin.

Huntington's disease   Choreic movements and dementia; variable age of onset.

juvenile glaucoma   Glaucoma developing by age eight.

Marfan's syndrome   Excessively long arms and legs with associated aneurysm of aorta. Abraham Lincoln is thought to have suffered this syndrome.

myokymia   Spontaneous muscle twitches.

myopia   Nearsightedness.

nail-patella syndrome   Dysplasia of the nails and absence of kneecaps.

neurofibromatosis   Presence of cafe-au-lait spots and numerous benign skin tumors.

osteogenesis imperfecta   Brittle bones resulting in frequent fractures, associated with loose jointedness, blue sclerae, and progressive deafness.

piebald trait   White forelock and white spotting over body.

polydactyly   Extra toes or fingers (figure 4.1a).

porphyria   Increased excretion of breakdown products of hemoglobin; increased sensitivity to sunlight.

ptosis   Drooping eyelids (figure 4.1b).

retinoblastoma   Tumors of the retina, resulting in blindness; fatal if eye is not removed.

woolly hair   Short, tightly curled and woolly hair.

From V. A. McKusick, *Mendelian Inheritance in Man,* 7th ed. Copyright 1986 Johns Hopkins University Press. Reprinted by permission.

The ethical questions raised by genetic technologies vary considerably with the specific genetic condition. As we discuss other diseases throughout the book, we will draw attention to other aspects of genetic counseling.

Autosomal dominant inheritance

At least 1,172 traits in humans have now been identified as the result of autosomal dominant inheritance (table 4.2), and 1,029 more traits are suspected. This is nearly twice the number known ten years ago. Pedigree analysis of dominant traits should be relatively simple. Affected offspring must have at least one affected parent (barring new mutations), and all children of homozygotes and half the children of heterozygotes should express the trait.

Achondroplasia

A well-documented autosomal dominant allele affects the growth centers of the long bones, causing a form of dwarfism called **achondroplasia.** The affected individual has a normal-sized torso, but very short limbs. Achondroplastic dwarfs average about four feet in height (figure 4.8). This condition affects less than one birth in ten thousand. Most cases are believed to result from spontaneous mutations, as most affected offspring are born to phenotypically normal parents. The homozygous condition is probably lethal. There is no treatment.

The Church of Jesus Christ of Latter-Day Saints (the Mormon church) maintains detailed genealogical records, useful in genetic studies. Achondroplasia is relatively common in several Mormon families, where achondroplastic dwarfs have routinely married and had children. In one family, seventy-six children were born of matings between normal (dd) and dwarf (Dd) individuals—a genetic testcross. The Mendelian prediction is that thirty-eight should be normal, thirty-eight dwarf. The actual result was that forty-two were normal, thirty-four dwarf. This is close enough to the expected values that the difference can be attributed to chance factors. A larger sample size would approximate more closely the 50% expectation of each type of child.

Neurofibromatosis

The play and film, *The Elephant Man,* described the life of John Merrick, a victim of **neurofibromatosis (NF).** The course of this disease, caused by an autosomal dominant allele, involves the development of hundreds of non-malignant tumors over the entire body. In combination with abnormal bone growth, the victim of NF may be grossly deformed.

This disease cannot be diagnosed prenatally and cannot be cured. Surgery is sometimes used to remove masses of tumors, especially those affecting nerves. About 20% of NF victims develop tumors of the spinal cord, which can cause deafness and blindness. Plastic surgery to correct the massive deformation is traumatic, and the results are seldom totally satisfactory. The onset and progress of the disease are unpredictable. Unfortunately, this relatively common condition afflicts over one hundred thousand Americans.

Nail-patella syndrome

Not all autosomal dominant traits have dramatic and serious effects. Carriers of the allele causing **nail-patella syndrome** have absent or malformed fingernails, toenails, and kneecaps (patellas). Some also have chronic inflammation of the kidneys. Otherwise, affected individuals exhibit no serious symptoms.

Earwax

Variation in the type of *cerumen* or *earwax* exemplifies a dominant Mendelian trait with innocuous phenotypic effects. Two types of cerumen occur in humans. Wet earwax, the result of a single dominant gene, W, is a sticky type of earwax found in about 85% of white Americans and 93% of black Americans. Dry earwax, genotype ww, produces a granular or scaly and non-adhesive earwax and is found in high frequency in Mongoloid populations—96% in northern Chinese, 92% in Koreans, and 85% in Japanese.

Such genetic differences are common among various racial groups, but the causes are often obscure. Wet earwax is associated with axillary (armpit) odor, which in Japan is considered a pathological condition requiring medical attention. In many mammals, including humans, scent is an important trait in the choice of a mate. Scientists have suggested that the high frequency of the allele for dry earwax among Mongoloids results from sexual selection—an historical preference for the aroma of people with dry earwax, or a distaste for the aroma of people with wet earwax. This

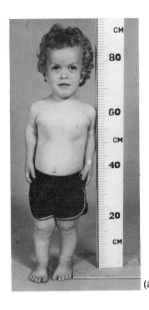

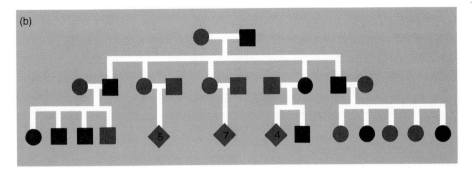

Figure 4.8 (a) An achondroplastic boy. (b) The hypothetical pedigree of a family affected by achondroplasia over three generations.
Source: (a) Dr. Judith Hall.

explanation suggests that the quality of earwax itself may have been of little selective importance. Rather, the allele producing dry earwax was selectively favored among Mongoloids because of one of its pleiotropic effects, absence of an offensive axillary odor.

The gene for cerumen type appears to exert a general effect on the apocrine glands, a type of sweat gland. The glands producing earwax and axillary odor are apocrine glands, as are the mammary (milk producing) glands of the breast. Yet another, more serious pleiotropic effect of the allele for dry earwax may involve breast cancer. Among Japanese women, breast cancer is about twice as common in women who have wet earwax as in women with dry earwax. Worldwide, the frequency and mortality rates for breast cancer are directly correlated with the frequency of the allele for wet earwax among all races. Although earwax type may be a relatively trivial trait, it is clear that the controlling gene probably has several important effects on phenotype and survival.

Anthropologists can often trace the origins of particular racial groups through genes that have well-defined racial differences. Native Americans are believed to have originated from Mongoloid ancestors who migrated across the Bering Straits thousands of years ago. Among full-blooded Navaho Indians of the American southwest, the frequency of the dry earwax phenotype is 70%, providing strong support for their origin from a Mongoloid stock. Among full-blooded Sioux Indians, the frequency is somewhat lower, about 47%. A possible explanation may be the fact that Sioux in the upper Mississippi Valley had contact with French trappers as early as 1700 (that is, they are not really full-blooded), while the relatively isolated Navaho have only recently come into contact with Caucasians.

Blood groups

A person's blood type is a phenotypic character that depends upon the types of molecules—called **antigens**—that occur on the surfaces of the red blood cells (RBCs). Although the ABO **blood group** is the best known, several dozen other blood groups have been defined in humans. The existence of so many blood groups indicates that the surface of a red blood cell is indeed crowded with proteins. It is now known that there are many "public" blood groups—antigens that occur on the surface of all

Table 4.3
ABO blood types and their inheritance in humans; antigen and antibody relations and permissible transfusions of red blood cells

Blood type		Antigen(s) on surface of red blood cells	Antibodies present in serum	Transfusion status	
Phenotype	Genotype(s)			*Can receive*	*Can donate to*
O*	OO	Neither A nor B	Anti-A, Anti-B	O	O, A, B, AB
A	AO or AA	A‡	Anti-B	A, O	A, AB
B	BO or BB	B	Anti-A	B, O	B, AB
AB†	AB	A, B	Neither anti-A nor anti-B	AB, A, B, O	AB

*Universal donor
†Universal recipient
‡Blood type A is actually more complex than consisting only of a single A antigen. There are two major type A antigens, designated A-1 and A-2. Eighty percent of persons with type A are of the A-1 genotype. Recently several other rare subgroups of the A antigen have been identified. Variants of the O and B groups are also known, but are quite rare. (Often a different allelic nomenclature is used for the ABO blood groups. The gene for the blood group is indicated by I, representing "isoagglutinin," the molecule involved in the agglutination (clumping) of RBCs in blood typing. Specific alleles are represented by superscripts, I^A, I^B, I^O.)

human RBCs, exhibiting no variation whatever. Some "private" blood groups, variants unique to only a few individuals, have also been found. Thus, the known blood groups almost certainly underestimate the immense diversity of molecules that occurs on each RBC.

The ABO blood group

The ABO blood group locus is characterized by three alleles, A, B, and O. O is recessive to both A and B, which are codominant to each other. A and B specify antigens on the surface of the RBC; O specifies no antigen. Although three primary alleles occur in the population as a whole, each individual possesses only two.

Discovery of the ABO group by Karl Landsteiner in 1900 depended not only upon the presence or absence of A and B antigens on the RBCs, but also upon the presence or absence of particular **antibodies** in the serum. Antibodies, a primary mechanism of defense for the body, combine with foreign proteins such as those on the surface of bacteria, neutralizing their pathological effects. Table 4.3 summarizes the combinations of antigens and antibodies of the four main ABO blood types.

Note that a person never has antibodies to his or her own antigens. For example, an individual of blood type A has antigen A, but anti-B antibodies. A person of blood type O has both anti-A and anti-B antibodies. If blood of different types is mixed, the antibodies combine with antigens of the same type (anti-A with antigen A, for example), causing a reaction called **agglutination.** Agglutination is visible to the naked eye (figure 4.9), and it was Landsteiner's observation of agglutination that resulted in discovery of the ABO blood group.

Most antibodies are produced only after a foreign antigen is introduced into the circulatory system. The antibodies of the ABO system, however, are present at birth. Landsteiner would not have made his fortunate discovery if the antibodies were not naturally occurring, and if all persons had the same ABO blood type.

Discovery of the ABO blood group was an important medical advance. During surgery or as the result of a serious wound, persons often need transfusions of blood. Prior to Landsteiner's work, transfusions had been used with variable results. Sometimes they were successful, but equally often the patient suffered a severe, sometimes fatal reaction to the transfusion.

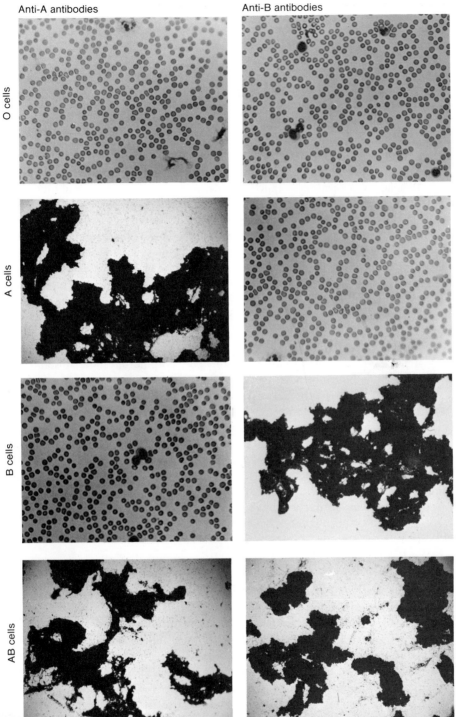

Anti-A antibodies   Anti-B antibodies

O cells

A cells

B cells

AB cells

Figure 4.9 The reaction of the four blood phenotypes when mixed with anti-A or anti-B antibodies. Agglutination (clumping) occurs when there is a match of the antigenic type of the blood with the antibody.
Source: Victor B. Eichler.

Current medical practice matches the blood type of the donor with that of the recipient prior to the transfusion. Usually, the recipient needs the oxygen carrying RBCs of the donor, so the critical aspect of the match is to insure that the donor's RBCs are not attacked by the antibodies in the serum of the recipient (table 4.3). Transfusion of type A blood into a type B person or type O person, for example, results in destruction of the type A cells, inducing shock in the recipient. Type B individuals are thus never given type A blood, and vice versa. Because type O persons have neither A nor B antigens, they cause no reaction in a recipient of any blood type and are considered *universal donors*. Conversely, a type AB person has neither anti-A nor anti-B antibodies and is a *universal recipient*. Donor-recipient ABO types are normally precisely matched, however, except in emergencies.

You might reasonably expect that transfusion of type O blood into a person of type B, for example, would have harmful consequences due to the presence of anti-B antibodies in the donor's serum. Usually this is not the case, however. The quantity of antibodies in the donor's serum is sufficiently diluted in the serum of the recipient to preclude much damage to the large volume of RBCs in the recipient. It is the far larger quantities of antibodies in the recipient's blood that cause injury.

Beyond its medical importance, the identification of the ABO and other blood groups has been of forensic (legal) importance. Because the ABO group shows strict Mendelian inheritance, blood types provide evidence in paternity suits. For example, a child of type AB cannot have a father of type O. In this case, the ABO group alone is sufficient to establish conclusively that an innocent man has been wrongfully accused. Frequently, however, ABO types provide inconclusive evidence. A type A man might be the father of a type O child, for example. The use of several blood groups can greatly increase the probability of a definitive finding of innocence. Evidence from blood groups can *never* prove paternity, although it can be highly suggestive.

Blood groups, like earwax, have been used to trace the history of particular racial groups. Many blood groups are highly polymorphic and show substantial differences among human populations. Considering again the origin of Native Americans, blood type B is extremely rare in most Native American tribes. Type B is totally absent from the Navaho, and occurs in about 4% of "full-blooded" Sioux. The occurrence of type B among Native Americans is believed to result from interbreeding with Caucasians. The combined frequency of the B and AB phenotypes among contemporary American whites is about 12%, thus supporting the hypothesis based on earwax frequencies that the Sioux have had a longer history of interbreeding with Caucasians than have the Navaho. Table 4.4 gives frequencies of ABO blood types in various human populations.

The MN blood group

Discovery of other blood groups has depended upon the use of antibodies not occurring naturally in humans. Landsteiner discovered the MN blood group by producing antibodies to human RBCs in rabbits, then using those antibodies to identify the antigens on the RBCs for which they were specific—that is, those that they caused to agglutinate. The MN blood group is controlled by two alleles, M and N, occurring at one genetic locus and expressed codominantly.

The MN blood group, of no medical importance, exhibits simple Mendelian inheritance. Because of codominant expression, however, two alleles at one locus produce three phenotypes (MM, MN, and NN) rather than the two phenotypes observed when one allele is dominant to the other. In the case of **codominance,** both genotypes and phenotypes from a mating of heterozygotes (MN × MN) occur in the ratio of 1:2:1 (simply a more precise expression of the 3:1 monohybrid ratio).

The MN group is inherited independently of the ABO group—that is, they are not linked. Mating of persons of blood types O,M and AB,MN produces four types of offspring in equal proportions A,M; B,M; A,MN; and B,MN (figure 4.10).

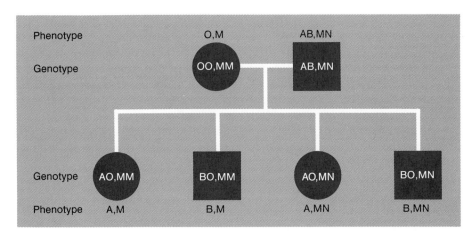

Figure 4.10 Four phenotypes result from a cross of a female of blood type O,M with a male of blood type AB,MN. Note that the male can produce four genetically different types of sperm, while the female produces only a single type of ovum.

Table 4.4
Frequencies of ABO blood types in human populations

| | Phenotype | | | |
Population	*O*	*A*	*B*	*AB*
Armenians	.289	.499	.132	.080
Eskimos (Greenland)	.472	.452	.059	.017
Austrians	.427	.391	.115	.066
Danes	.423	.434	.101	.042
Irish	.542	.323	.106	.029
French	.417	.453	.091	.039
Russians (Moscow)	.340	.377	.206	.077
Indians (Calcutta)	.304	.255	.375	.066
Vietnamese	.418	.218	.305	.059
Chinese	.439	.270	.233	.058
Japanese (Niigata)	.296	.369	.233	.102
Nigerians (Yoruba)	.515	.214	.232	.039
Amerindians (Lima)	.860	.140	.000	.000
Amerindians (Bolivia)	.931	.053	.016	.001
U.S. whites (St. Louis)	.453	.413	.099	.035
U.S. blacks (Iowa)	.491	.265	.201	.043

Source Table 18.3 from *An Introduction to Human Genetics*, Fourth Edition, by H. Eldon Sutton, copyright © 1988 by Harcourt Brace Jovanovich, Inc., reprinted by permission of the publisher.

The HLA antigens

The best studied system of cell surface antigens are those of the red blood cells, especially the ABO system. However, a complex set of antigens on the surface of most body cells, including the white blood cells, the **human leukocyte antigens (HLA),** is also of great medical importance. The individual HLA antigens are detected using a panel of antisera made to each of the known antigens. This complex system shows great Mendelian variation and governs the tissue rejection response, which will be discussed in chapter 11.

HLA antigens are the products of at least two genetic loci with major effects on the immune response and at least three loci with minor effects. All five loci are on the short arm of chromosome 6. The two major loci, designated HLA-A and HLA-B, are probably the most polymorphic loci yet identified in human populations. At least twenty-three codominant alleles encoding twenty-three unique antigens are known at the HLA-A locus, and another forty-seven alleles have been identified at the HLA-B locus.

Together, the A and B loci can generate an extraordinarily large number of unique genotypes. The A locus with twenty-three alleles generates 276 possible genotypes, while forty-seven alleles at the B locus generate 1,128 different genotypes. In combination, these two loci alone can produce 311,328 unique genotypes. At least forty-one other alleles are known at the three minor HLA loci, creating millions of unique HLA genotypes. Because not all alleles occur in equal frequency, in the population many genotypes are exceedingly rare. The single most common HLA genotype has a frequency of less than one in every hundred persons. Figure 4.11 illustrates the allelic diversity found within a family of six. Note that no two individuals share an HLA genotype.

HLA and disease

The widespread use of HLA typing has revealed striking correlations between the incidence or severity of many diseases and certain HLA phenotypes (table 4.5). One of the best-studied associations is between the B27 antigen and a type of arthritis known as ankylosing spondylitis. This HLA antigen occurs in only 8% of the British population, but in 90% of all British patients with ankylosing spondylitis. Of sixty relatives of ankylosing spondylitis patients tested, thirty-two had the B27 antigen, further evidence of a genetic association. Similar data have been obtained in Japan. Among healthy Japanese the B27 antigen is very rare, but over 90% of affected Japanese have this antigen.

Despite the strong association, the B27 antigen cannot be considered to cause ankylosing spondylitis. Not all persons with B27 develop the disease. Perhaps the B27 antigen increases susceptibility to another causal agent, such as a virus. If such an agent could be identified, genetically susceptible persons could be vaccinated or given early treatment.

Recently HLA antigens have been associated with susceptibility to leprosy. With routine tissue typing, children found to be at risk can be given drug treatment that may prevent or diminish the severity of this disease.

Athlete's foot, a nuisance disease, is caused by a fungus. Most persons occasionally develop this condition, but it is easily treated. A small group of persons, however, suffer chronic athlete's foot. Recent work has indicated that these individuals share a common HLA genotype, suggesting a genetic component or susceptibility to persistent fungal infection. As with ankylosing spondylitis and leprosy, understanding of the genetic connection can mean early screening and treatment of the disease.

HLA and forensic medicine

HLA antigens have become very useful in forensic medicine. Because the HLA system is so diverse, only a small fraction of the population shares any particular tissue type. While individual genetic uniqueness is a major barrier to transplant surgery, it also means that cell surface antigens provide each person with a "biological fingerprint" of great legal value.

As with ABO typing, HLA typing has had wide application in paternity suits; of which there are an estimated 200,000 in the United States each year. The larger number of HLA alleles, however, greatly increases the probability of excluding from paternity males accused unfairly. In a recent California case, for example, a woman accused a man of fathering her twins. ABO, MN, and HLA types were obtained, revealing that the putative father's HLA antigens matched that of one twin, but

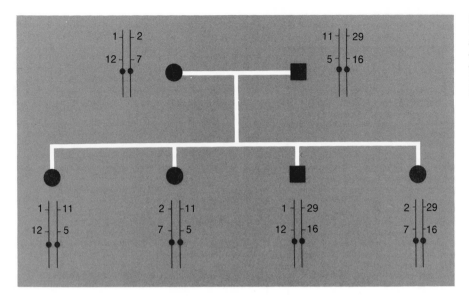

Figure 4.11 Inheritance of phenotypes at the HLA-A and HLA-B loci linked on chromosome 6. The numbers designate alleles at the A locus (top) and B locus (bottom).

Table 4.5
Some examples of HLA and disease associations*

Disease	Antigen	Frequency in patients (%)	Frequency in controls[†] (%)
Ankylosing spondylitis	B27	90	8
Celiac disease	B8	71	23
	A1	64	29
Insulin dependent diabetes	B8	37	22
	BW15	23	15
Multiple sclerosis	A3	36	26
	B7	34	24
	DW2	70	16
Psoriasis	B13	20	4
	B17	26	8
	B37	8	1
Ragweed hay fever (Ra5 sensitivity)	B7	50	19

*Data are based on Ryder et al. (1974) and McDevitt and Bodmer (1974).
[†]The population at large.

not of the other (figure 4.12). Subsequently, the mother admitted involvement of a second man. His HLA type indicated that he was the probable father of the second twin. These nonidentical twins were the seventh documented case of **superfecundation,** the simultaneous birth of children fathered by different men. While rare in humans, superfecundation is common in other mammalian species. It requires that a female have intercourse with several males within a few hours of ovulating more than one egg. Superfecundation was first documented in humans in 1810 when a woman gave birth to twins, one white and the other mulatto.

In the California case, neither ABO nor MN typing provided definitive information for establishing paternity. The mother, both twins, and both men were blood type A; the mother, one twin, and one man were MN, while the other twin and second man were type M, excluding neither male from paternity of either twin. The mother's HLA antigens were 2,44/11,27. Each twin had 2,44 from the mother. Additionally, one twin was 24,54 and the other 2,15. One man was 2,7/24,54, whereas

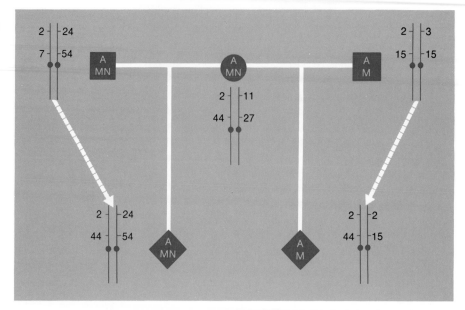

the other was 2,3/2,15. Using HLA types, each man was positively excluded from fathering one child, while being strongly implicated as father of the other. The frequency of the 24,54/2,15 genotype required by a third male who could have fathered both children was estimated to be one in 140,000.

This case also illustrates that while science can sometimes provide truth, it cannot enforce justice. After genetic analysis, one of the males accepted responsibility for *both* children, even though he was the biological father of only one of them. The mother dropped the paternity suit, and the second male was legally absolved of responsibility for the child he had fathered.

Autosomal linkage

Simultaneous examination of two or more autosomal traits in humans usually shows that they obey Mendel's Law of Independent Assortment. Knowledge that genes controlling specific traits are linked, however, can sometimes permit detailed analysis that might not be possible otherwise, such as precise identification of the genotype of both parents.

The genes for both nail-patella syndrome and the ABO blood group are located on chromosome 9 (figure 2.10). Figure 4.13 shows two pedigrees for ABO-NP phenotypes. In figure 4.13a the mother is blood type B and normal (nn) for NP syndrome. The father has NP syndrome (Nn) and is type A. The two sons have different phenotypes: the first has NP syndrome and is type B; the second is normal and type AB. What are the parental genotypes?

The genotype of the second son must be ABnn. He must have received an An gamete from his father and a Bn gamete from his mother. The first son must also have received Bn from his mother, while receiving ON from his father. Knowing the linkage groups, we are thus certain that one of the father's chromosomes is ON, the other An. One of the mother's chromosomes is Bn, and the other is either Bn or On.

The pedigree in figure 4.13b is more complex. The father is a doubly recessive homozygote, OOnn, producing only On gametes. Because two of the children are type O and two are normal for nail-patella, the mother must be heterozygous AO, as well as Nn. Which of her alleles occur on the same chromosome?

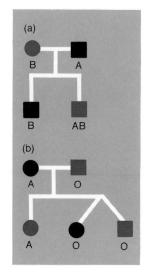

Figure 4.13 Pedigrees involving the linked genes, ABO and nail-patella. (Note that ABO phenotypes are given; individuals with darkened coloring are those who have nail-patella syndrome.) (a) A pedigree where there is no indication of crossing-over; (b) a pedigree where crossing-over must have occurred.

The first daughter received On from her father and thus received An from her mother. The mother's second chromosome is therefore ON. Union of a maternal ON gamete with a paternal On gamete would produce the second daughter. A crossover between the two maternal loci could then produce an On gamete, giving rise to the son. Pedigree analysis, however, often does not give a definitive answer. In this case, the limited pedigree is also consistent with two other types of maternal gametes, AN and On. If this were the case, what children could have resulted from maternal crossover?

Summary

1. Pedigree analysis is a common methodology of human genetics.
2. Genetic relatedness can be calculated between relatives and measures the proportion of shared alleles that are descended from a common ancestor. In general, relatedness decreases by a factor of one-half with each step in a relationship.
3. Consanguineous matings increase the probability of expressing recessive alleles.
4. The closer the relationship in a consanguineous mating, the more severe the effects, on average, of inbreeding depression. Incest taboos reflect societal recognition of the harmful effects of inbreeding.
5. Inborn errors of metabolism usually result from a deficiency or structural alteration of an enzyme in a metabolic or biochemical pathway.
6. Cystic fibrosis is the most common lethal genetic disease of American whites. Sickle-cell anemia is the most common lethal genetic disease of American blacks.
7. Frequencies of some polymorphic alleles differ greatly among populations. These genetic differences can be used to trace the history and origin of populations and races. For example, both cerumen type and ABO blood type have been used to show the origin of Native Americans from a Mongoloid stock, and to show the effects of Native Americans interbreeding with American whites.
8. Both red and white blood cells have many surface antigens that specify a variety of blood groups. Because these blood group phenotypes are inherited in a Mendelian fashion, they are often used to determine paternity. The more alleles that occur at a locus, the more useful it is in paternity testing. Many blood group alleles show associations with certain diseases.
9. Linkage relationships can sometimes be used to identify genotypes precisely.
10. Genetic counseling can help a family understand the genetic basis of disease and assist them in managing genetic conditions afflicting a family member. Often, genetic counselors can advise on the probability of recurrence of genetic diseases.

Key Words and Phrases

achondroplasia

agglutination

albinism

antibody

antigen

blood group

carrier

clone

codominance

coefficient of relationship

consanguineous

consanguineous mating

cystic fibrosis (CF)

dizygotic twins

genetic counseling

human leukocyte antigens (HLA)

inborn error of metabolism

inbreeding depression

incest taboo

index case

monozygotic twins

nail-patella syndrome (NP)

neurofibromatosis (NF)

pedigree

phenocopy

phenylketonuria (PKU)

pleiotropic

prenatal screening

sickle-cell anemia (SCA)

sickle-cell trait

superfecundation

therapeutic abortion

Questions

1. Why are humans so difficult to study genetically? What aspects of humans make them easier than some species to study genetically?

2. Could two albinos produce a normally-pigmented child? What genetic mechanism or mechanisms could produce such a result?

3. On the basis of ABO blood group evidence *alone,* should a type AB woman ever instigate a paternity suit? For the evidence to be most incriminating, what should be the ABO type of her child? If the woman were type O, would she be more or less likely to find ABO evidence in her favor than if she were type AB?

4. A man of type B is accused in a paternity suit of fathering a child, type A, by a woman of type O. Could he have been the father?

5. Suppose a woman of blood type B, M marries a man of type A, N. What possible ABO and MN types of children might they have?

6. Assume that eye color is controlled by one locus where allele B for brown eyes is dominant to allele b for blue eyes. If a female albino, genotype aaBB, marries a doubly heterozygous male, AaBb, what genotypes of children might they have? What will be the phenotypes associated with each genotype?

7. Why can the use of blood group evidence in paternity cases never *prove* that a particular male is the father of a child? How does the use of multiple blood groups increase the strength of such evidence in paternity cases?

8. If crossing-over occurred in the father in figure 4.11, what possible genotypes could be found among his children? If crossing-over occurred in both the mother and the father, how many different genotypes could be found among their offspring?

For More Information

Allison, A. C. 1956. "Sickle Cells and Evolution." *Scientific American,* 195:87–94.

Clarke, C. A. 1968. "The Prevention of Rhesus Babies." *Scientific American,* 219:46–66.

Friedman, M. J., and W. Trager. 1981. "The Biochemistry of Resistance to Malaria." *Scientific American,* 244:154–66.

Guillery, R. W. 1974. "Visual Pathways in Albinos." *Scientific American,* 230:44–62.

Hartl, D. L. 1985. *Our Uncertain Heritage: Genetics and Human Diversity,* second edition. Harper and Row, Pub. Inc., New York.

Kidd, K. K. 1987. "Phenylketonuria: Population Genetics of a Disease." *Nature,* 327:282–3.

Marx, J. 1985. "New Sickle Cell Test." *Science,* 230:1365 and 1350–54.

McKusick, V. A. 1986. *Mendelian Inheritance in Man,* seventh edition. The Johns Hopkins University Press, Baltimore.

Novitski, E. 1982. *Human Genetics,* second edition. Macmillan Publishing Co., New York.

White, R. 1986. "The Search for the Cystic Fibrosis Gene." *Science,* 234:1054–55.

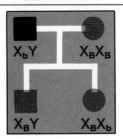

Genetics of sex

(a)

(b)

(c)

Figure 5.1 Sexual dimorphism in vertebrates: (a) Costa Rican frogs (the male is the smaller frog clasping the larger female from behind); (b) pair of wood ducks (the male is more decorated); (c) a male elephant seal defending his section of a mating beach. Source: (a) Roy McDiarmid; (b) Comstock; (c) Leonard Lee Rue III/Animals Animals. (See colorplate 6.)

In many species, the most obvious phenotypic variation is associated with the sexes, **sexual dimorphism** (meaning two forms). We seldom have difficulty distinguishing male humans and female humans, but in some species the differences between the sexes are dramatic. Male elephant seals, for example, are several times the size of females and bellow through a bulbous snout in defense of territory and harem (figure 5.1c). Males of many species possess elaborate ornamentation that appears to attract females (the peacock, for example). The sexes differ so strikingly in some fish that biologists first classified them as different species.

Sexual dimorphism is not universal among multicellular organisms, however. Many species of plants and invertebrate animals are **hermaphroditic** (from the combined names of the Greek god Hermes and goddess Aphrodite). In animals an hermaphrodite possesses both male and female gonads (testes and ovaries). The flowers of most plants, for example, have both male and female organs—each plant is both male and female. Hermaphrodites may be self-fertilizing, cross-fertilizing, or both. In some hermaphroditic annelid worms and snails, mating pairs exchange sperm, thus avoiding self-fertilization, which is the most deleterious form of consanguineous mating.

Some plants and most vertebrates have more or less clearly differentiated sexes. In some species, dimorphism is observable only in the gonads, the **primary sexual characters.** In the ginkgo tree, for example, the sexes can be distinguished only during flowering or when the female plants bear their foul-smelling fruit. Male ginkgoes are a very desirable ornamental tree, but the sexes cannot be distinguished until reproductive maturity is reached. Gardeners often must choose between chopping down a handsome shade tree or tolerating the annual nauseous stench when their promising sapling becomes a sexually mature female.

Traits other than the gonads that show sexual dimorphism are called **secondary sexual characters.** Examples in humans are the external genitalia; the internal reproductive tracts through which the gametes travel and in which the fetus develops in the female; patterns and coarseness of bodily and facial hair; patterns of fat deposition; hormonal balances; and size.

In many vertebrates, size is the only obviously dimorphic trait. Males, however, are not always the larger sex; perhaps more commonly, females are larger (figure 5.1a). In a few species such as the jacanas (a group of tropical aquatic birds), the female is not only larger but also gaudier than the male, and she maintains a harem of males whom she actively courts and defends. In many other species, the sexes are so externally similar that they can be told apart only by dissection and examination of the gonads.

Sex determination

The genetic mechanisms associated with sex determination are extremely variable and few generalizations apply to all species. However, in those organisms with sexual reproduction and some degree of sexual dimorphism, the following principles usually apply.

1. Like all phenotypic characters, sexual phenotypes are governed by the interaction of genes and the environment. The sexual phenotype involves many morphological, physiological, and behavioral traits, indicating the probable involvement of many different genes on many different chromosomes.

2. Nonetheless, sex determination in many species is associated with a major dimorphism in one pair of chromosomes, the *sex chromosomes*. The sex chromosomes, however, function very differently in different species. Thus, no one mechanism explains sex determination in all species.

3. The sex-determining genes direct the development of an individual into one of two pathways—male or female—leading to a sexual phenotype. In other words, each individual has the potential to develop into either sex. Gene mutations may cause an individual to develop into the sex opposite to that indicated by the chromosomes. Interruption of developmental pathways may cause an organism to display an ambiguous sexual phenotype.

4. In some species, sex is determined by a genetic response to external environmental factors. In some cases, when the environment changes, the individual changes sex. Controlling environmental factors may be either external or internal. Thus, some organisms change sex as they increase in age or size.

Chromosomal mechanisms of sex determination

By 1910, sexual phenotype had been shown to behave like a Mendelian character. Genes were hypothesized to occupy chromosomes (the Sutton-Boveri hypothesis), and a difference in one pair of chromosomes was shown to exist between the sexes in some species. Several persons—including Mendel himself in 1870—suggested that the 1:1 ratio of males to females at birth resembled the result of a testcross. This observation was essentially correct for humans, because the Y chromosome functions much like a dominant gene to produce maleness. However, other chromosomal mechanisms of sex determination have also been discovered.

The XX-XY system

As noted in chapter 2, many species have one pair of sex chromosomes in which the homologues may differ in structure. Among most placental mammals, females possess two similar sex chromosomes, called the X chromosomes. Males, in contrast, usually possess one X chromosome and one chromosome dissimilar in morphology from all others, the Y chromosome. In humans, females are XX for chromosome pair 23, and males are XY. Because the male's sex chromosomes are different, the male is called the **heterogametic sex;** the female is the **homogametic sex.** In humans the Y chromosome is smaller than the X (figures 2.8 and 2.9).

In humans the presence of the Y chromosome normally results in the male phenotype. Genes on the Y chromosome direct differentiation of the embryonic gonad to a testis, whose hormonal products then induce a male phenotype. Genes on the Y also control spermatogenesis. Thus, the presence of the Y chromosome determines maleness. Without a Y chromosome, the female phenotype develops.

A variation: the ZZ-ZW system

In birds and some insects, the male is the homogametic sex, and the female is heterogametic. In order not to confuse this type of sex determination with the **XX-XY** mechanism, the male chromosomes are labelled ZZ, the female ZW. Other than the reversal of homo- and heterogametic sexes, the **ZZ-ZW system** functions similarly to the **XX-XY system.**

The XX-XO system

In some species, the sexes have different numbers of chromosomes. The difference often involves the sex-determination mechanism. This phenomenon, common in insects, is called the **XX-XO system** (O indicates the absence of one sex chromosome). The female has two sex chromosomes just as in the XX-XY system, but the male has only one and is thus designated XO. In these species, the diploid number of the male is one less than that of the female as a result of the absence of one sex chromosome.

Sex determination by ploidy

In many species of hymenoptera (bees and wasps), sex is determined by the number of sets of chromosomes, or **ploidy,** of the individual. Females are diploid, producing normal haploid gametes via meiosis. Most eggs are fertilized by haploid sperm from males, but a few are not. The fertilized eggs become females, while the unfertilized eggs develop into haploid males (figure 5.2). Haploid individuals, of course, cannot undergo normal meiosis, so males produce gametes via mitosis.

Because males do not use meiosis for gamete production, all the sperm from one individual are genetically identical to each other and to the male parent. This has the interesting consequence of increasing the genetic relatedness of a male's daughters. Remember that in this system a male produces no sons—only daughters. On average, all daughters of one mated pair of bees share 75% of their genes, rather than the normal 50% relatedness of siblings of most species. The daughters are identical for all 50% of the genes received from their father, plus one-half of the 50% of the genes from their mother, an additional 25%, or a total of 75% genetic relatedness.

In some vertebrates, unusual degrees of ploidy are associated with sex. Some species of lizards consist only of triploid (3N) females and no males. In fact, males would be superfluous, because the females have become completely *parthenogenetic*—a triploid female produces a triploid egg, which undergoes complete development without being fertilized. Haploid male bees are also the result of **parthenogenesis.**

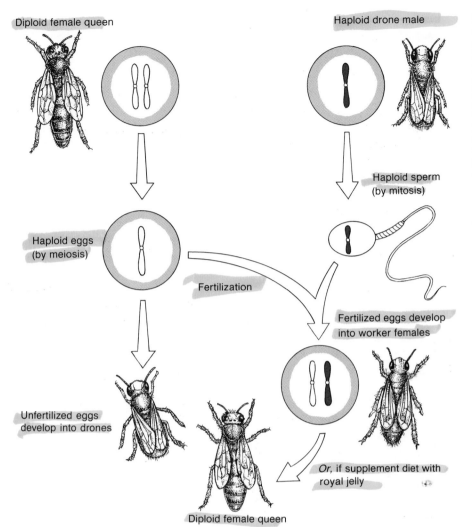

Diploid female queen

Haploid drone male

Haploid sperm
(by mitosis)

Haploid eggs
(by meiosis)

Fertilization

Fertilized eggs develop
into worker females

Unfertilized eggs
develop into drones

Or, if supplement diet with
royal jelly

Diploid female queen

Figure 5.2 Sex determination in honeybees: a haploid-diploid system. Females are diploid and males are haploid. The queen has been fed "royal jelly" and is larger than the average nonsexual worker females who do not reproduce. Unfertilized eggs develop into haploid males (drones) whose only function is to provide sperm for the queen.

Environmental sex determination

In some invertebrates and fish, sex is determined by the environment of the larva or juvenile. In the marine annelid worm, *Bonellia,* a larva may establish a parasitic relationship with an adult female, in which case it becomes a male. If it encounters no females, the larva develops into a female itself.

A similar phenomenon occurs in some deep-sea fish. In these species, the larva actually burrows into the side of a female and becomes little more than a parasitic testis, barely recognizable as a separate individual. These fish occur in very low numbers in the deep ocean. Such a mechanism probably maintains equal numbers of each sex and guarantees each mature individual a mate.

In other marine fish, all individuals are born as females and upon maturity become a member of one male's harem. Each male defends his harem and his territory, a coral outcrop. Females within a harem establish a dominance hierarchy. When the male dies, the highest-ranking female undergoes a sex change, becoming the male and assuming possession of the harem. Other species are known in which sex reversal occurs at a certain size or age, independent of dominance hierarchies.

Physical rather than biological factors may also influence sex determination. Incubation temperatures of eggs of some species of turtles, for example, determine sex. If eggs are incubated at relatively high temperatures, most produce females; if incubated at low temperatures, most become male (box 5.1).

Sex determination and sea turtle conservation

(a)

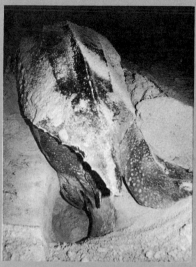
(b)

Figure 5.3 Female sea turtle (a) making nest preparatory to laying eggs; (b) covering nest after laying eggs. Source: (a and b) Frans Lanting.

Knowledge of the reproductive systems of animal and plant species has direct application to agriculture and conservation. Environmental sex determination, for example, is a relatively unusual phenomenon in nature, especially among vertebrate animals. It was first discovered in turtles in the 1960s, and now is also known in a number of egg-laying vertebrates. Temperature-dependent sex determination is crucial information in the fight to save some species of sea turtles (figure 5.3).

Sea turtles spend virtually all their lives at sea, except for one night each year when the females laboriously haul themselves onto a beach to dig nests and lay eggs, before returning to the sea. Because turtles are so vulnerable at this time, many have been taken as food by humans at turtle nesting areas throughout the tropics, and turtle populations declined dramatically in the past century.

Conservationists began collecting eggs from sites where endangered species nested. These eggs were incubated in captivity, but many of these early efforts were of limited success. Because the eggs were carefully incubated at only one temperature, all the artificially reared young were of one sex.

Since the discovery of temperature-dependent sex determination, conservationists now incubate eggs at several temperatures or in thermally fluctuating environments that mimic the variation found at nesting sites in nature. Thus, young of both sexes are produced, and the future of these species is more secure.

Sex linkage

Genes on the sex chromosomes are **sex linked;** a gene that is on only the X chromosome is **X linked;** one found only on the Y chromosome is **Y linked.** Homologous genes that occur on both the X and Y chromosomes are *partially sex linked*. They should exhibit the same pattern of inheritance as autosomal genes. No partially sex-linked genes have yet been identified in humans, although a small amount of meiotic pairing suggests that some may exist.

Table 5.1
X-linked traits in humans

cleft palate Incomplete cleft of the secondary palate.

color blindness, partial Deutan series, insensitivity to green; accounts for roughly 75% of color blindness.

color blindness, partial Protan series, insensitivity to red; accounts for roughly 25% of color blindness.

congenital night blindness Night blindness associated with nearsightedness.

deafness At least three separate X-linked forms of deafness have been described for congenital and progressive deafness.

Duchenne muscular dystrophy Onset of muscle wasting by age six, chair ridden by age twelve, and death usually by age twenty. A heart defect is associated with this X-linked form of muscular dystrophy (figure 5.7a).

ectodermal dysplasia Absence of teeth and sweat glands, accompanied by scanty body hair and early baldness (figure 5.13).

glucose-6–phosphate dehydrogenase deficiency Reduced activity of a key enzyme of glucose metabolism. Usually this condition is not serious, but when persons of this genotype ingest certain drugs, their red cells lyse, resulting in anemia. Removal of the offending substances brings complete recovery.

hemophilia A, classical hemophilia Defect in antihemophilic factor.

hemophilia B, Christmas disease Defect in Christmas factor, plasma thromboplastic component.

ichthyosis Rough, scaly skin; present from birth (figure 5.7c).

Lesch-Nyhan syndrome Enzyme deficiency resulting in mental retardation, palsy, and self-mutilating behavior, involving compulsive biting of fingers and lips (figure 5.7b).

ocular albinism Absence of pigment in the eye accompanied with rolling of the eyeball.

retinitis pigmentosa Degeneration of the retina of the eye.

testicular feminization syndrome XY males with female external genitalia, breast development, blind vagina, and abdominal or inguinal testes.

vitamin D resistant rickets A dominant X-linked mutation producing a defect in vitamin D metabolism. This disease is not cured by high doses of vitamin D (figure 5.7d).

Xg(A) blood group antigen A blood group antigen inherited as a dominant X-linked trait. This antigen is useful as a genetic marker for linkage studies and for determining where nondisjunction occurred in cases of X chromosome aneuploidy.

From V. A. McKusick, *Mendelian Inheritance in Man,* 7th ed. Copyright 1986 Johns Hopkins University Press. Reprinted by permission.

X linkage

During meiosis in human males, the X and Y chromosomes do not line up side by side in normal synapsis. Instead, they pair end to end. Because of dissimilarity in gross morphology as well as the absence of normal pairing, the X and Y chromosomes probably share few genes.

At least 124 genes have been positively assigned to the X chromosome (table 5.1), and 162 more are suspected of being X linked. Because so few genes are known on the Y chromosome, the terms "X linked" and "sex linked" are often used synonymously.

X-linked traits have a unique mode of inheritance because females have two doses of an X-linked gene, while males have only one. Males are thus **hemizygous** for X-linked traits. Dominant X-linked genes are always expressed in both sexes,

Figure 5.4 The crisscross pattern of inheritance of the sex chromosomes. The two X chromosomes of the female and the one X chromosome of the male are distinguished by their patterning; they can carry the same or different alleles of all the X-linked genes. The mother contributes one of her chromosomes to her sons, who will express any sex-linked genes on the X because they are hemizygous. The daughters receive an X from each parent and will not express their father's X-linked genes, unless they are dominant X-linked genes or unless they also receive the same X-linked recessive from their mother. The crisscross refers to the mother to son, and the father to daughter transmission pattern.

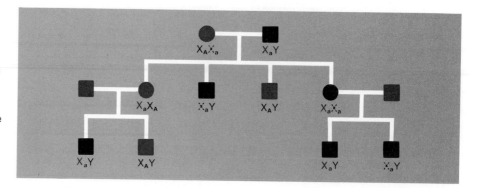

just as with autosomal traits. Because females have two X chromosomes, the frequency of dominant X-linked traits is always higher in females than in males. Such traits are expressed in both homozygous and heterozygous females.

The opposite is true for recessive alleles. Because they are hemizygous, males always express recessive X-linked alleles. Females, however, express recessive alleles only when they are homozygous. Thus the frequency of X-linked recessive traits is always lower in females than males.

Most X-linked genes have been discovered because they have recessive alleles with deleterious effects appearing in males. There are probably many common recessive traits whose effects are innocuous and thus escape the attention of geneticists.

Males transmit their X chromosome to every daughter, their Y chromosome to every son. Recessive X-linked traits thus show a pattern of inheritance in which the phenotype is usually expressed only in males of alternate generations. A male bearer transmits the recessive allele to a daughter, who does not express the allele because it occurs in the heterozygous condition. However, each of her male offspring has a 50% chance of receiving that allele and expressing the phenotype. The trait should thus appear in $\frac{1}{4}$ of her offspring ($\frac{1}{2}$ of her offspring are expected to be male, and $\frac{1}{2}$ of her sons receive the recessive allele: $\frac{1}{2} \times \frac{1}{2} = \frac{1}{4}$). The heterozygous female is a *carrier* of the allele. X-linked inheritance is often said to show a *crisscross* pattern of inheritance because of its pattern in pedigrees: the allele is transmitted male to female, female to male, but the trait is expressed only in males in alternate generations (figure 5.4).

Red-green color blindness

Many persons cannot see certain colors. The most common such defect is an inability to distinguish red from green. This condition, *partial* **color blindness,** affects more than five million American males. Two X-linked loci are now known to cause partial color blindness. One locus causes *deutan-type color blindness* (about 75% of the cases), while the other causes *protan-type color blindness*. Each gene affects different pigments in the retina, but both produce confused color perception. A far rarer condition is total color blindness, inherited as an autosomal recessive, where persons have no color perception at all.

Partial color blindness was recognized as an hereditary condition in the eighteenth century, but the mode of inheritance was not demonstrated until 1911. The recognition that genes are located on chromosomes soon led to association of partial color blindness and the X chromosome, both of which show the crisscross pattern of inheritance. Color blindness and hemophilia, both X linked, were the first genes in humans assigned to a specific chromosome. In 1986 the genes corresponding to the red, green, and blue visual pigments were isolated and sequenced.

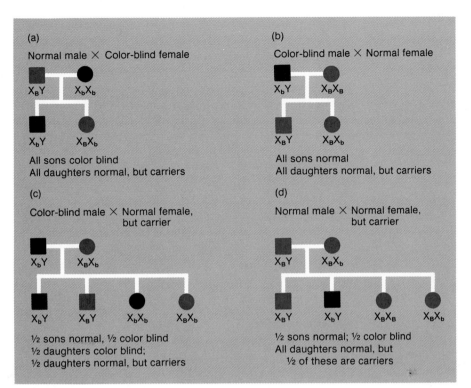

Figure 5.5 Four possible matings involving inheritance of the sex-linked recessive trait, partial color blindness: (a) normal male × color-blind female; (b) color-blind male × normal female; (c) color-blind male × normal female who is a carrier; (d) normal male × normal female who is a carrier.

(a)

Normal male × Color-blind female

X_BY — X_bX_b

X_bY X_BX_b

All sons color blind
All daughters normal, but carriers

(b)

Color-blind male × Normal female

X_bY — X_BX_B

X_BY X_BX_b

All sons normal
All daughters normal, but carriers

(c)

Color-blind male × Normal female, but carrier

X_bY — X_BX_b

X_bY X_BY X_bX_b X_BX_b

½ sons normal, ½ color blind
½ daughters color blind;
½ daughters normal, but carriers

(d)

Normal male × Normal female, but carrier

X_BY — X_BX_b

X_BY X_bY X_BX_B X_BX_b

½ sons normal; ½ color blind
All daughters normal, but
    ½ of these are carriers

Patterns of sex-linked inheritance do *not* conform to Mendel's Law of Parental Equivalence. Figure 5.5 shows the results of reciprocal matings of affected and unaffected parents. The mating of a normal male with a color-blind (recessive homozygous) female produces normal but heterozygous daughters. All sons have the disease (figure 5.5a). The reciprocal mating demonstrates crisscross inheritance. A color-blind male (hemizygous) mated with a normal (homozygous) female produces no affected offspring; the sons receive a normal X from their mother, while all the daughters are carriers (figure 5.5b). A color-blind male mated with a carrier female is analogous to a testcross: 50% of the offspring of both sexes are normal, 50% are color blind (figure 5.5c). Another possibility, a normal male mated with a carrier female, produces all normal female offspring, but 50% of male offspring are affected (figure 5.5d). The mating of two color-blind individuals results in all color-blind offspring, if they have the same type of color blindness.

Hemophilia

Hemophilia, also known as "bleeders' disease," is a condition characterized by the inability to form blood clots. An affected individual is in continual danger of bleeding to death from the smallest cut or from internal hemorrhaging after even a minor bruise. Rheumatism is a common complication due to bleeding within the joints.

At least three different genetic loci control different forms of hemophilia. The rarest is controlled by an autosomal recessive gene, but the other two result from recessive alleles at two X-linked loci. One of these, *Christmas disease,* or *hemophilia B,* comprises about 20% of all hemophilia and is caused by an abnormality or deficiency of plasma thromboplastin component.

Hemophilia A, classical hemophilia, is caused by an abnormality or deficiency in anti-hemophilic factor, AHF. Until recently, hemophilia A was untreatable; only about 25% of affected males reached age twenty-five. Treatment with AHF now results in a more normal life span, but the cost is high (six thousand dollars to ten thousand dollars per year).

The frequency of hemophiliacs is about one in ten thousand males, but much lower in females, perhaps one in one hundred million or less. A female hemophiliac can result only from the mating of a heterozygous female with an affected male. Such a mating is highly unlikely because, until recently, few male hemophiliacs survived long enough to reproduce. Females were also believed to die at the onset of menstruation, but some female hemophiliacs have now survived to maturity.

Hemophilia may have been understood as an inherited disease longer than any other genetic condition. The Talmud, the Hebrew book of law, noted that when excessive bleeding occurred during circumcision of two male infants of one mother, future male offspring were exempt. When sons of three sisters exhibited bleeding, sons of other sisters were also exempt. However, sons of brothers were not exempt, implying an understanding of the crisscross pattern of inheritance.

Hemophilia A has been called the "Royal disease" because it affected males in the royal families of Europe. Queen Victoria, a carrier of the hemophilia allele, had nine children (figure 5.6). Her eighth child, Leopold, was a hemophiliac who died at age thirty-one. Her other three sons were unaffected, so we know they did not receive the allele. One daughter had no children; her status as a carrier cannot be assessed. Two daughters had children, none of whom were hemophiliac, indicating that the mothers probably were not carriers. Two other daughters were carriers giving birth to hemophiliac sons.

The possible historical influence of hemophilia is tantalizing. Victoria's third child was Princess Alice, whose daughter Alix married Czar Nicholas II of Russia. The Czarina, Alix, had four daughters before giving birth to the long-awaited son, Alexis, heir to the Russian throne. Unfortunately, Alexis had the hemophilia allele, a legacy from his great-grandmother, Queen Victoria. Distressed over their son's condition, the Czar and Czarina turned to the monk Rasputin, a charlatan who claimed he could help Alexis. While affairs of state deteriorated, culminating in the Russian revolution, the Czar was preoccupied with the health of his son and manipulated by Rasputin.

Among Victoria's descendants, eight of twenty-five males in four generations were hemophiliac. Queen Victoria almost certainly received the gene for hemophilia A as the result of a mutation on the X chromosome she received from her father, Edward, Duke of Kent. He was fifty-two years old at the time of Victoria's birth, and such mutations may occur more frequently in the germ cells of older males.

More recently, a serious threat to victims of hemophilia has arisen due to their continuing dependence upon blood transfusions. Such transfusions are one means of contracting **acquired immune deficiency syndrome (AIDS),** and some hemophiliacs have in fact acquired AIDS in this way. Improved methods of screening donor blood for infection by the AIDS virus has reduced but not entirely eliminated this risk. Bioengineering techniques (chapter 13) offer the long-term hope that the defective gene(s) causing hemophilia can be treated directly. In the meantime, continued surveillance of donor blood supplies is required to protect hemophiliacs and all others requiring transfusions.

Duchenne muscular dystrophy

Most forms of muscular dystrophy are controlled by autosomal genes. **Duchenne muscular dystrophy (DMD),** however, is controlled by an X-linked recessive allele and primarily affects young males (figure 5.7a). Muscular deterioration begins between the ages of three and five years, by the time affected individuals reach their teens, they are confined to wheelchairs; and they die in their early twenties from atrophy of the respiratory muscles. Few affected males reproduce, so the condition is transmitted mainly by female carriers. Duchenne muscular dystrophy occurs in

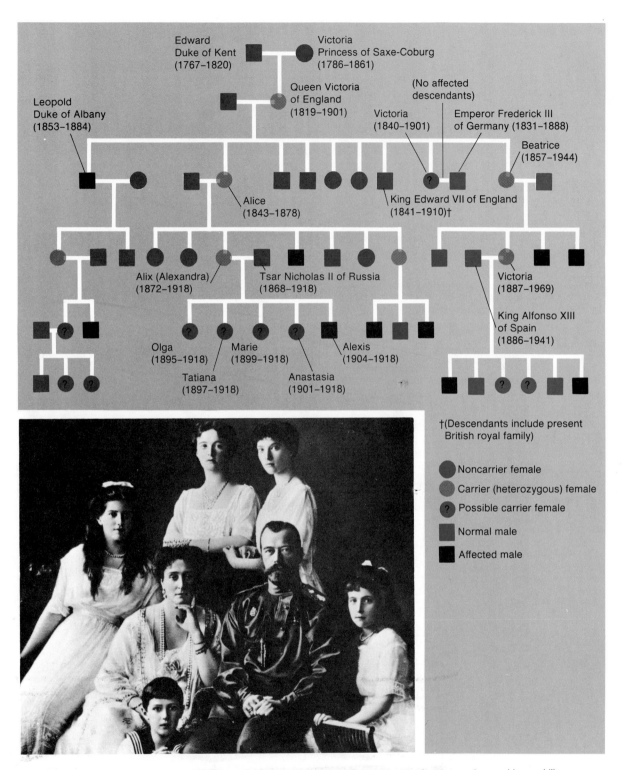

Figure 5.6 The Royal Hemophilia. Partial pedigree of the descendants of Queen Victoria showing carriers and hemophiliacs, as well as normals and possible carriers. The present royal family of Great Britain is free of the hemophiliac allele. The photograph is of Czar Nicholas II, the Czarina Alexandra, their four daughters, and their son Alexis, who had hemophilia.
Source: Historical Pictures Service, Chicago.

Figure 5.7 Examples of conditions caused by sex-linked genes. (a) Young boy exhibiting muscle deterioration typical of Duchenne muscular dystrophy; (b) a boy with Lesch-Nyhan syndrome, characterized by compulsive self-destructive behavior; (c) a boy afflicted with the skin disorder ichthyosis; (d) a mother who is heterozygous for the dominant vitamin D-resistant rickets and her children.
Source: (a) Stanley Willard; (b) with permission of W. L. Nyhan and N. A. Sakati. *Diagnostic Recognition of Genetic Disease:* Lea & Febiger, 1987, Philadelphia; (c) Lester V. Bergman and Associates, Inc.; (d) E. W. Lovrien.

(a)

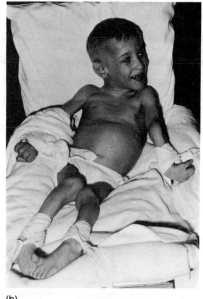

(b)

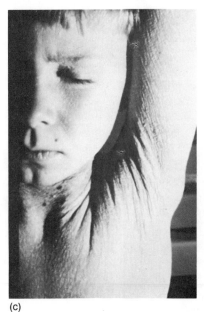

(c)

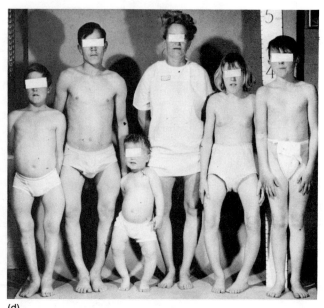

(d)

roughly one in every four thousand newborn males; perhaps one-third of these are due to new mutations. This is several times more frequent than the occurrence of hemophilia; other X-linked disorders are far rarer.

In 1986 the defective gene that causes Duchenne muscular dystrophy was isolated and cloned. This led to the identification in early 1988 of a protein (dystrophin) whose absence causes DMD. While this finding does not constitute a cure for DMD, it offers new research avenues that may eventually allow treatment. As with hemophilia, a longer-term hope may be direct modification of the defective DNA (see chapter 13).

Ichthyosis
Consanguineous matings can greatly increase the frequency of expression of X-linked traits in females. In consanguineous pedigrees containing X-linked recessive alleles, females have a high probability of carrying the X-linked allele, as they can receive

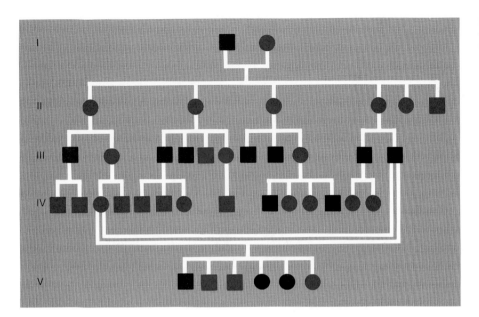

Figure 5.8 A pedigree demonstrating the occurrence of an X-linked recessive trait, ichthyosis, in females, as a result of consanguineous mating.

the allele from either parent. In turn, matings of carrier females and affected males (figure 5.5c) produce daughters and sons with an equal likelihood (50%) of being affected.

Ichthyosis is a disorder characterized by extreme dryness, roughness, and scaliness of the skin (*ichthy* meaning fish-like). One form, usually present at birth, is caused by an X-linked recessive allele (figure 5.7c). Figure 5.8 shows a pedigree of a family in which consanguineous mating resulted in expression of this X-linked trait in females.

The allele for ichthyosis first appeared in male I-1. His first four daughters all transmitted the allele, without expressing it themselves. Male III-11 mated with his first cousin, once removed, IV-3, who must have been a carrier, having received the ichthyosis allele through two generations of females. One of their three sons and two of their three daughters exhibited ichthyosis, a highly unlikely result without consanguineous mating.

Chicken feathers: ZZ-ZW linkage

Sex-linked patterns of inheritance in birds are identical to those in mammals, except that the female, not the male, is the hemizygous sex. A dominant Z-linked allele, designated S, has been discovered in chickens. Bearers of the S allele have silver-colored down at hatching. The alternate allele, s, produces a gold-down feather. This locus has been put to practical use by many chicken farmers.

For egg production, male chicks are of no value. In fact, males cost the farmer money, but the sexes cannot be distinguished until they reach sexual maturity. The S allele can be used to distinguish the sexes immediately after hatching, thus permitting farmers to dispatch males and raise only females (figure 5.9). If a hemizygous (SW) silver female is mated to a gold (ss) male, crisscross inheritance produces an obvious phenotypic distinction between the sexes. Female chicks are all gold (sW), while males are silver (Ss). Males can then be sorted at birth, saving the farmer the expense of raising them to maturity.

Figure 5.9 Feather color in chickens, a sex-linked trait used by poultry breeders to select breeding chickens. Down feather color is controlled by a pair of alternate alleles, S, producing a silver color down, and s, producing a gold color. Gold-feathered males are bred with silver-colored females. All male offspring will have silver down and all female offspring will have gold down. Thus breeders can select the females for raising, rather than raising all chicks. (See colorplate 7.)

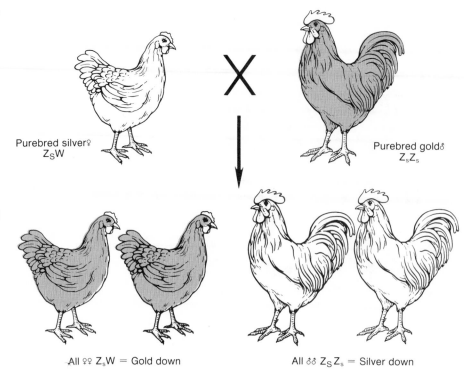

Purebred silver♀
Z_SW

Purebred gold♂
Z_sZ_s

All ♀♀ Z_sW = Gold down

All ♂♂ Z_SZ_s = Silver down

Y linkage

Genes that occur only on the Y and not the X chromosome are *Y linked* or *holandric* (meaning entirely male). Y-linked genes should have the simplest possible pattern of inheritance. Because only males have a Y chromosome, Y-linked traits should appear only in males. Because they are hemizygous, every Y-linked gene should be expressed. Each male contributes his one and only Y chromosome to every male offspring, so Y-linked traits should show strict father-to-son transmission.

The only genes known for certain to be Y-linked play an important role in differentiation of the testes in males (see chapter 6), and it is not surprising (it is hard to imagine otherwise) that the gene determining maleness is found on the Y chromosome. Remember that in mammals the presence of the Y chromosome is the crucial karyotypic factor in sex determination. Genes affecting gonadal differentiation, spermatogenesis, stature, and other aspects of maleness are likely to be on the Y chromosome.

X inactivation

It is sometimes possible to distinguish the phenotypes of individuals who are homozygous or heterozygous for autosomal dominant alleles. Consider a gene producing compound A. Individuals who are AA normally produce twice as much of this compound as individuals who are Aa. Thus, it is reasonable to ask if homozygous normal females produce twice as much gene product for X-linked genes as hemizygous normal males. Alternatively, are females homozygous for a deleterious X-linked allele affected twice as severely as affected males? The answer appears to be no: X-linked traits are usually phenotypically similar in males and females, due to a phenomenon known as **dosage compensation.**

Early in embryonic development in female mammals, one of the two X chromosomes is turned off or inactivated. In spite of possessing twice as many X chromosomes as a male, a female is functionally hemizygous at X-linked loci, just like males.

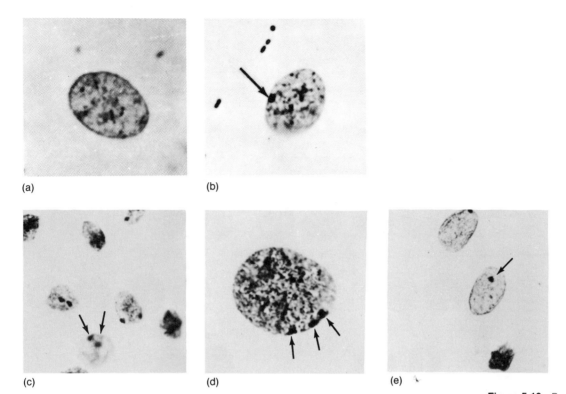

(a) (b) (c) (d) (e)

Figure 5.10 Barr bodies are darkly staining bodies seen in the nuclei of nondividing cells of normal females. They are inactivated X chromosomes. Turner females (XO) do not have Barr bodies, while Klinefelter males (XXY) have a single Barr body. XXX females would have two Barr bodies. (a) Normal male, no Barr body; (b) normal female, one Barr body; (c) XXX female, two Barr bodies; (d) XXXX female, three Barr bodies; (e) Klinefelter male, one Barr body.
Source: (a and b) Wilson and Foster, *Williams Textbook of Endocrinology,* 7th ed. © W. B. Saunders 1985; (c–e) National Jewish Hospital and Research Center.

Giemsa staining of somatic cells reveals **X inactivation** as a distinct sexual dimorphism. In cells of females, one darkly staining region is found at the edge of the nucleus (figure 5.10). This region, named a **Barr body** (after its discoverer, M. L. Barr), is the inactivated X chromosome. One X chromosome is necessary for normal development in both sexes, but if an individual (of either sex) has more than one, all but one are inactivated and visible in stained somatic cells as a Barr body. Thus, somatic cell nuclei of normal males have no Barr bodies, and those of normal females have one.

The hypothesis that all but one X chromosome are inactivated in each cell was proposed by geneticist Mary Lyon in 1961 and is known as the **Lyon hypothesis.** Crucial evidence for this hypothesis was provided by sexually aneuploid individuals. **Aneuploidy** (meaning not true number) refers to the possession of an abnormal number of chromosomes. Aneuploid individuals have the normal diploid number, *plus or minus* one or more chromosomes. Females lacking one X chromosome exhibit Turner's syndrome, designated 45, XO (for 45 chromosomes, with one X missing). Males with an extra X have Klinefelter's syndrome, designated 47, XXY. Cells from 45, XO females have no Barr bodies, while those from 47, XXX females have two; 47, XXY males have one Barr body, 48, XXXY males have two (figure 5.10).

Testing for the number of Barr bodies is simple and inexpensive compared to preparation of a complete karyotype. Examination of Barr bodies can be done easily to screen for sex-chromosome abnormalities. The mandatory "sex tests" that have been required of Olympic athletes include a count of Barr bodies in epithelial cells scraped from the inside of the cheek. Males disguised as females have no Barr bodies.

The inactivation of X chromosomes during development apparently occurs at random. Early in development, the maternally-derived X is inactivated in some cells, while the paternally-derived X is inactivated in others. Thereafter, descendants of

Figure 5.11 Barr body formation in human females. In about the third week of embryonic development, one of the two X chromosomes in each cell becomes inactivated to form a Barr body. All descendants of these cells have the same chromosome inactivated, so females are functional mosaics for their maternally-derived and paternally-derived X chromosomes. In the germinal tissues Barr bodies are not seen. Both X chromosomes may need to be functional in these cells.

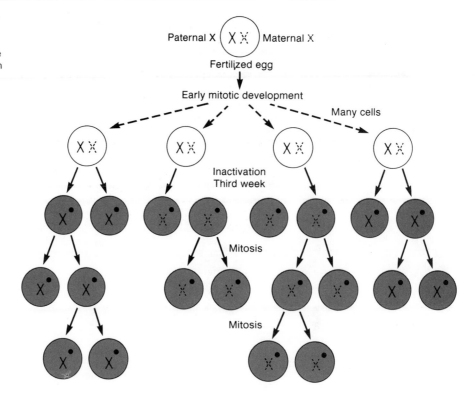

a particular cell have the same X inactivated (figure 5.11). If a female is heterozygous for an X-linked gene, she is **mosaic** for that trait. One of her X chromosomes is active in roughly half of her cells, while the second X is active in her other cells.

Calico cats exhibit mosaicism due to dosage compensation. Several loci control coat color in cats, but only one X-linked locus is involved in producing calico individuals. Two alleles occur at that locus, R and R′. In hemizygous males, R produces rust coat color, R′ black. In females, X inactivation produces clones of R-bearing rust fur intermixed with R′-bearing black fur—the calico or tortoiseshell cat (figure 5.12). Thus, almost all calico cats are females. The only male calico cats result from sex-chromosome aneuploidy. XXY males also undergo X inactivation, so an occasional calico male occurs.

When a female is heterozygous for a deleterious X-linked allele, the effects of the cell lines bearing the normal allele may compensate for the harmful effects of the cell lines bearing the deleterious allele. In females heterozygous for partial color blindness, for example, some cell clones in the retina are in fact color blind, but the presence of other normal clones results in normal color vision.

X inactivation in humans can sometimes be seen in females heterozygous for certain X-linked traits. In 1875 Darwin described *ectodermal dysplasia,* a condition now known to be X linked that involves the lack of some teeth and sweat glands. Heterozygous females show a mosaic of areas of the jaw with and without teeth, and patches of skin with and without sweat glands. In figure 5.13, the females in generation III are identical twins; they have developed from a single fertilized egg and are therefore genetically identical. However, due to the *random* inactivation of different X chromosomes during development, they show considerable difference in the location of patches of skin lacking sweat glands. X-chromosome inactivation is one example of a developmental process that can produce *phenotypic* differences in *genotypically* identical individuals. Thus, even clones may differ significantly.

Figure 5.12 A calico with patches of color resulting from random inactivation of X chromosomes bearing color determining genes in cells giving rise to hair. (See colorplate 8 for more examples.)
Source: Mike Chiaverina.

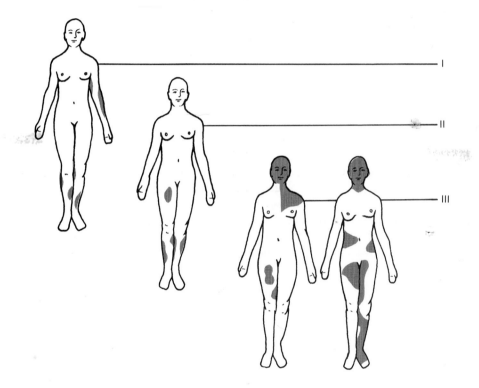

Figure 5.13 Mosaic phenotypes in three generations of females, all heterozygous for the X-linked gene for ectodermal dysplasia. Stippling indicates areas in which sweat glands are missing. The affected areas differ in each woman because the controlling gene is inactivated randomly, even in the identical twins in generation III.

Sex-influenced inheritance

Two types of inheritance exhibit sexually related effects, although neither is sex linked. **Sex-influenced** genes show differing patterns of expression in each sex—usually, the trait behaves as a dominant in one sex and a recessive in the other. Sex-influenced genes occur only on autosomes.

The best documented example of sex-influenced inheritance involves *pattern baldness*. Individuals expressing pattern baldness begin to lose their hair relatively early in life, often in their twenties. Affected individuals are not totally bald: a distinct rim of hair surrounds the head in patterns varying from person to person. This trait is controlled by an autosomal gene. The allele for pattern baldness is dominant in males, and heterozygous males therefore express pattern baldness. In females, however, the gene is recessive.

...ing
...orn
...ited
...somes
...arge horns in
...emales also carry
...s gene, but do not
express it.
Source: Leonard Lee Rue III/
Photo Researchers.

Sex-limited inheritance

Some autosomal genes are expressed only in one sex or the other and are thus **sex limited.** The most obvious examples are the external genitalia and other secondary sex characteristics. Only females, for example, possess fully developed breasts, and only males develop full beards. Nonetheless, the genes controlling the development of these characters occur in both sexes, although the traits are only expressed in one. In chapter 6, we describe how all individuals possess the potential for developing the phenotype of either sex, indicating clearly that all of us possess genes for traits of the opposite sex that are normally never expressed. The horns of male sheep and goats are a non-human example of sex-limited traits (figure 5.14).

Sex ratios

Because sex is determined by the Y chromosome, and because males produce X- and Y-bearing gametes in approximately equal numbers, Mendel's Law of Segregation predicts that the sexes should occur in equal proportions, or as commonly expressed, in a male:female ratio of 1:1. In most species most of the time, the number of males and females is about equal, but that is not always the case. In humans, for example, different **sex ratios** occur for different age groups. The **primary sex ratio** is the sex ratio at conception. It is extremely difficult to measure, but it certainly favors males. Estimates as high as 1.60:1.00 have been made. The **secondary sex ratio** (at birth) in white Americans is about 1.06:1.00, again favoring males.

Tertiary sex ratios may be measured at any time after birth. In humans, the sexes are about equally abundant around age twenty, but females increasingly outnumber males at later ages. The mean life expectancy of white females in the United States is about ten years longer than for males, which is reflected in the unbalanced sex ratio at later ages.

Several factors seem to affect the sex ratio. For unknown reasons, males of many species have a higher death rate than females at all ages. In humans, it has long been postulated that male hemizygosity for the X chromosome could be responsible. Any deleterious X-linked genes are expressed in males, but may be hidden in females, if recessive. That hypothesis has not been proven, however, and there is some evidence against it. Remember that in all female cells except oocytes, one of the X chromosomes is inactivated (the Barr body), so that deleterious genes on the active X should be expressed, just as in males. Further, in many species without dimorphic sex chromosomes, males still have a higher death rate.

Natural selection apparently favors equality in the sex ratio at the time of sexual maturity. The ratio is 1:1 in humans at about age twenty, for example. Thus, the proportion of males at birth is increased sufficiently that the higher male death rate produces a balanced sex ratio at about puberty.

Given the apparently equal production of X- and Y-bearing gametes by meiosis, how can the sexes occur in other than equal proportions? As usual, both environmental and genetic factors can alter the balance. As described earlier, in sea worms and deep-sea fish, juveniles develop into a female if they cannot find a female upon whom they can become parasitic males. Such a mechanism virtually guarantees one male per sexually mature female. As we noted earlier, the temperature at which eggs of some turtles are incubated alters the proportion of the sexes produced.

Several genetic factors can also alter sex ratios. Any sex-linked recessive lethal gene causes the death of hemizygous males, thus unbalancing the ratio in favor of females. Additionally, single genes are known in many species that cause unequal production of gametes with respect to different characteristics.

Finally, physical characteristics of X- and Y-bearing sperm may make a difference. Y sperm are lighter and thus may swim faster. If Y sperm regularly reach the egg first, the probability of male offspring should be increased. Dairy cattle breeders have attempted to use the difference in weight to separate the two types of sperm by centrifugation of bull semen. The heavier X-bearing sperm are then used for artificial insemination to increase the probability of conceiving milk-producing heifers. This technique has met with varied success.

Summary

1. In many species the sexes exhibit marked sexual dimorphism in both primary and secondary sexual characteristics.
2. Hermaphrodites possess sexual characteristics of both sexes.
3. Mechanisms of sex determination probably involve the interaction of environmental factors with many genes located on many chromosomes.
4. In many species a major chromosomal dimorphism is associated with the two sexes. The three most common chromosomal mechanisms of sex determination are the XX-XY, XX-XO, and ZZ-ZW systems.
5. In bees, wasps, and some vertebrates, the sexes exhibit different degrees of ploidy.
6. For some species, environmental factors exert primary control over the sex of an individual.
7. In species with an XX-XY mechanism, genes on the sex chromosomes may be X linked, Y linked, or partially sex linked.
8. Recessive X-linked traits show a crisscross pattern of inheritance. Female heterozygotes are carriers who pass the trait to 50% of their male offspring. Recessive X-linked traits are expressed far less commonly in females than males.
9. The only traits that seem likely to be Y linked are some that directly affect certain sexual dimorphisms.
10. In female mammals, dosage compensation occurs by the inactivation of one of the X chromosomes in all somatic cells. This is called the Lyon hypothesis, and its effects are seen clearly in female mosaics.

11. Sex-limited and sex-influenced traits are the result of genes on the autosomes. The expression of these genes is sexually dimorphic.
12. For most species, the sex ratio is about 1:1 most of the time, but many factors are known that can greatly alter this ratio.

Key Words and Phrases

acquired immune deficiency syndrome (AIDS)

aneuploidy

Barr body

color blindness

dosage compensation

Duchenne muscular dystrophy (DMD)

hemizygous

hemophilia

hermaphrodite

heterogametic sex

homogametic sex

Lyon hypothesis

mosaic

parthenogenesis

ploidy

primary sexual characters

secondary sexual characters

sex-influenced inheritance

sex-limited inheritance

sex linked

sex ratios: primary, secondary, tertiary

sexual dimorphism

X inactivation

X linked

XX-XO systems

XX-XY systems

Y linked

ZZ-ZW systems

Questions

1. A woman has parents with normal color vision, but her brother is color blind. What is her genotype?
2. A woman with normal color vision, whose father was color blind, marries a man with normal color vision. What proportion of her children will be color blind? If her husband had been color blind, what proportion of her children would be color blind?
3. A man who has protan-type color blindness marries a woman who has deutan-type color blindness. What phenotype or phenotypes will their offspring have?

4. A non-bald man marries a non-bald woman whose mother was bald. What is the probability that their first child will be bald?

5. Woolly hair is inherited as an autosomal dominant trait. If a woolly-haired man marries a woman with normal hair, what proportion of their sons will have woolly hair?

6. The diploid number in cats is 38. How many chromosomes would you expect in a female calico cat? In a male calico cat?

7. It is possible for an individual to be a mosaic of two or more cell lines. In an individual whose karyotype is mosaic 45, XO/47, XXY, how many Barr bodies would you predict to find in the somatic cells?

8. The size and shape of female breasts are sex-limited traits. A small-breasted mother produced a daughter with large breasts. To what do you attribute this?

For More Information

Arehart-Treichel, J. 1983. "Duchenne Muscular Dystrophy: A Cure in Sight?" *Science News,* 123:42–43.

Avers, C. J. 1974. *The Biology of Sex.* J. Wiley and Sons, New York.

Diamond, J. M. 1986. "Variation in Human Testis Size." *Nature,* 320:488–89.

Goodfellow, P. N. 1986. "Duchenne Muscular Dystrophy: Collaboration and Progress." *Nature,* 322:12.

Kolata, G. 1986. "Maleness Pinpointed on Y Chromosome." *Science,* 234:1076–77.

Kowles, R. V. 1985. *Genetics, Society, & Decisions.* Scott, Foresman & Co., Glenview, Illinois.

McKusick, V. A. 1965. "The Royal Hemophilia." *Scientific American,* 213:88–95.

McKusick, V. A. 1986. *Mendelian Inheritance in Man,* seventh edition. The Johns Hopkins University Press, Baltimore.

Miller, J. A. 1986. "The Genes behind Vision's Palette." *Science News,* 129:246.

Mittwoch, U. 1963. "Sex Differences in Cells." *Scientific American,* 209:54–70.

Molton, J. D. 1986. "Understanding Color Vision." *Nature,* 321:12–13.

Morse, G. 1984. "Why Is Sex?" *Science News,* 126:154–57.

Rushton, W. A. 1975. "Visual Pigments and Color Blindness." *Scientific American,* 232:64–74.

Stine, G. J. 1977. *Biosocial Genetics.* Macmillan Publishing Co., New York.

Human sexuality: genotype and phenotype

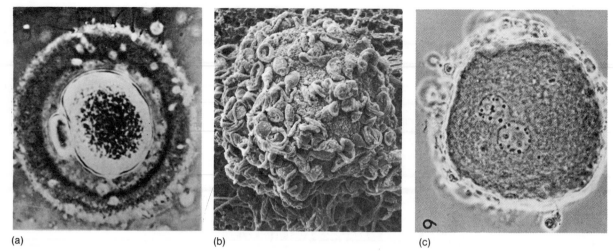

(a) (b) (c)

Figure 6.1 A human egg at the time of conception: (a) with the polar body visible; (b) covered with sperm, only one of which will achieve fertilization; (c) after the sperm has penetrated the egg, but prior to uniting with the egg nucleus (both sperm and egg nuclei are visible).
Source: (a) Dr. Landrum Shettles; (b) David Scharf/Peter Arnold; (c) Z. Dickman, T. H. Cleve, W. A. Bonney, and R. W. Noyes, "Anatomical Records," 152 (1965).

Human sexual function and identity is a complex mixture of genetic, morphological, physiological, and psychological traits. Understanding human sexuality thus requires knowledge of the contribution and relationships of these varied factors. In this chapter, we discuss aspects of human sexuality that demonstrate the complexity of the genetic and biological processes that control the sexual phenotype.

Human reproductive anatomy

The primary sexual structures are the *gonads,* testes in males and ovaries in females. Gonads of both sexes serve two functions: (1) production of haploid gametes and (2) production of hormones that control a variety of physical and physiological traits. Testes and ovaries share a common origin from undifferentiated gonadal tissue that appears in the abdomen of the developing fetus about one month after conception. By the end of the first three months of gestation, the embryo has begun to exhibit clear differentiation in secondary as well as primary sex structures.

Female anatomy

In an adult female, the **ovaries** are paired oval structures about $1\frac{1}{2}$ inches long. One is located on each side of the upper part of the pelvic cavity (figure 6.2), held in place by suspensory ligaments. Internally, each ovary contains about four hundred thousand *ovarian follicles,* which were formed during fetal development. Each follicle consists of one **primary oocyte** (meaning first egg cell), surrounded by supportive cells.

Following *menarche,* the onset of menstruation and puberty, one or two primary oocytes begin development per month, continuing meiosis and **oogenesis** to produce one egg or **ovum** per oocyte. A female thus ovulates only about four hundred to five hundred eggs throughout her lifetime, one to two per month for thirty-five to forty years. The remaining oocytes never develop, and ovulation ceases at *menopause.* The ovaries produce the sex hormones, **estrogen** and **progesterone,** throughout a female's lifetime.

The remainder of the female reproductive system is adapted for receiving the male penis and sperm, for the passage of sperm and egg, and for the development and birth of a fetus, should fertilization occur. Lying over each ovary is the fringed

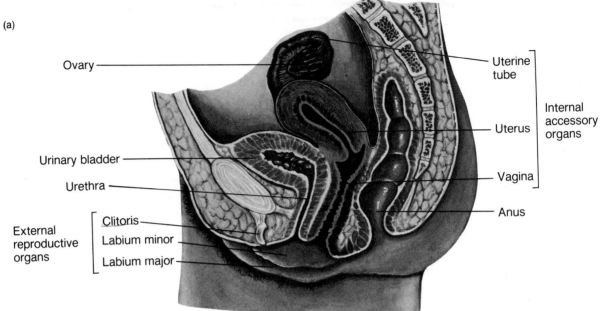

(a)

Ovary

Uterine tube

Internal accessory organs

Uterus

Urinary bladder

Urethra

Vagina

Anus

External reproductive organs

Clitoris

Labium minor

Labium major

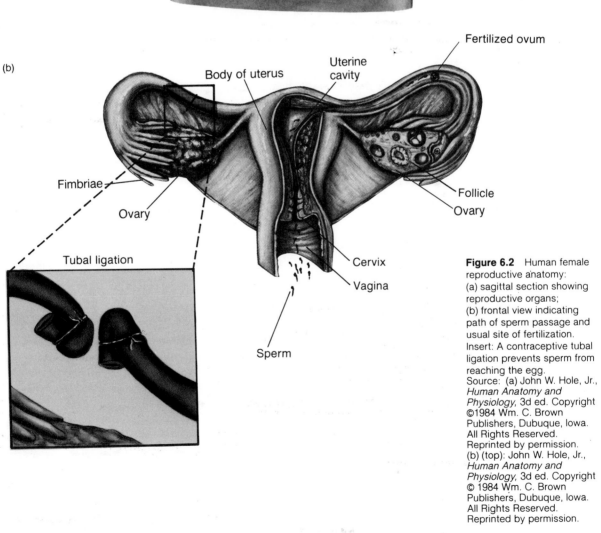

(b)

Body of uterus

Uterine cavity

Fertilized ovum

Fimbriae

Ovary

Follicle

Ovary

Tubal ligation

Cervix

Vagina

Sperm

Figure 6.2 Human female reproductive anatomy: (a) sagittal section showing reproductive organs; (b) frontal view indicating path of sperm passage and usual site of fertilization. Insert: A contraceptive tubal ligation prevents sperm from reaching the egg.
Source: (a) John W. Hole, Jr., *Human Anatomy and Physiology,* 3d ed. Copyright ©1984 Wm. C. Brown Publishers, Dubuque, Iowa. All Rights Reserved. Reprinted by permission. (b) (top): John W. Hole, Jr., *Human Anatomy and Physiology,* 3d ed. Copyright © 1984 Wm. C. Brown Publishers, Dubuque, Iowa. All Rights Reserved. Reprinted by permission.

entrance to the *fallopian tube* or *oviduct*. At ovulation, an ovum is released by one of the ovaries and enters the fallopian tube. If sperm are present, fertilization normally occurs in the fallopian tube. If sperm do not reach the egg within about twenty-four hours after ovulation, the egg dies.

The ovum (figure 6.1), whether fertilized or not, travels down the fallopian tube and enters the *uterus*. The uterus is a muscular, hollow organ located low in the pelvic cavity. The fallopian tubes enter on each side near the broad upper end. If the egg has been fertilized, it implants in the highly vascularized lining of the uterus, the *endometrium*. If the egg has not been fertilized, the endometrium is sloughed off about once every twenty-eight days as part of the menstrual flow.

Implantation occurs about seven to eight days after fertilization. The uterine wall grows to surround the developing zygote, and as the zygote also grows, tissues of both zygote and mother form a complex, intertwined structure called the **placenta.** The placenta provides a structure for the passage of nutrients from the mother to the fetus and of waste products from the fetus to the mother. The placenta also serves as a filter, isolating the fetus from the larger molecules in the mother's circulatory system. Nonetheless, the fetus is not perfectly protected. Toxins such as alcohol and nicotine readily cross the placenta, and some of the mother's own proteins and antibodies may also cross the placenta, damaging the fetus as in the case of Rh incompatibility.

The lower, constricted portion of the uterus, the *cervix*, opens into the *vagina*, a tubular passageway from the uterus to the surface of the body. The vagina serves not only as a birth canal, but also as a passageway for menstrual flow and, in the opposite direction, for sperm.

Externally, the opening of the vagina is surrounded by two sets of folds of skin and adipose tissue, the *labia majora* (the external lips) and the *labia minora* (the internal lips). Where the labia minora join anteriorly is the small, cylindrical *clitoris*. The clitoris is homologous to the male penis in structure and function. It is erectile tissue, densely innervated and functioning in tactile stimulation during intercourse. Located within the folds of the labia minora are two sets of glands, which produce mucus for lubrication during intercourse.

Male anatomy

The male reproductive system (figure 6.3) does not have to fulfill as many functions as the female system. The male system is designed primarily for the production of the male gametes (the sperm) in the testes and the transport of sperm to the female reproductive tract.

Like the ovaries, the **testes** originate high in the pelvic cavity, but during fetal development they descend into the *scrotum* (box 6.1). Internally, each testis is divided into a series of *lobules*. Within each lobule are long, convoluted *seminiferous tubules*. The walls of the seminiferous tubules are the site of **spermatogenesis,** and the immense surface area on the interior of the seminiferous tubules allows production of millions and millions of sperm (figure 6.4). Interstitial cells between the seminiferous tubules secrete **testosterone,** the male sex hormone.

Developing sperm cells are released into the lumen, or hollow interior, of the seminiferous tubules, where they travel to the *epididymis* on the surface of the testis. The epididymis is a long, coiled tube in which sperm are stored as they complete maturation.

Sufficient sexual stimulation results in *ejaculation*, the expulsion of mature sperm and accessory fluids. The sperm are forced through the system by the rhythmic contraction of muscles in the walls of the epididymis and other portions of the male reproductive tract. The sperm themselves have long tails for "swimming" that begins only after deposition in the female reproductive tract.

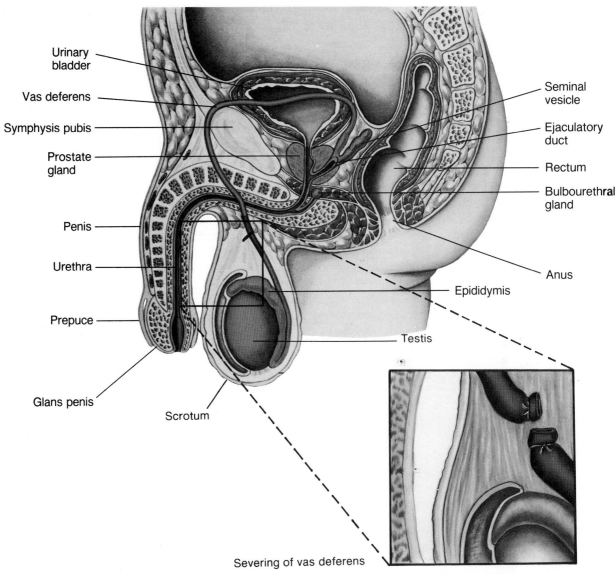

Urinary
bladder

Vas deferens

Symphysis pubis

Prostate
gland

Penis

Urethra

Prepuce

Glans penis

Scrotum

Seminal
vesicle

Ejaculatory
duct

Rectum

Bulbourethral
gland

Anus

Epididymis

Testis

Severing of vas deferens

Figure 6.3 Human male
reproductive system in
sagittal section. Insert:
Contraceptive vasectomy
prevents sperm transport.
Source: (top) John W. Hole,
Jr., *Human Anatomy and
Physiology,* 3d ed. Copyright
© 1984 Wm. C. Brown
Publishers, Dubuque, Iowa.
All Rights Reserved.
Reprinted by permission.

As the epididymis contracts, sperm are forced into the *vas deferens,* which passes through the *inguinal canal* and up into the lower pelvic cavity. Each of the vasa deferentia actually loops over the pubic bone, one passing on each side of the urinary bladder. They then enter the *prostate gland,* within which they join the *urethra.* The urethra, the tube that drains the urinary bladder, also passes directly through the center of the prostate. Beyond the prostate, the products of the urinary and reproductive systems share a common passageway, which is still called the urethra. In females, the urethra is always separate from the reproductive tract, although it opens within the space between the labia minora.

The urethra passes through the length of the *penis.* The shaft of the penis consists of three parallel columns that, during sexual excitement, become engorged (filled with blood), resulting in *erection.* The head or *glans* is enlarged and richly innervated. The opening of the urethra is through the glans.

The ejaculate, called **semen,** consists not only of sperm cells, but also the products of several *accessory glands.* The *seminal vesicles* open into the vasa deferentia just as they enter into the prostate. They secrete a viscous alkaline fluid that comprises over half the volume of the semen and provide nutrients for the sperm. The

The testes

The testes illustrate the delicacy and precision of the reproductive process. Male gametogenesis requires a temperature of about 35°C (95°F), several degrees lower than core body temperature. Therefore, the testes are suspended from the body in a pouch, allowing sufficient cooling for successful gametogenesis. The walls of the scrotum contain the *dartos muscle,* which contracts when the temperature is cold, warming the testes, and relaxes when the temperature is hot, cooling them. Failure of the testes to descend during gestation is called *cryptorchidism,* and it results in sterility due to lack of gametogenesis. Cryptorchidism can be treated easily by hormone administration or surgery, if necessary.

However, the location of the delicate testes in an exposed sac leaves them subject to injury. Because males are capable of continuous reproductive activity following puberty, the testes remain vulnerable throughout life.

In humans the *inguinal canal* normally closes permanently following descent of the testes. However, it may open as a result of trauma, causing an *inguinal hernia.* Some humans, however, actively attempt to keep the canal open throughout life. In Japan, the parents of future sumo wrestlers massage their sons' testes from birth, passing the testes into and out of the abdomen. This maintains an open canal, and when the youths begin wrestling, the scrotum is tightly bound, forcing the testes into the abdomen and protecting them from injury.

Unlike humans, most mammals reproduce only seasonally, often just once a year. In males of many species, the testes are retracted into the abdominal cavity during most of the year, descending into the scrotum only for the few weeks of annual reproductive activity.

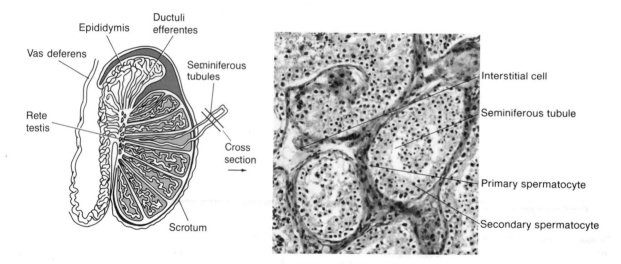

Figure 6.4 Cross section through a testis. Insert: Details of a single seminiferous tubule. Source: (insert) Gary Retherford.

prostate gland also secretes an alkaline fluid into the semen. The environment within the epididymis is acid, primarily due to the waste products of the developing sperm cells, and the sperm are relatively inactive as long as they remain in acid conditions. The alkaline secretions of the seminal vesicle and prostate thus activate the sperm at the time of ejaculation.

The vagina is also an acid environment that sperm alone could not tolerate, but the alkaline seminal fluid produces an environment in the female reproductive tract more favorable for the sperm. The ducts of the two bulbourethral or Cowper's glands enter the urethra just below the prostate. These glands secrete a small amount of alkaline fluid that lubricates the glans of the penis prior to sexual intercourse.

Each ejaculate contains about two to six milliliters of semen and about fifty to one hundred million sperm per ml of semen. Below a concentration of twenty million sperm per ml, a male is usually sterile. Medical attention has recently focused on a significant decline in the average **sperm count** of American males. Over the past fifty years, the average sperm count has decreased from about ninety to one hundred million per ml in 1929 to sixty to seventy million per ml in 1979. About 23% of all American males have a sperm count less than twenty million per ml, the level of functional sterility.

The causes of this extraordinary decline are unknown. Present hypotheses suggest that several chemicals widely used in our society may be responsible: the pesticide DDT; the industrial chemicals called PCBs; and dioxin, a highly toxic contaminant of many herbicides. All of these chemicals are now found in seminal fluid, and dioxin has been implicated in causing a variety of birth defects.

Gametogenesis

The process of gamete formation involves not only meiosis but also the maturation of the products of meiotic division. There are significant differences between spermatogenesis and oogenesis, so we discuss them separately.

Spermatogenesis

Male gametogenesis begins with *spermatogonia,* precursor cells just inside the wall of the seminiferous tubules. Each diploid spermatogonium divides mitotically to produce one daughter cell spermatogonium that remains on the periphery of the tubule as a precursor cell and one daughter cell that is pushed toward the interior of the tubule. As the diploid daughter cell moves inward, it undergoes a developmental change to become a **primary spermatocyte,** in which meiosis begins (figure 6.5).

In the primary spermatocyte, replication of the chromosomes occurs prior to meiosis I. Meiosis I produces two daughter or **secondary spermatocytes** that are haploid with respect to the number of centromeres, but retain chromosomes in the duplicated condition. The secondary spermatocytes then undergo meiosis II, producing a total of four haploid **spermatids** of two kinds, X bearing and Y bearing. The spermatids are then released into the lumen of the seminiferous tubules, where they move to the epididymis for completion of *spermiogenesis,* becoming mature **spermatozoa** (literally, "seed animals") (figure 6.6).

Mature spermatozoa are highly specialized cells, each consisting simply of head, body, and tail, but almost no cytoplasm. The head contains the chromosomes that the father contributes to the zygote if the sperm achieves successful fertilization (the odds are slight). The body contains rows of mitochondria that provide energy for the long journey the sperm will attempt to the upper reaches of the oviduct—a swim of 180 millimeters (seven inches) for an organism only .06 mm long (.0023 inches), or almost three thousand times its body length, equivalent to a swim of three miles for an adult human. The tail, a long, whiplike structure, propels the sperm.

The rate of spermatogenesis is extraordinary. The production of mature spermatozoa from a spermatogonium requires about two months. Each ejaculate contains several hundred million sperm. Thus, spermatogenesis is a relatively continuous process producing billions of sperm throughout a male's lifetime.

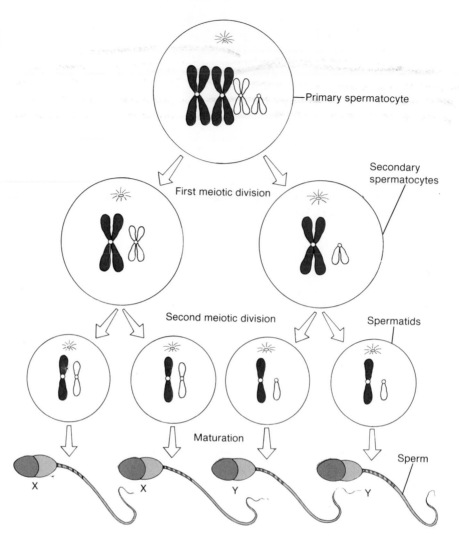

Oogenesis

Development of mature ova also involves meiosis, but the timing and results differ considerably from spermatogenesis. Recall that at birth all the potential eggs a female will ever possess are already present in ovarian follicles. Each follicle consists of numerous supportive cells surrounding one oogonium, which may become an ovum.

Prior to birth, each oogonium develops into a *primary oocyte,* which replicates its chromosomes and begins meiosis I. However, development stops near the end of prophase I, and the oocytes remain in this condition from birth until the onset of puberty at about age twelve. Then, one or two oocytes per month become active and resume meiosis. In contrast to the millions of sperm that mature daily in a male, a female produces only four hundred to five hundred mature ova throughout her entire reproductive life span. The remaining million or more oocytes simply degenerate.

When a follicle resumes development, meiosis I continues from the stage of prophase I where it ceased, and the first meiotic division is completed (figure 6.7). However, in contrast to spermatogenesis, the two daughter products are not of equal size. Both contain a duplicated haploid set of chromosomes, but the division of cytoplasmic contents is highly unequal. One daughter cell, the *secondary oocyte,* receives virtually all the cytoplasm from the primary oocyte, while the other, the *first* **polar body,** receives essentially none.

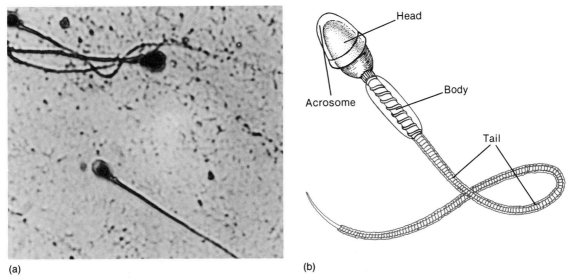

(a)   (b)

Figure 6.6 Morphology of human sperm: (a) photo of sperm; (b) diagrammatic interpretation of sperm.
Source: (a) Kirk Kreutzig/Photo Graphics. (b) John W. Hole, Jr., *Human Anatomy and Physiology,* 4th ed. Copyright © 1987 Wm. C. Brown Publishers, Dubuque, Iowa. All Rights Reserved. Reprinted by permission.

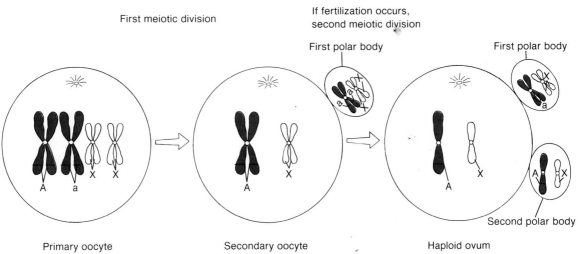

Figure 6.7 Oogenesis in a human female. Primary oocytes have the normal diploid complement of forty-six chromosomes, twenty-two pairs of autosomes plus XX. These are represented by a single pair of shaded chromosomes for the autosomes and a second pair of unshaded X chromosomes. (See colorplate 9.)

The follicle matures at the end of the first meiotic division, and the secondary oocyte, polar body, and some surrounding cells are released by the ovary, or *ovulated.* The secondary oocyte moves into and down one of the fallopian tubes. If it encounters a sufficiently large number of sperm, fertilization occurs (figure 6.1). Although only a single sperm penetrates the ovum, the presence of many sperm within twenty-four hours after ovulation is required for penetration to occur.

Meiosis II begins only *after* fertilization. The polar body may or may not divide. The secondary oocyte undergoes equatorial division, and again the cytoplasm is divided unequally. One daughter cell, the *second polar body,* receives essentially no cytoplasm, while the other, the ovum, retains all the cytoplasm. The ovum, of course, has already been fertilized, and it continues development into a zygote, embryo, and fetus. The two or three remaining polar bodies degenerate.

The results of meiosis are thus very different for males and females. Male meiosis produces four haploid and equal (but not genetically identical) gametes. Female meiosis produces only one mature ovum, plus two or three polar bodies of no further function. Female meiosis is not completed unless fertilization occurs.

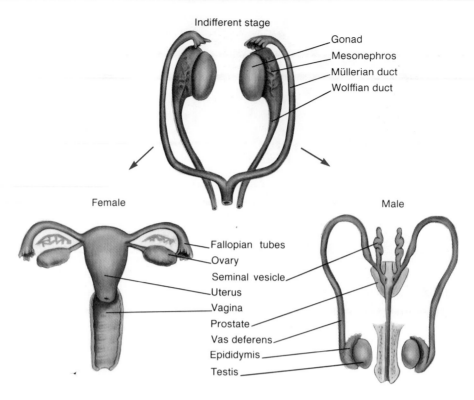

Figure 6.8 Differentiation of internal reproductive organs in human males and females. Every fetus passes through an indifferent stage of reproductive anatomy. As the fetus subsequently becomes either male or female, some structures develop further while others regress entirely.

Development of the sexual phenotype

For at least the first month of gestation, human embryos exhibit no observable sexual dimorphism. At this stage, the embryonic reproductive system has three components: (1) undifferentiated gonads; (2) two systems of genital ducts, *Müllerian ducts* and *Wolffian ducts;* and (3) externally, a set of genital folds (figure 6.8).

Control of the male phenotype

Differentiation begins earliest in the male embryo when, at about one month of age, the gonads begin to differentiate to testes. Although the process is only partially understood, the link between the XY genotype and the first step in sexual differentiation seems to depend primarily upon one or more genes on the Y chromosome that induce testis differentiation. The gene(s) concerned is called **testis-determining factor (TDF)** and has been mapped to the short arm of the human Y chromosome. Another Y-linked gene is located on the long arm of the Y chromosome and produces a characteristic male cell marker, the **H-Y antigen,** whose presence is essential for normal spermatogenesis. The H label refers to its discovery as a determinant of tissue, or *histological,* incompatibility between the sexes; the Y label refers to its location on the Y chromosome.

The next step of phenotypic development depends upon the products of the testes. The testes secrete two controlling hormones, Müllerian inhibiting substance and testosterone. Müllerian inhibiting substance causes the regression of the Müllerian duct system. Testosterone induces *virilization* or differentiation of embryonic tissue to masculine structures. In the embryo, this results in alteration of the Wolffian duct to produce the epididymides, vasa deferentia, and seminal vesicles. Primordial cells

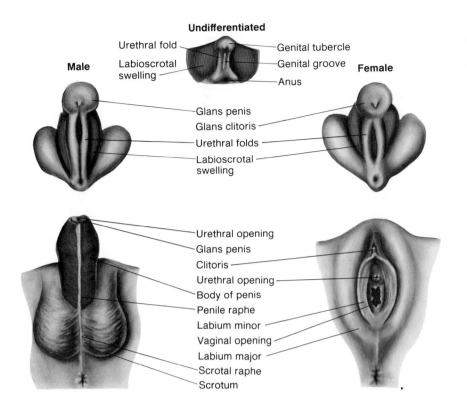

Undifferentiated

Urethral fold — — Genital tubercle

Male — Labioscrotal swelling — — Genital groove **Female**

— Anus

— Glans penis

— Glans clitoris

— Urethral folds

— Labioscrotal swelling

— Urethral opening

— Glans penis

— Clitoris

— Urethral opening

— Body of penis

— Penile raphe

— Labium minor

— Vaginal opening

— Labium major

— Scrotal raphe

— Scrotum

Figure 6.9 Differentiation of external reproductive organs in human males and females.

differentiate to produce the prostate gland and Cowper's glands. Genital folds elongate and form the penis, while swellings surrounding the folds enlarge to form the two halves of the scrotum (figure 6.9). Descent of the testes into the scrotum is usually complete by birth.

Testosterone performs a variety of important functions throughout the life of a male. In addition to its function in sexual differentiation, testosterone is continually required for maintenance of secondary male structures. Testosterone regulates bone growth; muscle development; fat deposition; sperm production; sexual behavior; patterns of facial, axillary, body, and pubic hair; and enlargement of the thyroid cartilage that results in deepening of the voice.

Control of the female phenotype

Female sexual differentiation begins somewhat later than in the male, becoming apparent by about the second month of development. In the absence of the TDF gene, the primitive gonads develop into ovaries. In the absence of testosterone and Müllerian inhibiting substance, the Wolffian ducts degenerate. The Müllerian ducts develop into the fallopian tubes, uterus, and part of the vagina (figure 6.8). The genital tubercle, which gave rise to the male glans, develops into the clitoris. The genital folds, which became the shaft of the penis, become the labia minora; and the genital swellings, precursors to the scrotum, become the labia majora (figure 6.9). The primordial cells that became the prostate develop into Skene's glands; and the primordia of the male Cowper's glands become Bartholin's glands.

Thus, a broad set of homologies exists between male and female sexual characteristics. Without the action of the TDF gene and testosterone, the fetus "automatically" develops a female phenotype. With them, it develops a male phenotype. The male phenotype is thus an *inducible* condition, the female *noninducible*. Individuals of both karyotypic sexes possess the capacity to develop the phenotype of either sex.

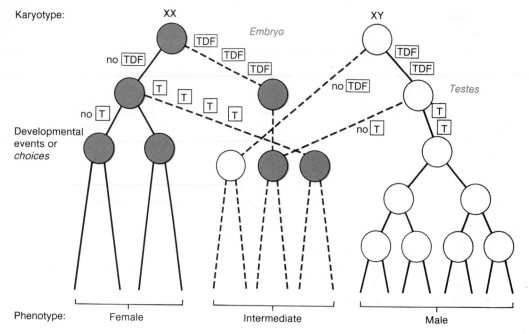

Figure 6.10 Development of the sexual phenotype. An individual will follow one of many possible developmental pathways, depending upon the person's genotype and environmental interactions. Both sexes encompass a wide range of "normal" phenotypes, but some individuals follow pathways that produce sexually intermediate phenotypes. T = Testosterone produced; TDF = TDF gene present.

Development may therefore be viewed as a sequence of diverging pathways. In the case of the sexual phenotype, the presence of the TDF gene commits an individual to the male pathway. The subsequent masculinization of tissues represents the response of each of those tissues to developmental signals—a continuation down the pathway to maleness (figure 6.10). In some cases, genetic defects result in the inability of tissues to respond to appropriate signals. In the next section, several genetic conditions are discussed that represent a redirection of a developmental pathway due to the inability to respond to testosterone.

Genetic sexual disorders

Many genetically based variations on normal sexual development have been discovered. We discuss some of these because they illuminate the mechanism of sex determination and the definition of sex.

Single gene disorders

Although the chromosomal basis of sex determination is well known, single genes are sometimes able to cause massive alteration of the sex phenotype.

Testicular feminization

Bearers of this rare syndrome possess a normal male karyotype (designated *46, XY* for 46 chromosomes and the normal XY male sex chromosomes). However, they possess internal testes and no other internal reproductive organs. Externally, they are phenotypically well-developed females—so well developed, in fact, that they are often characterized as voluptuous. Their psychological orientation is clearly female. They usually come to medical attention because of their failure to begin menstruation as teenagers.

The cause of this condition appears to be a recessive mutation at an X-linked locus designated Tfm. Persons with an XY karyotype and the Tfm mutation produce a normal H-Y antigen, which induces differentiation of the gonad to testes.

Hormone production by the testes is normal, and Müllerian inhibiting substance results in degeneration of the Müllerian ducts, which is normal. However, the Tfm mutation makes the tissues non-receptive to the effects of testosterone. As a result, all tissues respond as if testosterone were not present, differentiating into female secondary structures. The Wolffian ducts therefore regress, leaving no internal reproductive organs at all. The **testicular feminization** syndrome illustrates the extraordinary effect only one locus can have on the sexual phenotype.

Guevodoces

True **hermaphrodites** possess both types of primary sex organs, testes and ovaries. Some humans possess both ovarian and testicular tissue, but both types are almost never functional. **Pseudohermaphrodites** have only one type of gonad or none at all, but they show ambiguous genitalia, i.e., indeterminate secondary characteristics or secondary characteristics of both sexes. In the village of Salinas in the Dominican Republic, twenty-four pseudohermaphrodites have been detected showing a 46,XY karyotype. These individuals, called *guevodoces,* are born with ambiguous external genitalia—the scrotum appears to be labia, a blind vaginal pouch is present, and the penis resembles a clitoris. Until recently, these children were raised as girls.

At puberty, however, they show a distinct virilization of the external secondary sex structures. The voice deepens, the muscular development is masculine, and the apparent clitoris enlarges to become a penis. (Guevodoces means, literally, "penis at twelve.") The individuals become fully functional as males, their psychological orientation is masculine, and they are fertile.

This condition results from an autosomal recessive allele regulating the use of testosterone. Testosterone acts directly on the Wolffian ducts, but before causing virilization of the external genitalia it must be biochemically altered to a related chemical, *dihydrotestosterone.* A male homozygous for the recessive allele controlling the enzyme that catalyzes testosterone to dihydrotestosterone does not show virilization of the external genitalia. Apparently, the effect of testosterone alone is sufficient to induce virilization of the genitalia at puberty. Again, as with testicular feminization, one gene exerts massive influence over the sex phenotype.

The high frequency of guevodoces in this village in the Dominican Republic is the result, as might be expected, of consanguineous matings. Salinas is an isolated village of forty-three hundred persons, yet this rare condition has appeared in twenty-four males, twenty-three of whom trace their ancestry through twelve families back to one woman.

Chimeras and mosaics

Rarely, an individual is found who possesses both ovarian and testicular tissue and is thus a true hermaphrodite. Usually, these individuals are identified at birth because of ambiguous genitalia. Cytological and histological examination usually reveals that their tissues consist of two different cell types. These persons have two distinct karyotypes, one per cell line. Such an occurrence can be explained if the individual resulted from the fusion, early during development, of two or more different zygotes. Such an individual is called a **chimera.** Alternatively, an hermaphrodite can arise as a result of early mitotic nondisjunction (see below) of an XY or an XXY embryo, producing a **mosaic** of cell lines, XO/XY, XX/XXY, and so on.

Most chimeras are discovered because the fusing zygotes are of different sexes. The karyotype is designated chi46,XX/46,XY. Chimeras resulting from fusion of zygotes of the same sex are discovered far more rarely because they do not exhibit obvious defects, but they are probably equally common.

Chimeras and mosaics may occur in a variety of ways. In some instances, a polar body is fertilized by a sperm at the same time the ovum is being fertilized by a different sperm. If one of the sperm is X bearing and one Y bearing, the resulting

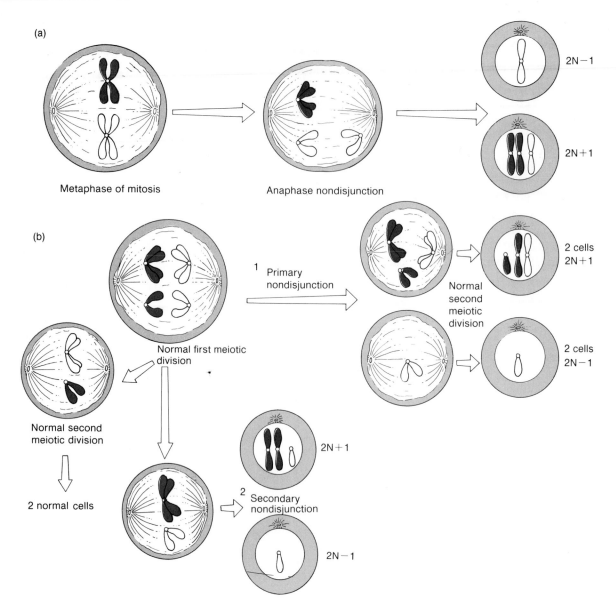

(a)

Metaphase of mitosis

Anaphase nondisjunction

2N−1

2N+1

(b)

1 Primary
nondisjunction

Normal first meiotic
division

Normal
second
meiotic
division

2 cells
2N+1

2 cells
2N−1

Normal second
meiotic division

2 normal cells

2 Secondary
nondisjunction

2N+1

2N−1

Figure 6.11 Nondisjunction during cell division leading to chromosomal aneuploidy: (a) mitotic nondisjunction occurs with centromere misdivision at anaphase; (b) meiotic nondisjunction may occur at either anaphase I or anaphase II, with differing results.

zygotes are of opposite sexes. Subsequent fusion of those zygotes would give rise to one individual bearing two different genetic types of cells. While chi46,XX/46,XY is most commonly identified, other chimeras discovered include: chi45,XO/46,XY; chi46,XX/47,XXY; and chi45,XO/46,XY/47,XYY.

Chromosomal disorders

Several syndromes have been discovered that result from **aneuploidy** of the sex chromosomes. Aneuploidy results from an error during cell division called *nondisjunction* (figure 6.11). As the chromosomes segregate during anaphase, one pair of homologues does not separate and divide equally. Instead, one daughter cell receives both members of an homologous pair, while the other daughter receives none. Nondisjunction usually involves only one pair of chromosomes at a time.

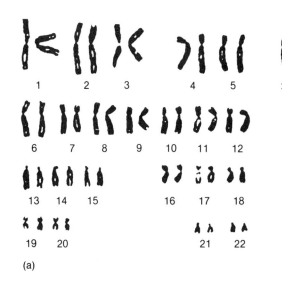

(a)

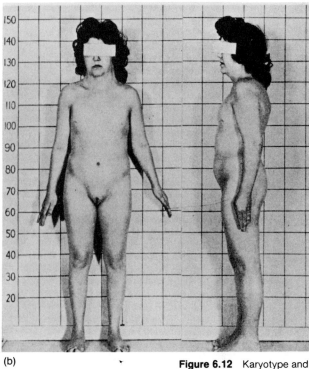

(b)

Figure 6.12 Karyotype and individual with Turner's syndrome.
Source: (a) A. Redding and K. Hirschhorn, "A Guide to Human Chromosome Defects," D. Bergsma (ed.) New York: The National Foundation–March of Dimes, BD:OAS IV (4) 1968; (b) M. Bartalos and T. A. Baramski: *Medical Cytogenetics,* © 1967 Williams & Wilkins Co.

Nondisjunction can occur during either meiotic division. If it occurs during the first division, it is called *primary nondisjunction* (figure 6.11b1). None of the gametes produced by primary nondisjunction is normal: two possess an extra member of the affected pair of homologues, two have none of that pair. *Secondary nondisjunction* (figure 6.11b2) occurs during the second meiotic division, producing two normal gametes, one gamete with one extra chromosome, and one gamete with a missing chromosome.

Most nondisjunctions affecting autosomes are lethal, especially those involving loss of an autosome. However, a few aneuploid karyotypes are viable, and those affecting autosomes are discussed in chapter 10. A number of aneuploidies of the sex chromosomes also are viable.

Turner's syndrome—XO

About one in every two thousand female births is deficient in one X chromosome, symbolized 45,XO (figure 6.12). Phenotypically, the individual is female, but the ovaries are poorly developed, the individual is short and has a webbed neck, and the secondary sexual characteristics are incompletely developed.

Turner's females have no Barr bodies. The inactivation of one X chromosome in normal female cells might lead to the expectation that loss of one would have limited effect. In fact, X inactivation may explain why *monosomy* (meaning one chromosome) for the X is the only monosomy that is viable at all in humans. However, X chromosomes are not inactivated in female germinal tissue; both Xs are required for female gametogenesis. Therefore, XO females are sterile.

Turner's females usually arise from meiotic nondisjunction during gametogenesis (figure 6.13). They can also arise from mitotic nondisjunction (figure 6.11a) during early embryonic development. In this case, they are a mosaic of XX and XO tissues. Because Turner's females, like males, are hemizygous for the X chromosome, they show an increased frequency of expression of X-linked traits.

Figure 6.13 Examples of how meiotic nondisjunction can give rise to persons with Turner's or Klinefelter's syndromes: (a) primary or secondary nondisjunction in females produces eggs with two X chromosomes or no X chromosomes leading to XXX, XXY, XO, or YO fertilized eggs; YO is lethal; (b) primary nondisjunction in males leads to sperm with no sex chromosomes, or with both an X and Y chromosome giving XXY and XO fertilized eggs. Secondary nondisjunction produces sperm with no sex chromosomes, with XX and with YY; these sperm produce XXX, XYY, and XO fertilized eggs.

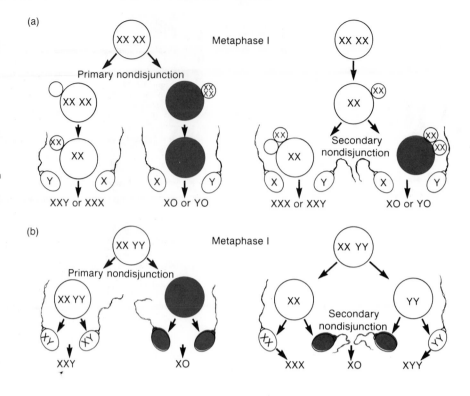

Klinefelter's syndrome—XXY

Approximately one of every five hundred males born has an extra X chromosome, a 47,XXY karyotype (figure 6.14). Because the Y is male determining, these individuals show a male phenotype. However, they tend to have longer limbs than normal, and some of the secondary sex characteristics are feminized. They usually have reduced intelligence.

Klinefelter's males are also usually sterile. One Barr body is present in somatic cells of XXY males (remember, there is always one less Barr body than there are X chromosomes). The presence of two Xs is associated with sterility in XXY males. Nonetheless, the male phenotype of Klinefelter's males is convincing evidence that the Y chromosome—and not the dosage of X chromosomes—is the controlling factor for maleness in humans.

XYY males

It is estimated that one male in a thousand has an extra Y chromosome and is thus 47,XYY. Phenotypically, XYY males are fertile and normally developed, with the single exception that they tend to be somewhat taller on average than the normal male population. These individuals result from the same type of meiotic nondisjunction that causes Turner's syndrome (figure 6.13). Nondisjunction of the Y chromosome during the second meiotic division produces one O gamete (i.e., no sex chromosome) and one YY gamete. Fertilization by the first gamete causes Turner's syndrome, XO, while fertilization by the second gamete produces an XYY male.

Attention has focused on XYY males in recent years because of the discovery that they are disproportionately represented among white inmates of mental/penal institutions. XYY males are twenty times more likely than normal XY males to be incarcerated in such an institution; about 2% of all inmates appear to be XYY.

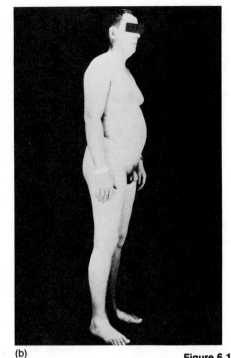

(a)

(b)

Figure 6.14 Karyotype and individual with Klinefelter's syndrome.
Source: (a) A. Redding and K. Hirschhorn, "A Guide to Human Chromosome Defects," D. Bergsma (ed.) New York: The National Foundation–March of Dimes, BD:OAS IV (4) 1968; (b) F. A. Davis Co. and Dr. R. H. Kampmeier.

This fact led to the hypothesis that XYY males are more aggressive than XY males and thus tend to criminal behavior. While XYY males may have an increased probability of incarceration, the causes of this phenomenon have yet to be determined (box 6.2). No proof has been found that an extra Y chromosome leads to criminal behavior. Until more definitive data are found, it is best to remember that the vast majority of XYY males lead apparently normal lives.

Other sex chromosome aneuploidies

Females have been discovered who are triplo-X (XXX), tetra-X (XXXX), and penta-X(XXXXX). Triplo-X females are relatively normal phenotypically, while tetra-X and penta-X individuals exhibit mental retardation. Triplo-X and tetra-X females are often fertile. All females with extra X chromosomes show extra Barr bodies—two Barr bodies for XXX individuals, three for XXXX, and four for XXXXX.

Males showing extreme symptoms of Klinefelter's syndrome have been discovered who are 48,XXXY or 49,XXXXY. They are almost always mentally retarded. Also known are 48,XXYY and 49,XXXYY males. They tend to be tall, more aggressive than normal, and mentally deficient. Like females with extra X chromosomes, their somatic cells have one less Barr body than the number of X chromosomes.

Gender

Sex or *gender* is a complex phenomenon. Numerous variations have been observed in both karyotypic and phenotypic measures of sex. In some instances, an individual cannot be unambiguously classified as either male or female. The causes of these anomalies are complex and may include genetic, hormonal, and environmental factors. Testicular feminization, for example, involves a clearly karyotypic male exhibiting a clearly female external phenotype. Which sex is the individual? The answer is less relevant than an understanding of the factors that may cause this condition. There is no legal definition of sex.

BOX 6.2

A long-term study of XYY males

In 1968 a long-term study of male infants with karyotypic abnormalities was begun at the Boston Hospital for Women. The goal was to identify XXY (Klinefelter's) and XYY males, then to study their development and behavior. Klinefelter's males present no ethical problem. They are often mentally retarded, prone to institutionalization, and therefore clearly require special care.

The case of XYY males, however, provides a genuine dilemma. About one male in a thousand is XYY, and most lead perfectly normal lives. The primary effect of the condition is that many XYY males are unusually tall. Concern has been raised that XYY males may be prone to aggressive or antisocial behavior, but no conclusive evidence yet supports that contention. The main evidence indicating such a possibility is that the risk of an XYY male being incarcerated in a hospital for the criminally insane is twenty times greater than the risk for an XY male.

The Boston study was designed, in part, to determine if XYY males were in fact prone to antisocial behavior. The program was terminated in 1975, however, due to public opposition. Many people believed that identifying a child as XYY would create a self-fulfilling prophecy: because the child would be labeled from birth as potentially antisocial, he would be treated differently—and the different treatment, *not* the extra Y chromosome, would produce aberrant behavior. As with the IQ controversy, separating nature from nurture is extraordinarily difficult.

The ethical dilemma is obvious. If an XYY male is not genetically prone to aggressive behavior, then identification serves no positive purpose. Instead, the stigma of being labeled genetically different could promote undesirable behavior. Even parents would be unlikely to avoid treating an XYY child differently. Normal childhood rowdiness would always be under suspicion as overaggressive behavior in an XYY son. Parents might then punish the XYY child with undue severity, eventually producing the undesirable behavior.

On the other hand, if XYY children are indeed subject to producing antisocial behavior, their identification at birth could permit treatment. For example, XYY males might experience greater than normal reading and learning disabilities, frustration over which could in turn generate overtly aggressive behavior. Knowledge that a child is XYY could enable parents and teachers to provide special education—including much patience and understanding—that could ultimately lead to normal development. In this case, failure to identify XYY males at birth could actually *deprive* them of a normal life.

Other possibilities also exist. The increased height of XYY males may elicit unusual reactions to them. The effect of differences in treatment due to height, combined with different treatment due to, for example, learning disabilities, might also produce antisocial behavior. While the opponents of the Boston study may be correct that identification of XYYs could create a self-fulfilling prophecy of aggressive behavior, the only way that the truth will ever be determined is by long-term studies such as the Boston group was attempting.

The determination of gender is further complicated by consideration of psychological factors. An individual's perception of his or her sexual identity is referred to as _gender identity_. In most cases, gender identity agrees with karyotypic and phenotypic sex, but it need not. Some karyotypic and phenotypic males _feel_ as if they are females, and vice versa. Obviously, such internal conflict can cause serious emotional and psychological problems.

A person whose psychological identity is at odds with his or her phenotypic identity is called a *transsexual*. While karyotypic sex cannot be altered, phenotypic sex can. Since the mid-1950s transsexual surgery has been performed to alter phenotypic sex to correspond to psychological identity. Normally, six or more months of hormone treatments are given prior to surgery. Administration of large doses of testosterone, for example, results in reduction in size of breasts, redistribution of fat to enlarge shoulders and reduce hip size, growth of facial and body hair, and deepening of the voice. Doses of estrogen have an opposite effect, and even prior to surgery, hormone treatment can cause major changes in secondary sex structures.

Transsexual surgery is less complicated for the male to female transition. The testes and the penis are removed, and the tissue from the penis and the scrotum are used to create an artificial vagina. In female to male surgery, the breasts, ovaries, and uterus are removed. A scrotum and, sometimes, a penis are constructed. The extent to which such surgery can affect psychological adjustment and happiness is a matter of ongoing debate. A stigma is still attached to individuals who have undergone such surgery.

Yet another complication arises with respect to sexual preference or *object choice*. Most persons develop love and sexual relationships with persons of the opposite sex and are thus *heterosexual*. The object choice of an *homosexual* is a person of the same sex; while *bisexuals* may choose a person of either sex. The causes of sexual preferences, like gender identity, are poorly understood. As with transsexuals, society only partially accepts individuals who exhibit non-heterosexual behavior.

Knowledge that a zygote of either XX or XY karyotype may differentiate either to male or to female phenotypes has helped science redefine the meaning of maleness and femaleness. Testicular feminization demonstrates that an individual can be psychologically healthy with a sexual orientation different from karyotypic sex. Guevodoces successfully change psychological orientation midway through life—they are normal little girls before virilization, normal adult males afterward. Such information persuasively argues the importance of developmental factors in producing sexually stereotyped behaviors. The definition of gender will continue to increase in complexity as the many factors governing sexual identity and differentiation continue to be elucidated.

Summary

1. The human embryo possesses primordial structures that may differentiate into the sexual structures of either sex. Both the internal and external reproductive structures of males and females are homologous.
2. TDF, the product of a gene or set of genes on the Y chromosome, appears to govern the process of sexual differentiation. It induces differentiation of embryonic gonadal tissue to testes. Females do not produce TDF, and in its absence the gonads differentiate to ovaries.
3. If testes are formed, they produce testosterone, which induces formation of male secondary sexual structures. In the absence of testosterone, female secondary characteristics develop.
4. Spermatogenesis results in the formation of four spermatozoa from each primary spermatocyte. Two of the four carry a Y chromosome, two carry an X chromosome.
5. Oogenesis results in the formation of only one mature ovum from each primary oocyte, plus two or three polar bodies.
6. A male produces billions of sperm throughout his reproductive life span. A female produces only a few hundred mature ova.
7. Expression of the sexual phenotype may be altered as the result of mutations involving a single gene on either an autosome or a sex chromosome. Syndromes having major phenotypic effects result from aneuploidy of the sex chromosomes.
8. X-chromosome aneuploidies are always reflected in the number of Barr bodies in somatic cells.

9. Chimerism is the result of fusion of two or more zygotes and may cause true hermaphroditism.
10. Nondisjunction of chromosomes occurs during anaphase of cell division and results in aneuploid gametes and mosaic individuals.
11. Gender is a phenomenon involving the complex interaction of genotype and phenotype.

Key Words and Phrases

aneuploidy	ovum	sperm count
chimera	placenta	spermatid
estrogen	polar body	spermatogenesis
hermaphrodite	primary oocyte	spermatozoon
H-Y antigen	primary spermatocyte	testicular feminization
Klinefelter's syndrome	progesterone	testis
mosaic	pseudohermaphrodite	testis determining factor (TDF)
oogenesis	secondary spermatocyte	
ovary	semen	testosterone
		Turner's syndrome

Questions

1. Why is a woman able to conceive only one day of each month?
2. What would you infer about the potential fertility of an adult Japanese sumo wrestler?
3. What are the functions of the accessory glands to the reproductive tract of each sex?
4. A normally pigmented female whose mother was an albino is a carrier for Lesch-Nyhan syndrome (Ll). She is married to an albino male. Considering these two traits, how many different types of eggs can she produce? What is the probability of the first child being an albino with Lesch-Nyhan syndrome?
5. A color-blind woman marries a normally visioned man, and they have a color-blind daughter. How do you explain this? How might this same couple produce a child with normal vision?

For More Information

Beckwith, J. 1975. "The Harvard XYY Study." *Science,* 187:298.

Chandra, H. S. 1985. "Is Human X Chromosome Inactivation a Sex-Determining Device?" *Proc. Nat'l. Acad. Sci.,* 82:6947–9.

Diamond, J. M. 1986. "Variation in Human Testis Size." *Nature,* 320:488–9.

Epel, D. 1977. "The Program of Fertilization." *Scientific American,* 237:128–40.

German, J. 1970. "Studying Human Chromosomes Today." *American Scientist,* 58:182–201.

Kolata, G. 1986. "Maleness Pinpointed on Y Chromosome." *Science,* 234:1076–7.

Miller, J. A. 1986. "X Chromosomes: Too Few and Too Many." *Science News,* 129:358.

Naftolin, F. 1981. "Understanding the Bases of Sex Differences." *Science,* 213:1263–4.

Ohno, S. 1980. "Two Major Regulatory Genes for Mammalian Sex Determination and Differentiation." *Genetica,* 52/53:267–73.

Roberts, L. 1988. "Zeroing in on the Sex Switch." *Science,* 239:21–3.

Segal, S. J. 1974. "The Physiology of Human Reproduction." *Scientific American,* 231:52–62.

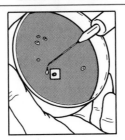

CHAPTER SEVEN

Reproduction: technologies and choices

Birth Technologies
Artificial Insemination
Sperm Banks
Surrogate Motherhood
In Vitro Fertilization
Practical problems of IVF / Box 7.1 Birth technologies have revolutionized animal husbandry / Ethical problems of IVF
Prenatal Diagnosis
Amniocentesis
Chorion Villus Sampling
Fetoscopy
Use of Prenatal Diagnosis
Risk / Cost / Use of information / Criteria for prenatal diagnosis
Bioethics and Therapeutic Abortion
Utilitarian or Absolutist?
Societal Standards

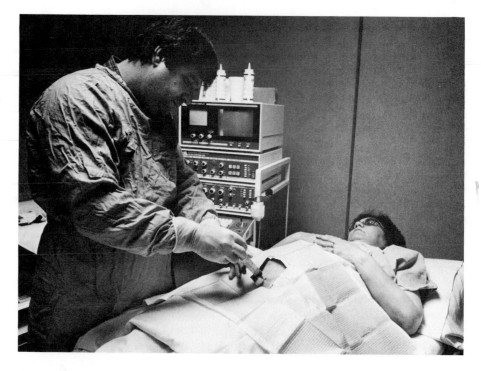

Figure 7.1 A pregnant woman having a sample of amniotic fluid withdrawn for genetic testing of her fetus. To avoid damage to the fetus, the fetus is visualized on the adjacent ultra-sound equipment (figure 7.5a) prior to the doctor inserting the needle.
Source: Robert Goldstein/Photo Researchers.

Understanding the details of human reproductive physiology has led to the development of a variety of techniques that allows medical manipulation of conception, gestation, and health of the fetus (figure 7.1). In this chapter we consider some of these techniques and their genetic and ethical relationships.

Birth technologies

Techniques for enhancing fertility are an important aspect of contemporary medical practice. Some of these procedures have received considerable attention in the public media, often being incorrectly labeled as a form of genetic engineering. They are not genetic engineering because the genomes of neither parent nor child are affected by such procedures. We discuss them here because reproduction is so closely linked to other aspects of genetic health.

Artificial insemination

Artificial insemination (AI) refers to the introduction of human sperm into the female vagina and cervical canal by means other than normal sexual intercourse. The first recorded use of AI in humans was in 1790. AI is most often used by couples when the husband is sterile, but the wife is fertile. In the past the doctor would arrange for a suitable anonymous sperm donor, often a medical student. The donor male collects the semen by masturbation, and the semen is then introduced into the woman's reproductive tract by the doctor. It is estimated that 1% of all children born in the United States were conceived by AI.

The probability of successful fertilization is increased by careful monitoring of the woman's ovulatory cycle, so the sperm can be introduced at the most opportune time. Generally, healthy women conceive after one to four AI attempts. In some cases of AI the sperm donor is the husband, who may have a low sperm count. Combining semen from several ejaculations can increase the sperm count and increase the probability of fertilization. This is now possible because of the technique of *sperm banking.*

Sperm banks

Human sperm can be maintained indefinitely at ultracold temperatures in **sperm banks,** just as can the sperm of other animals. Sperm banks were established commercially in the United States in 1970. A fresh sample of semen is usually distributed among several ampules and then frozen in liquid nitrogen ($-196°$ C).

Most commonly, semen for artificial insemination has been donated anonymously by medical students. More recently, men undergoing sterilization by vasectomy have placed samples in sperm banks as a form of fertility insurance, in the event they later wish to reproduce. In the 1960s, geneticist H. J. Muller advocated that AI not rely upon randomly donated and chosen sperm. Instead, he urged that sperm should be collected from potentially high-quality donors and not be used until after the donor's death, so that the donor could be evaluated properly. The crux of such a program would be the criteria used to evaluate donors; understandably, agreement on qualities of donors would be difficult to achieve in a diverse democratic society. Muller contended, however, that children conceived by AI should be given every possible opportunity for a high-quality life.

In 1980 a California businessman established an exclusive sperm bank. All donors were required to have exceptional IQs, and Nobel prizewinners were particularly solicited. Several Nobel laureates have contributed thus far, all anonymously except William Shockley, a cowinner of the 1956 Nobel Prize for Physics. Shockley was in his seventies at the time of his deposit, a factor that might run counter to Muller's high-quality criteria, as mutations in the germ cells of both males and females accumulate with age (see chapter 10.).

Sperm from this exclusive bank are made available only to selected intelligent, healthy females under thirty-five years of age who are married to sterile males. The sperm bank was advertised in the bulletin of MENSA, an organization of members whose IQs are among the top 2% of the nation. In 1982 the first child born to a woman using the exclusive sperm bank was announced.

Although it is a private citizen's right to open a selective sperm bank, such an act implies both that there is a high genetic component to human intelligence and that extremely high intelligence is desirable. The first implication is a matter of intense debate (see chapter 9), and the second is a matter of personal values.

Surrogate motherhood

When a wife is sterile, her husband's semen can be used to fertilize the ovum of a **surrogate mother** by artificial insemination. Following birth, the child is given to the couple to raise as their own. The surrogate or genetic mother (the media often say "biological mother") is usually paid a large fee for her services and for surrendering her legal rights to the child; ten thousand dollars is a common amount. The surrogate mother's medical expenses plus all legal fees are also paid by the couple receiving the child.

Surrogate motherhood is now practiced openly in a number of countries around the world. (It has probably been practiced secretly in many societies throughout human history.) However, surrogate motherhood is still not common, receives much media attention, and has yet to achieve full acceptance by society. Many people object to a mother surrendering her own child under any circumstances and especially doing so for money. One case in Britain received lurid publicity because the contracting parties decided to avoid the medical expenses of artificial insemination and used more traditional methods of conception. This fact was revealed by the surrogate mother in an article she wrote for a daily newspaper. She received almost as much money for the article as for being a surrogate mother.

A new specialty of law has developed to deal with the many problems created by surrogate maternity. Some surrogate mothers, for example, have changed their minds following birth and wished to keep the child. If the child is deformed or genetically abnormal (Turner's syndrome, for example), has the surrogate mother fulfilled her part of the contract, or must the child be normal? What assurances can be made that the AI sperm actually fertilized the egg, and not the sperm of the husband of the surrogate mother? Contracts negotiated between the couple and the surrogate mother attempt to specify unambiguously the rights of all parties, but it will likely be decades before most legal ramifications of surrogate motherhood are fully resolved.

A case that received much publicity in the United States illustrates how complex and soul-wrenching surrogate motherhood can become. A couple who had no children due to the wife's infertility contracted with a married woman to be the surrogate mother of the husband's child. Following artificial insemination, the contracted surrogate became pregnant and gave birth to a child with a serious defect, *hydrocephalus*, an accumulation of fluid in the center of the brain that causes an enlarged skull and compressed brain. The contracting couple then refused to accept the child, not on the grounds that the child was defective, but that the husband was not the father of the child—in other words, that conception had not occurred as a result of the artificial insemination. Subsequent paternity analysis showed that the contracting husband was indeed not the father of the child; the results supported the probability that the surrogate's husband had fathered the child. The surrogate mother then accepted responsibility for the child.

Most people find such conflicts appalling and feel particular sympathy for the innocent child at the center of the debate. Although this particular case was resolved by a successful paternity test, such will not always be true. Furthermore, this case does not set a legal precedent for the rejection of a child *because* it is defective, as that was not the issue.

In vitro fertilization

Some women are sterile because their oviducts are blocked. Although they ovulate normally, the blocked passageway precludes access of sperm to the ovum. The technique of *in vitro fertilization* (figure 7.2) can often provide an effective cure for sterility for such women.

In theory, the technique is simple. A physician collects freshly ovulated ova from the woman. These eggs are then fertilized in a dish in the laboratory—*in vitro* (in glass)—using the husband's sperm or that of a donor if the husband is sterile. About a week after fertilization, all viable embryos are placed in the mother's uterus. If implantation occurs, a normal pregnancy usually follows.

The technique was first developed for use in animal husbandry over the past several decades (box 7.1), but was first successfully applied to humans in 1978. Doctors in Britain and Australia have pioneered the use of these techniques on humans, and now hundreds of babies resulting from *in vitro* fertilization have been born worldwide. It is estimated that one child is born each day from this technique. Nearly all the children have been healthy, and some cases have produced twins. The technology is rapidly advancing; some embryos have been stored in ultracold embryo banks for as long as six months prior to successful implantation.

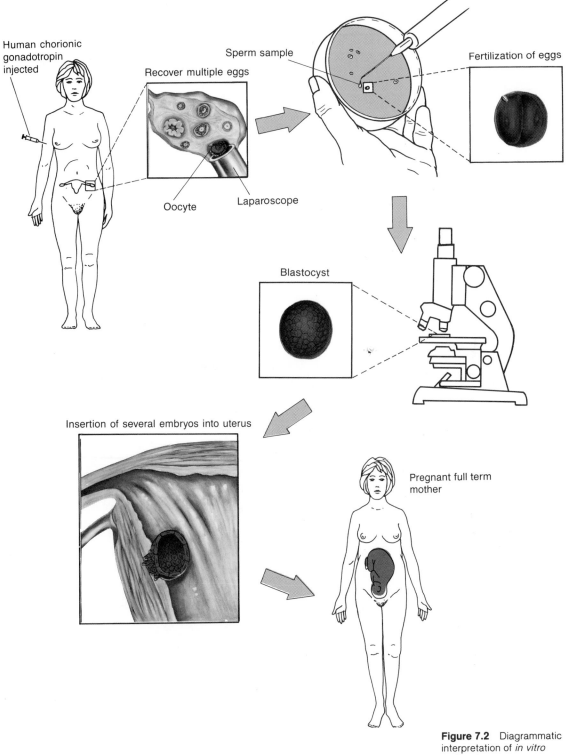

Human chorionic
gonadotropin
injected

Recover multiple eggs

Sperm sample

Fertilization of eggs

Oocyte

Laparoscope

Blastocyst

Insertion of several embryos into uterus

Pregnant full term
mother

Figure 7.2 Diagrammatic
interpretation of *in vitro*
fertilization.

BOX 7.1

Birth technologies have revolutionized animal husbandry

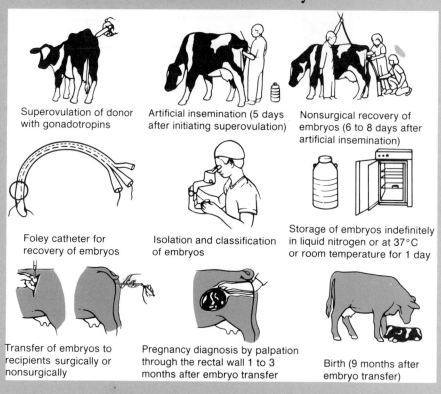

Figure 7.3 *In vitro* fertilization as practiced on cattle.

The spread of useful new traits such as increased fertility or the propagation of new breeds of domestic livestock has traditionally been a slow process. The normal reproductive rates of most animal species allow new characteristics to be spread through existing herds only over many generations, usually involving five, ten, or more years. Beginning in the 1950s, the techniques of **superovulation** and **embryo transfer** have been developed to increase the reproductive rates of valuable livestock (figure 7.3). Since the 1970s, these techniques have been used to respond to market factors and to improve the genetic quality of many herds. They have also served as a laboratory for development of IVF techniques for humans.

The popular media usually apply the provocative term "test-tube babies" to these children, but it is a misnomer. Gestation does not occur in a test tube. Other than the brief period of fertilization and early development, usually less than a week, gestation occurs totally within the uterus of the mother, just as in other pregnancies.

Although IVF is a technique of great importance to animal husbandry, its application to human fertility is fraught with both practical and ethical problems.

Practical problems of IVF
IVF has only limited application to curing infertility. The success rate is low (less than 10%), the treatment involves risks, and it is expensive.

BOX 7.1 (*Continued*)

Figure 7.4 A "genetic" mother and her ten offspring produced in one breeding season by superovulation and embryo transfer. Some of the ten surrogate mothers are shown on the left. Source: George Seidel.

The techniques used for animal IVF are very similar to those used for human IVF; the primary difference is that animal IVF does not require external fertilization. Injections of gonadotropic hormones cause a cow to superovulate, i.e., to ovulate many eggs rather than just one. Shortly after ovulation, the cow is inseminated, either naturally or artificially, with the sperm of a high-quality bull. About a week after insemination, the fertilized embryos are collected from the oviducts prior to implantation in the uterus and stored in liquid nitrogen ($-196°$ C).

Later, the frozen embryos are thawed and implanted in other cows, the *surrogate mothers*. Surrogate mothers are less valuable as breeding stock than the genetic mother. They may, for example, produce less meat or milk than females used as the genetic mothers, but they are physiologically well suited to the rigors of pregnancy and birth. About 50% of all embryos implanted in surrogates produce viable offspring. Recalling that more than half of all human embryos abort spontaneously in the first trimester, the success of this technique is remarkably good. All ten calves in figure 7.4 are the progeny of a single superovulation, and as many as twenty calves have been produced from a single superovulated cow.

Since its commercial application in the 1970s, this technology has produced tens of thousands of calves worldwide. Superovulation and embryo transfer are also widely used in the husbandry of horses, swine, goats, sheep, and rabbits. Frozen embryos of these species can be temporarily implanted in a rabbit and transported worldwide with relative ease. An entire herd of cattle can be readily introduced from one country to another in just a few small rabbits, without worry of transmitting disease and at relatively low cost.

IVF is suitable only for treating women with blocked oviducts, about 2% of all cases of infertility in women. Although IVF does allow some of these women to have children, it does not cure the blocked ovary. Thus, IVF must be used for each subsequent pregnancy attempt. Furthermore, fertilization is seldom achieved on the first attempt. Six to twelve monthly attempts are often required before pregnancy occurs. Each attempt involves hormone treatment (of unknown risk) and the risk of minor surgery. Also, spontaneous abortions are as common as with a normal pregnancy.

Financial costs for IVF are high. Each attempt may cost ten thousand dollars, and women requiring multiple attempts may spend fifty thousand to one hundred thousand dollars, with less than a 10% chance of success.

Other practical fears have not been realized. Many persons were concerned that the procedure might induce a high incidence of birth defects, particularly in embryos that had been frozen. Also, the low success rate has led physicians to introduce multiple embryos to the uterus in the hope of achieving a single implantation and pregnancy. Obviously, this has the risk of producing two or more children per pregnancy, an occurrence not welcomed by all parents. Thus far, neither birth defects nor multiple births appears to occur at higher than normal frequencies.

Ethical problems of IVF

Two major categories of ethical questions arise from IVF, the first from the great expense and low success rate of the technique. The costs virtually guarantee that only the wealthy will have access to IVF, unless public funds are provided to make IVF widely available. In the United States, IVF is undertaken at private clinics and not otherwise subsidized. It thus can be criticized as another example of the disparity in health care between the wealthy and the poor.

More broadly, the high cost of IVF has been examined in terms of the intended effect—the treatment of infertility. Because IVF is applicable to only a small percentage of infertile women and because it has a very low success rate, some groups observe that the primary beneficiaries of the technique are medical professionals who collect the large fees and receive public adulation. These groups argue that the money spent on IVF could be used more efficiently to conduct research and prevention programs on the numerous sexually transmitted diseases that cause infertility. The results of such research would benefit a greater number of people, including the poor who currently receive no benefits from IVF procedures. Because such research is less well paid and publicized, medical researchers may have less incentive to pursue it than they do to practice IVF.

A second ethical constraint on IVF is the problem of "excess" embryos. Women undergoing IVF receive human chorionic gonadotropin, a hormone that controls ovulation (so the ova can be collected at the proper time) and induces superovulation. This allows many eggs to be fertilized and multiple embryos to be introduced to the uterus. Extra embryos are also frozen for later use.

The fate of these frozen embryos or other embryos not implanted is the source of controversy. What should be done with them, particularly after the woman either becomes pregnant or stops IVF treatment? Those who believe that all human life is sacred and begins at conception view disposal of these embryos as a form of abortion or, in their words, murder. Other people take a more utilitarian view, believing that termination of the embryos is appropriate and that under appropriate controls the embryos may be used for medical experimentation. Views on both sides are very strongly held and expressed.

Federal government regulations on research on fetuses incorporate the recommendations of the National Commission for the Protection of Human Subjects of Biomedical and Behavioral Research, published in 1978. These guidelines allow the use of aborted fetuses for medical research if (1) they are less than twenty weeks of gestational age, (2) the research does not alter their lifespans, (3) the knowledge is important and cannot be gained by other means, (4) suitable preliminary experiments have been conducted on other animals, and (5) the research has been approved by appropriate review boards.

Although the notion of experimentation on aborted fetuses is unpleasant, fetuses are required for certain types of research. Furthermore, such research has saved thousands of lives. Research in the 1940s using spontaneously aborted fetuses led to the development of the polio vaccine. Today, a procedure called *fetal-cell surgery* is increasingly used as an experimental treatment for many diseases, including diabetes, Parkinson's disease, and the blood disease thalassemia. This technique uses treated fetal cells to replace defective tissues in adults because the fetal

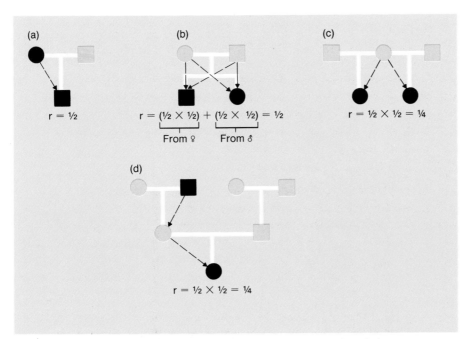

Plate 5 Determination of coefficients of relationship (r) for (a) parent-child, (b) full siblings, (c) half siblings, and (d) grandparent-grandchild. Note that the coefficient decreases by 1/2 for each step in the chain of relationship, and that, as in (b), two pathways of relatedness may connect one individual.

Plate 6 Sexual dimorphism in vertebrates: (a) Costa Rican frogs (the male is the smaller frog clasping the larger female from behind); (b) pair of wood ducks (the male is more decorated); (c) a male elephant seal defending his section of a mating beach. Source: (a) Roy McDiarmid; (b) Breck P. Kent/Animals Animals; (c) Dale and Marion Zimmerman/Animals Animals.

(a)

(b)

(c)

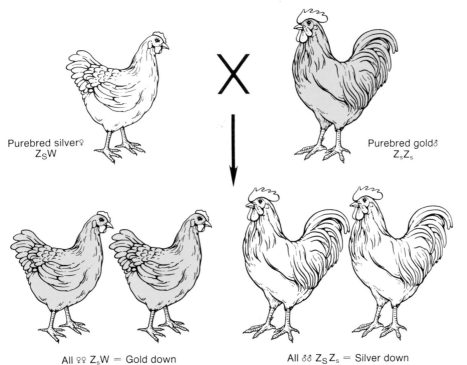

Plate 7 Feather color in chickens, a sex-linked trait used by poultry breeders to select breeding chickens. Down feather color is controlled by a pair of alternate alleles, S, producing a silver color down, and s, producing a gold color. Gold-feathered males are bred with silver-colored females. All male offspring will have silver down and all female offspring will have gold down. Thus breeders can select the females for raising, rather than raising all chicks.

Purebred silver♀
$Z_S W$

Purebred gold♂
$Z_s Z_s$

All ♀♀ $Z_s W$ = Gold down

All ♂♂ $Z_S Z_s$ = Silver down

Plate 8 Calico cats demonstrating the patches of color resulting from random inactivation of X chromosomes bearing color determining genes in cells giving rise to hair.
Source: Ron Kimball.

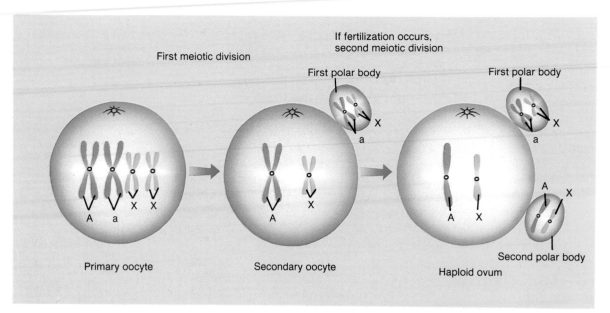

First meiotic division

If fertilization occurs,
second meiotic division

First polar body

First polar body

X
a

X
a

A
X

A X

Primary oocyte

Secondary oocyte

Haploid ovum

Second polar body

Plate 9 Oogenesis in a human female. Primary oocytes have the normal diploid complement of forty-six chromosomes, twenty-two pairs of autosomes plus XX. These are represented by a single pair of shaded chromosomes for the autosomes and a second pair of unshaded X chromosomes.

cells may not be rejected by the recipient as cells from an adult would be (see chapter 11 for discussion of transplant problems). In the face of considerable criticism, Dr. Robert Edwards, who, along with Dr. Patrick Steptoe, developed IVF techniques on humans, justified his work by the many advances in medical research likely to result from the ability to control and manipulate human embryos at the earliest stages.

Nonetheless, controversy over the use of IVF embryos will doubtless continue for many decades. The unpredictability of the course of this controversy is shown by a case in Australia in the mid-1980s involving a wealthy couple undergoing IVF treatment. The couple had yet to achieve a successful implantation when they were both killed in an accident, leaving some embryos frozen in the IVF laboratory. A suit was filed on behalf of the embryos as the rightful and only heirs to the wealth of the couple, presumably to require that they be implanted in a surrogate mother. A court ruled against the suit, however, and the embryos were allowed to die.

Prenatal diagnosis

Diseases of genetic origin are a major human health problem. Some 3%–5% of all humans, perhaps two hundred million persons, are seriously affected by deleterious genetic conditions (table 7.1). As the genetic basis for diseases such as cancer becomes better known, this estimate is likely to increase substantially.

Diagnostic techniques such as chromosome analysis, genetic-marker tests using restriction enzymes (RFLPs, see chapter 13), and direct gene probes for certain genetic diseases now permit identification of individuals carrying certain deleterious genes. Because of rapid advances in these techniques, in the coming decade we may be able to diagnose many, perhaps most, common genetic diseases and, also, identify the carriers for these conditions.

In this section we discuss briefly the primary techniques now available for sampling tissues for genetic analysis from fetuses *in utero*. Results of prenatal screening allow parents to make informed choices regarding the future of their child.

Amniocentesis

Developed in the 1960s as a test for maternal-fetal Rh incompatibility, **amniocentesis** is a procedure that samples the amniotic fluid surrounding the fetus. A needle is inserted through the mother's body wall and uterus, and about ten to twenty milliliters of fluid are withdrawn (figure 7.5). Amniotic fluid contains fetal secretions and living epidermal cells from the digestive and respiratory tracts. The living cells are grown in laboratory culture and then subjected to both karyotypic analysis and biochemical tests. Over sixty biochemical defects can now be identified as well as several dozen genes by direct analysis of the DNA (see chapter 13).

Timing of amniocentesis is crucial. Before the fifteenth week of pregnancy there is insufficient amniotic fluid to permit safe extraction for culturing. It takes two to three weeks to grow cells and culture chromosomes for karyotype analysis and an additional four to six weeks to grow enough cells for successful biochemical tests. Therapeutic abortion is legal in most states only through week twenty-eight. If results of the first amniocentesis are inconclusive, little time is available for a second testing. Furthermore, danger to the mother during abortion increases with duration of the pregnancy. According to the U.S. National Academy of Science Institute of Medicine, the risk of maternal death due to legal abortion in the first trimester is roughly 1.8 per one hundred thousand. Abortions following amniocentesis are performed in the second trimester when the risk of death has risen to 12.2 per one hundred thousand, still slightly lower than the risk during a noncaesarian live birth of 12.5 per one hundred thousand.

Table 7.1
The incidence of different categories of genetic diseases

Type of condition	Frequency per thousand
Chromosome abnormalities	6.9
Defects caused by single genes of major effect	
Autosomal dominant	1.9–2.6
Autosomal recessive	2.2–2.5
Sex linked	0.8–2.0
Subtotal	11.8–14.0
Congenital malformations*	(19.0–22.0)
Complex disorders†	(7.0–10.0)
Total genetic contribution (approximately)	(38.0–46.0)

Source *Report of the Task Group on WHO Human Genetic Programme Health for All by the Year 2000: The Contribution of Human Genetics.* WHO Report HMG/TG/81.3.1981.

*One-half of malformations such as spina bifida, congenital heart disease, and cleft lip with or without cleft palate are assumed to be genetic in origin.

†One-third of disorders such as schizophrenia and diabetes mellitus are estimated to be genetic.

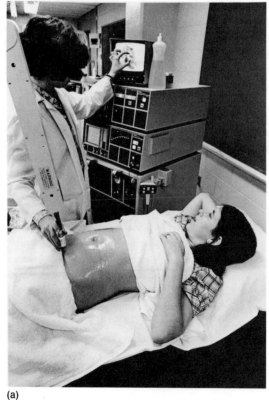

Figure 7.5 Amniocentesis. (a) Ultra-sound analysis is used to visualize the fetus and determine its position, to allow safe sampling of amniotic fluid. Ultra-sound analysis can also be used to monitor the growth of the fetus during pregnancy. (b) A needle and syringe are used to withdraw amniotic fluid, which is then centrifuged to separate fetal cells and fluid. The cells are cultured for karyotypic and biochemical analysis; some biochemical tests can be made directly on the amniotic fluid.
Source: (a) Will McIntyre/ Photo Researchers.

(a)

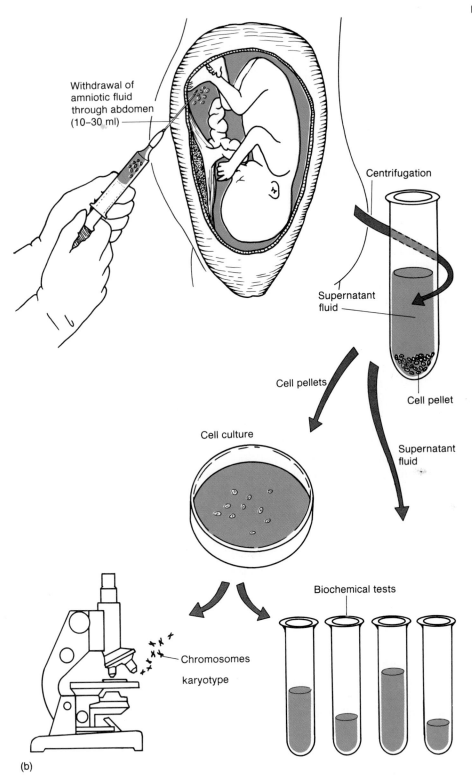

Withdrawal of amniotic fluid through abdomen (10–30 ml)

Centrifugation

Supernatant fluid

Cell pellets

Cell pellet

Supernatant fluid

Cell culture

Chromosomes

karyotype

Biochemical tests

(b)

Figure 7.5 (*continued*)

The most common conditions detected by amniocentesis are Down's syndrome (chapter 10; about five thousand cases annually, identified by the presence of an extra chromosome 21) and neural tube closure defects (about six thousand to eight thousand cases per year, detected by the presence of elevated levels of alphafeto-protein in the amniotic fluid). The latter condition results in anencephaly (absence of the brain and spinal cord) and spina bifida (an open spine, due to failure of normal closure of the spinal cord during embryonic development).

Chorion villus sampling

More recently **chorion villus sampling** has been introduced in the United States as an alternative to amniocentesis. Chorion villus sampling can be performed as early as the eighth week of pregnancy, and results can be obtained within twenty-four hours. This technique is painless. A plastic tube is inserted through the entrance to the womb, and a tiny piece of the chorion snipped off. Karyotypes can be prepared directly from mitotic cells in this tissue. Numerous enzymes can be assayed from homogenates of the tissue, eliminating the standard three to four week period needed to grow sufficient material from amniotic fluid to perform the same analyses. Chorion villus sampling gives the pregnant woman information about the fetus within twenty-four hours of the test and permits a first-trimester abortion if one is warranted. Chorionic villus sampling currently is being used successfully in China, England, France, Italy, and Russia, saving many women weeks of waiting for the tests results of amniocentesis.

Fetoscopy

Fetoscopy is a technique introduced in the early 1980s that allows a physician to visualize a fetal blood vessel *in utero*. A needle can then be inserted into the vessel to obtain a sample of fetal blood that can be examined for genetic conditions such as hemophilia. Additionally, intravenously administered therapies rely upon fetoscopy. For example, vitamin deficiencies, which can cause mental retardation, can now be treated prenatally; in addition, blood transfusions can be administered to fetuses.

Any technique allowing visualization of the fetus—fetoscopy or ultrasound, for example—can also be used to diagnose morphological defects prenatally. Sometimes, as in the case of blocked ureters, prenatal surgery may allow the fetus to be saved.

Use of prenatal diagnosis

Who should have prenatal diagnosis? At first glance, widespread screening of all pregnant mothers for the dozens of genetic conditions that can now be detected prenatally might seem a good idea. Prevention of all genetic diseases would be an obvious benefit to society, but three factors presently preclude the application of these techniques to all pregnant women: risk, cost, and use of the information.

Risk

Amniocentesis and chorion villus sampling are associated with a number of risks. Amniocentesis may lead to leakage of amniotic fluid. In addition, the needle may cause injury to the fetus, and lung and limb problems sometimes result. Both amniocentesis and chorion villus sampling can lead to infections; the spontaneous abortion rate is about 1%, slightly less than the rate of occurrence of serious genetic anomalies. No such risks are associated with ultra-sound observation of the fetus.

Cost

Prenatal sampling of tissues is usually an out-patient procedure requiring approximately one hour in a clinic. Total costs for ultrasound, amniocentesis or chorion villus sampling, karyotype analysis, and biochemical tests are usually about one thousand dollars or more. A single biochemical test may cost several hundred dollars.

Use of information

Although many genetic diseases can be diagnosed (and the number should continue to increase rapidly in the coming decade), many cannot be treated. The most common use of diagnostic techniques is to detect serious genetic conditions such as Tay-Sachs disease (table 4.1) or major chromosomal anomalies (chapter 10) and to provide information upon which a decision for therapeutic abortion can be based.

However, routine prenatal diagnosis would also discover many genetic conditions that have only minor effects or whose expressions are variable or uncertain. For example, karyotypic analysis would regularly reveal XYY males, but as discussed in box 6.2, XYY males present a dilemma. The condition cannot be treated, and XYY males usually lead normal lives, although there is some possibility that they may be prone to aggressive behavior. In addition, the *knowledge* that an individual is XYY may itself induce the feared aggressive behavior.

Thus, a spectrum of genetic conditions exists, some causing such serious problems that abortion seems the reasonable choice, but many others in which diagnosis itself may compound or even create problems. This consideration, alone and in conjunction with risk and cost factors, argues strongly that wide-scale, indiscriminate prenatal screening is inappropriate. Future reductions in risk and cost combined with the ability to treat most genetic conditions might warrant a different view, but for today's society such screening cannot be justified.

Criteria for prenatal diagnosis

In general, and with a certain amount of latitude on the part of the physician, prenatal diagnosis is recommended and permitted for cases in which all of the following criteria are satisfied:

1. *The fetus is known to be at special risk for certain genetic conditions.* If a family has a history of a serious genetic disease such as hemophilia or sickle-cell anemia, prenatal diagnosis is recommended. This is particularly true if a mother has already given birth to an affected child, and thus the parents are known to be carrying deleterious genes. Other special risk groups might include women taking certain mutagenic or teratogenic drugs such as thalidomide, women contracting rubella during pregnancy, or women over thirty-five, who have an increasing probability of giving birth to children with Down's syndrome.

2. *The genetic condition under screening is serious.* Because the most common result of a positive finding following prenatal screening is therapeutic abortion, screening for nonserious conditions is pointless. Screening for blood type, as an extreme example, is of no merit.

    While such a requirement (that the genetic condition under screening be serious) might not need explicit statement, at least one example suggests otherwise. In India, abortions are strictly limited and are legal only when the life or health of the mother is threatened. A recent study of abortions in India revealed that over 99.9% of abortions—7,997 of 8,000 cases, in this instance—involved female fetuses. Clearly, prenatal diagnosis of sex is practiced on a large scale, and the results are used illegally to pursue an ancient social prejudice.

Other societies also have biases against female children. In China, where government policy imposes penalties for families with more than one child, female infanticide is sometimes claimed to be a common practice. In Western societies, more prospective parents state preferences for male babies than for females. Although physicians in Western societies will not conduct prenatal diagnosis to determine the sex of the fetus alone, diagnosis of sex is used for serious X-linked diseases. If the parents are known carriers or are at risk of being carriers for hemophilia or Lesch-Nyhan syndrome (table 5.1), for example, a sex test is used; male fetuses can be aborted, and females can be permitted to continue gestation.

3. *There is no treatment for the disease.* If the condition is treatable, therapeutic abortion cannot be justified, and the parents gain little from knowing in advance of their child's illness. Such knowledge may cause needless worry during pregnancy. Screening for PKU, for example, would not be permitted.

4. *The parents must be willing to terminate the pregnancy following a positive diagnosis.* Again, diagnosis for its own sake has limited virtue and may cause needless psychological trauma. If the parents are not willing to abort an affected fetus, the risk of the diagnostic procedure is added to the risk of genetic disease with little or no benefit.

In some cases, willingness to terminate the pregnancy is not a requirement for prenatal diagnosis. Some physicians believe that knowledge that an unborn child is affected can help those parents unwilling to abort, by allowing them to make both emotional and physical preparations prior to birth, which may aid them in caring for their seriously affected child.

Bioethics and therapeutic abortion

In this chapter and in chapter 4, we discuss **therapeutic abortion** as one response to knowledge that a fetus is at risk of or suffering from a serious genetic disease. For many persons the decision to abort a fetus affected by a serious genetic condition such as Lesch-Nyhan syndrome is straightforward. These people support such a decision because it has many consequences that they see as beneficial to all involved—the fetus, the parents, the family, and society at large. The fetus is spared a short, painful life. The family is spared the overwhelming burden of disruption, emotional anguish, and medical expenses. Society is spared the high financial cost of caring for the child if the family cannot do so.

Others, however, find abortion unacceptable under any circumstance. They often support their views on the grounds that abortion is murder and that murder is never acceptable.

The conflicting views expressed over the suitability of abortion are referred to as a philosophical *dilemma,* a choice between alternate courses of action that each can be fully justified by its proponents. A dilemma poses an apparently insolvable predicament. The study of human conduct, especially in terms of distinguishing good and evil, lies within the philosophical field of *ethics.* Difficult choices between conflicting behaviors are said to be *moral decisions.*

Advances in biological knowledge have led to the development of a subdiscipline of ethics called *bioethics.* Bioethics deals with moral decisions arising from choices now available to humans because of new biological technologies and information. Many of these choices relate to human health and medicine, such as the decision to abort or to allow gestation to continue. (Other bioethical decisions arise from our treatment of the environment—for example, what limits should be set on exploitation of nature?)

Utilitarian or absolutist?

Bioethics cannot always determine what is right or wrong. Instead, it analyzes conflicting choices presented by new biological technologies within an ethical framework. Furthermore, the rapid development of new medical technologies is outracing our ability to conduct bioethical analysis and achieve a societal consensus on the use of new techniques. Thus, although bioethics can help us identify and understand the issues involved in a particular case, the final choice must be made by the individual based on his or her personal values. The two most common types of justification of bioethical choices are utilitarian and absolutist.

In the case of therapeutic abortion, if the decision to abort is based on a desire to spare the fetus, family, and society needless pain and suffering, bioethicists view the decision as a *utilitarian* choice. That is, the consequences of the decision are believed to justify the choice. Aborting the fetus reduces suffering, while refusing to abort increases suffering.

The alternative view—that abortion is unacceptable because it is murder—is an *absolutist* ethical stance. This view is based on the belief that some moral laws are always true and should never be violated. In this case, the moral law would be that murder is evil.

The absolutist position is often associated with a religious viewpoint or with moral absolutes, which may be derived from apparently self-evident laws of nature. A moral absolute such as "suffering is wrong" might be viewed as obvious to any civilized person, for example, and could lead to an absolutist justification for abortion. In this view, abortion would be ethically required when a fetus has been diagnosed as affected by serious genetic disease because it is a moral imperative to reduce suffering whenever possible.

Equally, a utilitarian view could support an antiabortion position. A utilitarian might argue that abortion contradicts the Hippocratic Oath (which requires physicians to save lives), undermining the ethical basis of medical conduct. Or, a utilitarian might support abortion only if public funds are provided, making abortion as available to the poor as to the wealthy, presently not the case in the United States. Otherwise, the system is unfair and discriminatory against the poor.

Societal standards

It is difficult to comprehend and balance the many aspects of bioethical choices, but the issues are pressing and decisions must be made. Most people will confront bioethical choices during their lives. Abortion, for example, is widely used in Western societies for both health and social reasons, and most families are likely to have a member confronted with this choice at some time.

Fortunately, society gives some guidance with respect to accepted standards of conduct. These guidelines may take the form of generally accepted religious codes or they may take the form of laws, legal limitations on behavior. Laws, of course, work only so long as they embody widely accepted social and ethical standards, and prohibition of both abortion and alcohol are examples of laws that were widely ignored because they violated this precept.

The U.S. law on abortion derives from the Supreme Court decision of *Roe v. Wade* of 1973. This ruling states that decisions on abortion in the first trimester of pregnancy are the sole right of the pregnant woman in consultation with her physician. Abortions later in the pregnancy are legal if they threaten the woman's health, and in the third trimester the state may intervene to protect the life of the fetus. Subsequent decisions have specifically prohibited the father of the child from interfering with the woman's right to decide, and parents of young unmarried mothers also have no right to interfere.

The role of the law is central in social consensus on ethical matters, but the law is usually conservative (in the general rather than political sense) in its approach. That is, the law responds to changing values only *after* the change. The law usually reflects social consensus at the time a specific law originated, often decades ago, and is subsequently only changed when a new consensus is achieved.

The rapid advance in technology emphasizes the conservative nature of the legal process and the need for an increasingly informed citizenry. Legal codes can only change and stabilize when a consensus is reached. For this to occur, the average citizen must be knowledgeable about new technologies and their ethical implications.

The present diversity of views regarding abortion makes it highly unlikely that a social consensus will be achieved in the near future. In fact, a number of subsidiary issues to *Roe v. Wade* remain to be resolved, for example, the appropriateness of public funding to support abortion. Additionally, many groups object to *Roe v. Wade* and continue to seek constitutional changes to prohibit abortion. And, as medical technology permits survival of fetuses earlier and earlier in pregnancy, the precise rights of fetuses and the determination of when a fetus is to be legally declared a human being are likely to be sources of increasingly fierce debate.

Summary

1. Artificial insemination, introduction of sperm into the uterus and cervical canal by other than natural means, is widely used when a male is sterile or has a low sperm count, but his wife is fertile.
2. Surrogate motherhood is sometimes used when a woman is infertile. Her husband's semen is used to artificially inseminate the surrogate, who surrenders her rights to the child upon birth. However, the numerous legal and ethical complications of this technique have limited its acceptance.
3. *In vitro* fertilization is a treatment for female infertility due to blocked oviducts. While increasingly widely used, the low success rate, high cost, and ethical problems related to use of excess embryos pose unresolved problems.
4. Prenatal diagnosis refers to a set of technologies that allow the genetic health of the fetus to be examined prior to birth. Diseases that can be assayed in chromosomal or biochemical tests can be diagnosed from amniotic fluid via amniocentesis or from samples of the chorionic villi.
5. Prenatal diagnosis is costly and risky and is not used on all pregnant women. It is applied most often when the fetus is known to be at risk for treatable serious genetic conditions and the parents are willing to terminate the pregnancy following a positive diagnosis.
6. Therapeutic abortion exemplifies the type of bioethical dilemma that many people now confront. Compelling arguments both for and against the suitability of therapeutic abortion can be advanced from many different viewpoints.
7. The two most common bases for bioethical decisions are utilitarian and absolutist arguments. Bioethical decisions must be based on our personal value systems.

amniocentesis

artificial insemination

chorion villus sampling

embryo transfer

fetoscopy

in vitro fertilization

prenatal screening

sperm bank

superovulation

surrogate mother

therapeutic abortion

Questions

1. Would you recommend that every pregnant woman undergo amniocentesis? Why or why not? For which groups of pregnant women would amniocentesis be most strongly recommended?
2. Considering *in vitro* fertilization: what happens to a fertilized ovum in the first week following conception in a normal pregnancy, i.e., where is the embryo and what is it doing?
3. How does H. J. Muller's concept of an incubator female differ from surrogate motherhood as it is now practiced?

For More Information

Beardsley, T. 1985. "Problems of Prenatal Testing." *Nature,* 314:211.

Broad, W. J. 1980. "A Bank for Nobel Sperm." *Science,* 207:1326–27.

Brody, B. 1975. *Abortion and the Sanctity of Human Life: A Philosophical view.* MIT Press, Cambridge, Massachusetts.

Feinberg, J. 1980. "Abortion." In *Matters of Life and Death,* ed. T. Regan. Random House, New York.

Harron, F.; J. Burnside; and T. Beauchamp. 1983. *Health and Human Values: A Guide to Making Your Own Decisions.* Yale University Press, New Haven.

Kuhse, H., and P. Singer. 1985. *Should the Baby Live? The Problem of Handicapped Infants.* Oxford University Press, Oxford.

Miller, J. A. 1986. "New Molecular Analysis for Genetic Disorder." *Science News,* 129:84.

Orkin, S. H. 1982. "Genetic Diagnosis of the Fetus." *Nature,* 296:202–3.

Rao, R. 1986. "Move to Stop Sex-Test Abortion." *Nature,* 324:202.

Reich, W. T., ed. 1978. *The Encyclopedia of Bioethics.* The Free Press, New York.

Rensberger, B. 1981. "Tinkering with Life." *Science 81,* November:45.

Saltus, R. 1986. "Biotech Firms Compete in Genetic Diagnosis." *Science,* 234:1318–20.

Shurkin, J. 1983. "Operating on the Unborn." *Science 83,* May:71–74.

SECTION III

GENETIC INFORMATION AND EXPRESSION

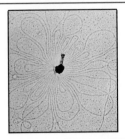

Informational macromolecules

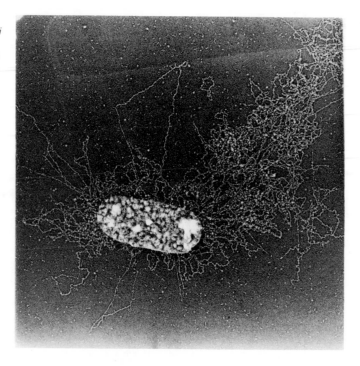

Figure 8.1 An electron micrograph of a cell of *E. coli* that has been treated with mild detergent to disrupt the cell wall and release its DNA (magnification = 20,000×). The small circle of DNA at 11:00 is a plasmid that the cell contained, demonstrating the contrast in size of genomic and plasmid DNA. (See colorplate 10.)
Source: Dr. Jack Griffith.

Like most good science, Mendel's work answered some questions, but also raised new ones. Through careful study of phenotypic characters, Mendel described a genetic mechanism and demonstrated the existence of discrete genes underlying phenotypic characters. He also demonstrated that those genes passed unchanged between succeeding generations. Mendel's work, however, shed no light on the physical and chemical nature of the gene. For decades after Mendel's work, many geneticists concentrated on establishing the chemical identity of the genetic material, as well as describing the chemical structure that could explain the Mendelian behavior of genes.

The search for the genetic material

We now know that the hereditary material for all species is **nucleic acid.** In most species it occurs in the form of deoxyribonucleic acid, DNA; for a few viruses, it occurs as ribonucleic acid, RNA. The primary functions of nucleic acid relate to the hereditary process: the storage and transmission of information; and the expression of hereditary information as proteins. Less than a century ago, however, **protein** rather than nucleic acid was the prime candidate for the hereditary molecule.

The structure of protein

Proteins were a good guess for the genetic substance. About 66% of the weight of the human body is water; 47% of the remainder (16% of total weight) is protein. Proteins fulfill a variety of crucial functions for the living organism. Of particular importance, one category of proteins—called **enzymes**—serves to *catalyse* the thousands of necessary biochemical reactions in the body. Catalysts accelerate chemical reactions, permitting them to occur in a fraction of a second.

Some proteins function to provide physical support, both for individual cells and for the entire organism. Keratin is a structural protein that is the primary component of skin and hair; tubulin forms part of the "skeletal" or supportive system of an individual cell. Muscle tissue consists of protein specialized for movement. *Antibodies* are proteins that protect the body from bacteria and other foreign substances. Hemoglobin is a protein specialized for the transport of gases in the blood.

Some *hormones* are proteins that provide communication among some tissues of the body. A brief introduction to the chemical structure of proteins can help illustrate how proteins fulfill so many different functions, and why they once seemed the most likely chemical to bear hereditary information.

Proteins consist of long sequences or chains of molecules known as **amino acids** (figure 8.2). Twenty amino acids are commonly used in the proteins of most species. Humans are able to synthesize twelve of these twenty amino acids, while the remaining eight *essential* amino acids must be supplied from the diet.

Two amino acids are bound together by a **peptide bond** to form a molecule known as a dipeptide. A chain of many amino acids is known as a **polypeptide.** Proteins may consist of only one polypeptide chain, several polypeptide chains, or polypeptide chains combined with other molecules. Hemoglobin, for example, is a protein consisting of four polypeptide chains, each bound to an iron-containing molecule (figure 8.2).

The sequence of amino acids in a polypeptide chain is called its **primary structure.** Most proteins consist of a single polypeptide chain one hundred to five hundred amino acids long, but some are considerably longer. Because a primary characteristic of an hereditary molecule must be the capacity to encode large amounts of information, it seemed possible that the sequence of amino acids in a long protein could easily fulfill this requirement. (As an analogy, the English language—which is quite complex and carries enormous amounts of information—is constructed from an alphabet of only twenty-six letters.)

The number of different possible sequences of amino acids in a protein is 20^n, where n is the number of amino acids in the protein. For a polypeptide only four amino acids long, for example, 160,000 different messages are possible; ten amino acids can produce 10,240,000,000,000 unique messages. For a protein one hundred amino acids long, the number of messages is unimaginably large. Biochemists of the nineteenth century knew that such a large molecule could possess sufficient variety of structure to fulfill the many functions of proteins, as well as encode the immense amount of hereditary information needed to direct the biochemistry of life.

The complexity of protein structure does not end with its primary structure. Commonly, the proteins in a polypeptide chain coil into a spiral or *helix* (from the Latin word for ivy), which is the *secondary structure* of a protein (figure 8.2). In turn, the helix is twisted and folded precisely, the *tertiary structure,* into a complex, three-dimensional conformation that is crucial to the protein's function. Finally, many proteins consist of linked chains of two or more polypeptides; the arrangement of multiple polypeptides provides such proteins with a *quaternary structure.*

Nuclein

In 1869—only a few years after the publication of Mendel's work, but decades before the recognition of its importance—a Swiss biochemist, Johann Friedrich Miescher, studied the nuclei of pus cells. Miescher isolated from the pus cells a phosphorus rich, nonprotein material that he called *nuclein,* named for its site of origin. (The suffix *-in* usually identifies a chemical as protein.) Nuclein proved to be a nucleic acid, **DNA.**

Work by Miescher and other scientists subsequently demonstrated that the nuclei of cells of all species contain nuclein, and that nuclein was an important constituent of chromosomes. Quantitative analysis of chromosomes showed that they consisted of about 60% protein and 40% nuclein (DNA)—quite reasonably suggesting to scientists at the time that protein was probably the genetic material. Further evidence supporting this view was the fact that proteins are composed of about twenty types of amino acids, while DNA has only six types of molecular components. In other words, the hereditary alphabet seemed to be substantially larger in proteins. Thus, it seemed likely that the apparently more complex molecules (proteins) should be the carriers of the genetic message.

Table 8.1

DNA content of selected vertebrate somatic and sperm cells (DNA measured in picograms = 10^{-12} gram—one-trillionth of a gram)*

Organism	2N cells	Sperm
Cow	6.6	3.3
Human	6.4	3.2
Chicken	2.6	1.3
Frog	15.0	7.5

*Most mammals have DNA amounts similar to those found in humans. The maximum DNA content of an animal cell is found in the salamander *Amphiuma*, which has 168.0 picograms. A *Drosophila* cell has 0.2 picograms.

Amounts of DNA and protein

Description of the meiotic behavior of chromosomes supported their role in heredity. The diploid number was constant in the somatic cells of all members of a species, but was halved in the germinal cells, returning to the diploid condition only after fertilization.

Careful measurements of the amounts of DNA and protein in somatic and sperm cells again indicated the likely hereditary role of DNA, but not protein (table 8.1). Within a species, the amount of protein in a sperm cell varied considerably. Furthermore, the amounts of protein in sperm and somatic cells bore no simple relationship to each other. On the other hand, the amount of DNA in sperm was relatively constant. Additionally, the amount of DNA in sperm cells was exactly one half that in somatic cells.

Transformation

An English microbiologist, F. Griffith, began the sequence of experiments that eventually led to the establishment of DNA as the genetic material. Griffith showed that bacteria of one strain could incorporate genes from a second strain. This phenomenon caused the first strain of bacteria to display phenotypic traits of the second, a process called **transformation** (box 8.1). Transformation is now a widely practiced procedure forming the foundation for the field of *biotechnology,* but in Griffith's time, transformation was regarded as absolutely astonishing.

Subsequent elegant work in 1943 by three American biologists, O. T. Avery, C. M. MacLeod, and M. McCarty, separated the contents of bacterial cells of one strain into four basic biochemical components: DNA, proteins, lipids, and carbohydrates. Each of the four components was mixed with a separate bacterial culture of a different strain. Only the DNA was capable of transforming the bacterial culture; none of the other biochemical components could do so.

Analogous experiments with viruses in the early 1950s by A. D. Hershey and M. Chase also gave results pointing exclusively to DNA as the genetic material (see box 8.2). Viruses are perfect experimental organisms for such a study, as they are composed of only two components, DNA surrounded by a protein coat. When the DNA and protein were labelled with unique radioactive isotopes, only the labelled DNA was shown to be capable of infecting bacteria and producing more viruses. The viral protein alone was unable to direct the production of more viruses.

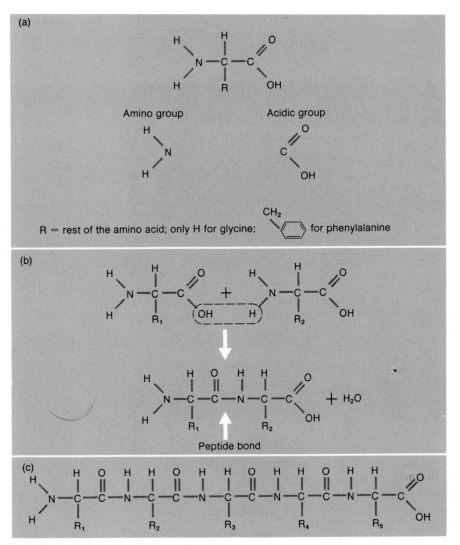

(a)

Amino group

Acidic group

R = rest of the amino acid; only H for glycine; CH_2—◯ for phenylalanine

(b)

+

Peptide bond

+ H_2O

(c)

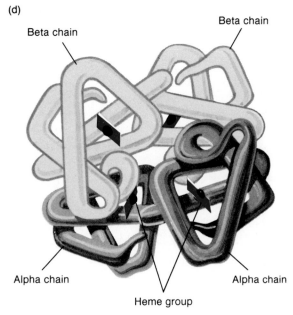

(d)

Beta chain

Beta chain

Alpha chain

Alpha chain

Heme group

Figure 8.2 The primary structure of proteins: (a) amino acids, the components of polypeptides; (b) the formation of the peptide bond between two amino acids; (c) a polypeptide chain of amino acids held together by peptide bonds; (d) the protein molecule hemoglobin, which transports oxygen to all body tissues. The primary structure is the sequence of the amino acids in the alpha chain and in the beta chain. The secondary, tertiary, and quaternary structures reflect the coiling and folding of the chains and their association into a tetrameric molecule with the oxygen containing heme groups.
Source: (d) Stuart Ira Fox, *Human Physiology*, 2d ed. Copyright © 1987 Wm. C. Brown Publishers, Dubuque, Iowa. All Rights Reserved. Reprinted by permission.

BOX 8.1

Outline of Griffith's experiments demonstrating transformation

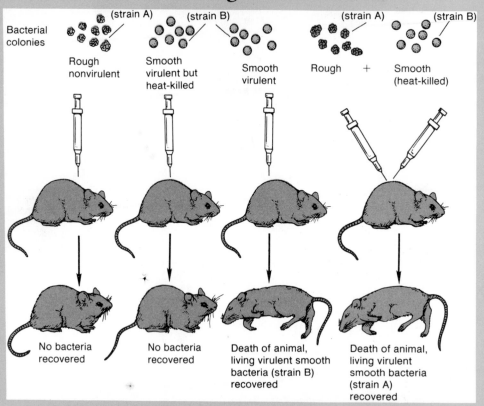

Figure 8.3 Schematic representation of Griffith's experiments with the bacterium *Diplococcus pneumoniae* and mice.

Griffith studied two pneumonia-causing strains of the bacterium *Diplococcus*. When grown *in vitro* (meaning *in glass*) in a laboratory culture flask or on an agar plate, these virulent strains, are smooth in appearance. Mutants of both strains, however, appear rough when grown in culture because they lack the normal bacterial capsule, nor do these mutant (rough) strains cause pneumonia (figure 8.3).

Smooth bacteria, which cause pneumonia in mice, can be recovered from mice who have died from pneumonia. When rough or smooth bacteria are heated, a process that kills most strains of bacteria, and injected into mice, all the mice survive, and no bacteria can be recovered from the mice. When rough bacteria of strain A were injected into mice simultaneously with heat-killed smooth bacteria of strain B, the mice died of pneumonia, and smooth bacteria of strain A were recovered. Thus, inside the mice, the live (rough) strain A bacteria were astonishingly transformed to smooth (virulent) bacteria. We now know this transformation occurred because the live strain A bacteria incorporated DNA coding for the production of the smooth capsule from the dead strain B bacteria. The rough strain A bacteria were transformed to smooth, pneumonia-causing bacteria, resulting in death of the mice and bacteria capable of producing a bacterial capsule.

BOX 8.2

DNA is the genetic material—evidence from viruses

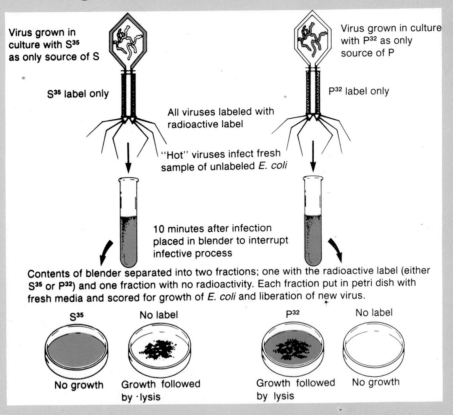

Virus grown in culture with S³⁵ as only source of S

S³⁵ label only

Virus grown in culture with P³² as only source of P

P³² label only

All viruses labeled with radioactive label

"Hot" viruses infect fresh sample of unlabeled *E. coli*

10 minutes after infection placed in blender to interrupt infective process

Contents of blender separated into two fractions; one with the radioactive label (either S³⁵ or P³²) and one fraction with no radioactivity. Each fraction put in petri dish with fresh media and scored for growth of *E. coli* and liberation of new virus.

S³⁵	No label	P³²	No label
No growth	Growth followed by ·lysis	Growth followed by lysis	No growth

Figure 8.4 Schematic interpretation of the Hershey–Chase "blender experiments" that confirmed that DNA is the hereditary material. (See colorplate 11.)

In 1952, A. D. Hershey and M. Chase presented even more convincing evidence that DNA was indeed the genetic material. Hershey and Chase took advantage of the simple morphology and life history of a virus to compare the roles of DNA and protein in viral reproduction (see chapter 11; figure 11.8).

Both DNA and protein contain the elements most characteristic of living organisms—carbon, hydrogen, oxygen, and nitrogen. They differ in one major respect, however: sulfur occurs only in protein and not DNA, while phosphorus occurs only in DNA and not protein. Hershey and Chase began their experiments by growing cultures of the virus T-2 (figure 8.5a) in suspensions of *E. coli* rich in one of two radioactive isotopes, sulfur-35 (S³⁵) or phosphorus-32 (P³²). All of the T-2 derived from these cultures thus had either the DNA labeled with P³² or the protein coats labeled with S³⁵.

Next, an unlabeled culture of *E. coli* was infected with S³⁵ labeled T-2, but growth was permitted to proceed for only about ten minutes, half the lytic cycle of T-2. Growth was interrupted by placing the mixture in a blender, which separated the empty protein coat (the *viral ghost*), from the surface of the bacterial cells. The mixture was then separated into two fractions, one containing radioactively labeled sulfur and one with no label. The two fractions were placed in separate cultures and permitted to resume growth. The culture containing the S³⁵ fraction showed no growth and produced no T-2. Examination showed that it contained only the viral ghosts (figure 8.4).

On the other hand, if *E. coli* were infected with P³²-labeled T-2 and the experiment repeated, the phosphorus-bearing fraction contained *E. coli* and the DNA of the T-2 that had been injected into the bacteria. Hence the fraction continued to grow, ceasing only when the T-2 completed their lytic cycle, killing all the bacteria. Thus, phosphorus-containing DNA had been transmitted between generations of bacteriophages. These results excluded protein as the genetic material.

The race to discover the structure of DNA

By the early 1950s scientists in several laboratories were intensely involved in a *de facto* race to determine the chemical structure of the hereditary material. Although evidence strongly supported DNA, knowledge of the precise structure of the molecule was necessary to understand how it could fulfill the exacting requirements of inheritance.

Some aspects of the components and structure of DNA were well known. *Guanine* had been isolated from crystallized bird feces (guano) in 1844, and *thymine* had been purified from calf thymus, also in the nineteenth century. Erwin Chargaff, a biochemist at Columbia University, had demonstrated that in DNA the amount of adenine always equalled the amount of thymine, and the amount of cytosine always equalled the amount of guanine—relationships now known as *Chargaff's rules*.

Such famous scientists as Linus Pauling actively pursued the structure of DNA, but two young and relatively unknown scientists made the discovery. James Watson had received his Ph.D. from the University of Indiana at the age of twenty-three after being refused admission to graduate school at Harvard University and the California Institute of Technology. In 1951 he arrived at Cambridge University, where he began his now-famous collaboration with Francis Crick.

Crick had received his university degree in physics in 1938, but was working towards a Ph.D. in biology when he met Watson. A key piece of the puzzle was supplied by X-ray crystallographic studies of DNA by Rosalind Franklin and Maurice Wilkins. Using the X-ray studies in conjunction with previously published work, Watson and Crick deduced the probable structure of DNA, publishing their model in a short paper in 1953. Watson, Crick, and Wilkins were awarded the 1962 Nobel Prize for Physiology or Medicine for their historic discovery.

Although the X-ray crystallography that provided crucial evidence was actually done by Rosalind Franklin, she was not awarded the Nobel Prize because of several technical restrictions. No more than three persons can share one prize, and they must all be living. Franklin died of cancer in 1958. Nonetheless, the role of Rosalind Franklin in solving the structure of DNA has become a matter of historical debate. She had an uneasy professional relationship with Wilkins, who apparently showed her work to Watson and Crick without her permission. There may thus be some irony in Wilkins' receipt of the Nobel prize. Very different accounts of Franklin's role are given by Anne Sayre (Franklin's biographer) and by James Watson.

The chemical structure of DNA

By the early 1950s the accumulating evidence favored nucleic acids as the hereditary material. However, proof of their role depended upon determination of a structure for DNA sufficient to permit it to fulfill the difficult functions of an hereditary molecule. The search for the structure of DNA is one of the best known and most fascinating detective stories in scientific history (box 8.3). It involved some of the most famous scientists of the time, but the race was won by two relatively unknown scientists—James Watson and Francis Crick (figure 1.1c). The discovery of the structure of DNA won them (and biophysicist Maurice Wilkins) the 1962 Nobel Prize for Physiology or Medicine and a permanent place in the history of science.

The hereditary molecule could not be a simple one; it must fulfill at least four functions. The hereditary molecule must be able (1) to contain all the needed information to direct the development and life of an organism; (2) to transmit that information to all parts of the body, translating the information into a usable form in all the tissues; (3) to transmit that information to its descendants with high accuracy; and (4) to be mutable, in spite of the normally precise transmission of information—that is, it must permit a small number of changes to occur routinely, creating genetic variability in a population.

In 1953, Watson and Crick published a short paper in the journal *Nature* describing a model for the structure of DNA, now known as the **Watson-Crick model.** At the conclusion of this paper, they appended a sentence understating the importance of their hypothesis: "It has not escaped our notice that the [structure] we have postulated immediately suggests a possible copying mechanism for the genetic material." Indeed, the Watson-Crick model of DNA, verified and elaborated by many subsequent studies, is the foundation for modern molecular genetics. Within a decade after the Watson-Crick hypothesis, DNA was shown to fulfill all requirements of the hereditary molecule.

The Watson-Crick model describes DNA as an extremely long, two-stranded molecule twisted like a rope into the shape of a **double helix** (figure 8.5). The basic building block of DNA is a molecule called a *nucleotide*. Each nucleotide has a short backbone of a small sugar molecule linked to a phosphate group. The five-carbon sugar is known as **deoxyribose.**

Within a DNA molecule, the deoxyribose and phosphate groups show no variation. Long chains of thousands of alternating sugar and phosphate groups from thousands of nucleotides are linked together, forming two parallel backbones. The two backbones are joined by the third component of each nucleotide, a **nitrogenous base.** One of four different nitrogenous bases is attached to each of the sugar molecules in every nucleotide: *adenine* (A), *cytosine* (C), *guanine* (G), or *thymine* (T). These four bases comprise the only letters in the genetic alphabet, and the major variation between the DNA of different individuals is in the sequence of bases along the length of the molecule.

On one side, each nitrogenous base is attached to a sugar molecule in the long chain of alternating sugars and phosphates. On the other side, each base attaches to a nitrogenous base extending from a sugar on the parallel strand. Thus, the two strands are bound at each sugar by two nitrogenous bases, bridging the gap between the parallel chains much like rungs of a ladder. The entire molecule is twisted into a two-stranded or double helix, just as if a long ladder had been twisted about its longitudinal axis.

The nitrogenous bases of one strand do not join randomly with those of the parallel strand. Rather, only two pairings are possible, adenine with thymine, and cytosine with guanine, A-T and C-G. Thus, one strand is *complementary* to the other (**complementary base pairing**), explaining Chargaff's rules (see box 8.3). The sequence of nucleotide bases on one strand reflects the sequence of bases on the other.

The sequence of bases and their complementary pairing on opposing strands are the keys to the function of DNA. The sequence of bases encodes hereditary information, while specific pairing provides the mechanism for accurate replication and transmission of this information between cells of one individual and between members of consecutive generations.

It was immediately apparent that DNA possessed the capacity to meet the first requirement of an hereditary molecule—that it be capable of encoding an immense amount of information. The number of unique messages that can be contained by a DNA molecule is 4^n, where n is the number of nucleotide pairs in the DNA. A DNA molecule only 4 base pairs long can encode 256 unique messages; 10 pairs can encode 1,048,576 messages; 100 pairs can encode over 10^{60} unique messages. The human genome contains over three billion nucleotide pairs. Thus, even with an alphabet of only four letters, the information-bearing capacity of DNA is far more than sufficient to encode all the information for life.

The Watson-Crick model resolved the search for the hereditary material. Each chromosome is now known to consist of immensely long chains of DNA surrounded by protein. While the protein probably serves an important function in the expression of the genetic message, it is DNA that bears the message. The Watson-Crick

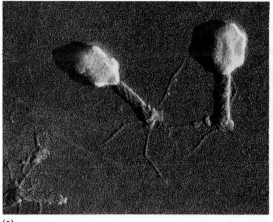

(a)

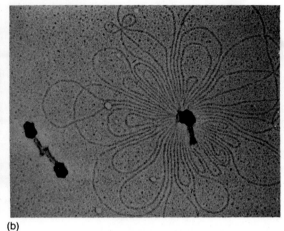

(b)

Figure 8.5 An electron micrograph of the phage T-2, (a) intact and (b) with its DNA liberated from the viral head by osmotic shock; (c) Watson-Crick model of DNA; (d) a schematic representation of the spiralling ladder of the DNA molecule held together by base pairing between adenine(A) and thymine(T), cytosine(C) and guanine(G); (e) an enlarged section indicating how the sugar and phosphate molecules form the ladder backbones of the DNA. For more detailed information on the chemistry and conformation of DNA, consult an advanced genetics text.
Source: (a) Dr. Michael Wurtz/Biozentrum der Universität Basel, Abteilung Mikrobiologie; (b) Biology Media/Photo Researchers; (c) Courtesy of Dr. Alexander Rich; (d and e) E. Peter Volpe, *Biology and Human Concerns,* 3d ed. Copyright © 1983 Wm. C. Brown Publishers, Dubuque, Iowa. All Rights Reserved. Reprinted by permission.

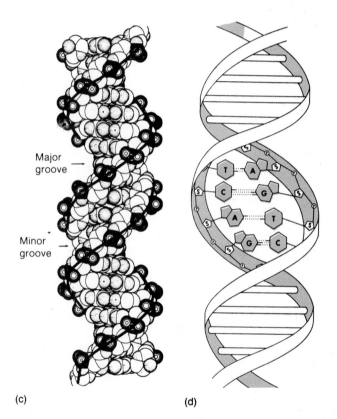

Major groove

Minor groove

(c)

(d)

model also moved the science of genetics into the realm of molecular biology. Geneticists no longer studied only the phenotypic expression of genes whose location and nature were unknown. Instead, geneticists began to study the genetic message itself.

RNA

RNA (ribonucleic acid) resembles DNA in structure and also plays a major role in the hereditary mechanism. Three structural differences distinguish RNA from DNA (figure 8.6). First, RNA contains **ribose,** a five-carbon sugar with one more oxygen atom than *deoxyribose.* Second, RNA is usually single rather than double stranded, although it is also helical in shape. Third, the nucleotide base thymine does not occur in RNA. Rather, it is replaced by a similar base, *uracil* (U).

Figure 8.5 *(continued)*

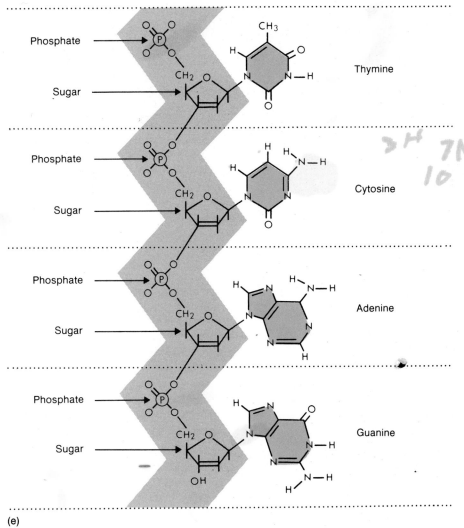

(e)

In some viruses, RNA rather than DNA serves as the hereditary molecule. Because of its analogous structure, it obviously has the same information-carrying capacity as DNA. However, all organisms have three forms of RNA, all of which are crucial to genetic function. All are transcribed directly from nuclear DNA, and they function as crucial intermediaries in the process of protein formation. They will be considered in detail.

DNA and chromosomes

Every cell in the body contains an identical set of chromosomes. In most species, each chromosome consists of one very long molecule of DNA. In eukaryotes, the DNA is surrounded and enclosed by several types of nuclear proteins, especially histones. Chromosomes containing only one copy of the DNA molecule are *unineme*. Some tissues of some species have *multineme* chromosomes (chromosomes consisting of many identical copies of DNA arranged in parallel), but little is known about the use of such an arrangement.

The DNA in a chromosome is remarkably long. If unravelled, the DNA from a normal-sized chromosome would stretch several inches, many times the diameter of the cell itself. Thus, DNA can be packaged into a cell only as a result of extreme coiling. One of the functions of histones is probably to organize the dense coiling of DNA.

Figure 8.6 A generalized structure for the informational macromolecule RNA has a single stranded sugar-phosphate backbone and four nitrogen bases. Note the sugar, ribose (X), differs from the deoxyribose of DNA, and the base uracil is found in place of the base thymine that occurs in DNA. In portions of RNA where base pairing can occur, adenine pairs with uracil and cytosine pairs with guanine.

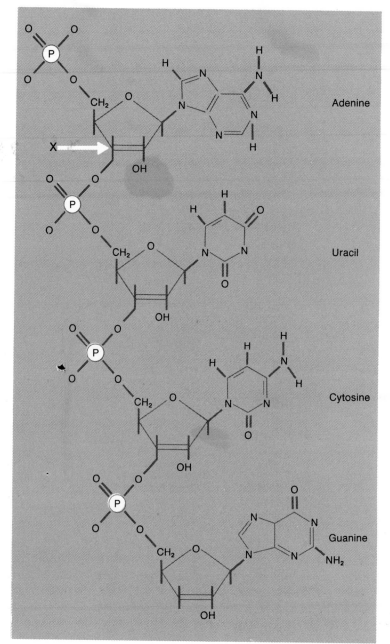

Both meiosis and mitosis produce daughter cells that contain genetic information derived from the parent cell, and both involve one episode of DNA duplication. Recall that replication of DNA occurs during the S or synthesis phase of the cell cycle—not at the time of cell division. Thus, at the beginning of cell division each chromosome consists of two identical double helices, joined at a single centromere.

Throughout most of the life of the cell, however, chromosomes are far more diffuse than during the brief period of cell division. It is probable that the absence of coiling during interphase is crucial to the function of DNA during the vegetative portion of the cell's life.

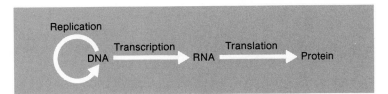

Figure 8.7 The central dogma as originally envisioned by Francis Crick. Contemporary understanding of genetic function has modified but not fundamentally changed Crick's model (see figure 8.17).

Function of DNA

The Watson-Crick model of DNA is a cornerstone of modern biology. In the few decades since the discovery of the chemical structure of the gene, molecular genetics has become one of the most intensely exciting and rapidly advancing fields of modern science. Since 1953, a large body of knowledge has accumulated that verifies the double-helical nature of DNA and describes in considerable detail how this structure permits DNA to fulfill necessary hereditary functions.

The central dogma

In 1958 Francis Crick described the modern concept of the relationship between genotype and phenotype as the **central dogma** (figure 8.7). Expression of the hereditary information contained in DNA requires several steps. The genetic message is first copied precisely to form **messenger RNA**—the process of **transcription.** In eukaryotes, messenger RNA then moves from the nucleus to the cytoplasm, where its encoded information is read precisely by both **ribosomal RNA** and **transfer RNA** to assemble proteins—the process of **translation.** The proteins themselves, particularly in the forms of enzymes, mediate the subsequent processes of phenotypic development.

Geneticists of the late nineteenth and early twentieth centuries were correct that proteins were intimately involved with heredity, but they were not correct in believing that proteins actually were the hereditary molecule. The sequence of amino acids in a protein is itself determined by the sequence of nucleotide bases in DNA. Like DNA, protein is an **informational macromolecule;** but it is not the controlling molecule. Genetic information flows from DNA to proteins, and not vice versa.

The central dogma also describes the flow of genetic information between generations. In the cell nuclei DNA replicates itself during interphase. The precise duplication and division of chromosomes during mitosis insure that each cell contains a set of genetic instructions identical to those in the newly fertilized zygote. Meiotic processes govern the recombination and independent assortment of the genetic material in germinal cells, with each gamete receiving a complete haploid set of chromosomes.

Replication

Complementary base pairing was recognized by Watson and Crick as a mechanism that permits precise replication of the DNA molecule. As a result of complementary pairing, each strand of a DNA molecule can direct the synthesis of a complete new molecule of DNA.

During replication, each strand of DNA serves as a *template* for the formation of a new parallel and complementary strand. As in most biochemical processes, the replication of DNA is controlled by a series of enzymes with precise functions. One type of enzyme breaks the bonds between complementary nucleotide pairs—in effect,

Figure 8.8 Semi-conservative replication in DNA.

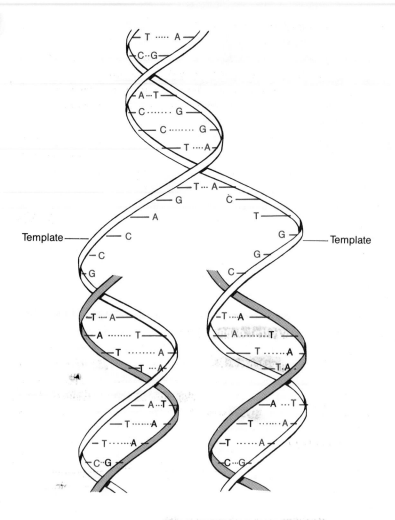

the halves of the DNA molecule separate much like the opening of a zipper (figure 8.8). As the two strands unwind, nucleotide bases are added to each, complementary to the nucleotides of the original strands. Each single strand thus provides the pattern for the formation of a new complete double helix, each identical to the original DNA molecule.

Each daughter DNA molecule contains one strand derived from the parent and one strand assembled from free nucleotides. Since each of us arose from a single fertilized egg, it is possible (but highly unlikely) that one strand of DNA from each of our parents is still in some cell of our body, even after numerous mitotic divisions. This type of replication, in which each daughter double helix receives one complete parental strand of DNA, is called **semi-conservative replication.**

The genetic code

The variable sequence of nucleotides in DNA immediately suggested that nucleotides comprise the letters in a genetic alphabet. Therefore a dictionary was necessary to understand this new language. Many geneticists immediately focused on decoding the genetic message.

DNA and proteins have analogous structures. Both are comprised of long chains of molecular subunits—nucleotides in the case of DNA, and amino acids in the case of proteins. The simplest model is that a one-to-one correspondence exists between the two sequences, that is, that they are **colinear.** In 1958, Crick proposed colinearity as an hypothesis to explain the structure of proteins: the sequence of nucleotides in DNA directly specifies the sequence of amino acids in a protein. Subsequent work has verified this hypothesis.

Because the DNA alphabet consists of only four different bases (letters), each base alone, a one-letter word, is obviously not sufficient to specify uniquely twenty different amino acids. Assuming a common word length for all amino acids and a four-letter alphabet, the number of amino acids that could be specified by a word of n letters is 4^n. Thus, a dictionary of two-letter words would also be inadequate: $4^2 = 16$. However, the use of three letters to identify an amino acid provides $4^3 = 64$ possibilities, far more than enough words to encode uniquely all amino acids.

Working on the hypothesis that the dictionary contained three-letter words, M. W. Nirenberg broke the **genetic code** in the early 1960s. Nirenberg added all of the protein synthesizing apparatus of *E. coli* to a test tube, as well as all twenty amino acids. He then synthesized an artificial genetic message (messenger RNA) consisting only of uracil: U-U-U-U-. . . . He subsequently examined the mixture for the presence of polypeptides and found only one type, a polypeptide consisting of phenylalanines and no other amino acids. He concluded that the messenger RNA sequence UUU encodes the message or word for phenylalanine. (The RNA sequence UUU mirrors the DNA sequence AAA. Messenger RNA is discussed later in this chapter.) Similar experiments showed that the RNA message AAA encoded the amino acid lysine; CCC encoded proline; and GGG encoded glycine.

Further detailed work soon elaborated the entire genetic code (table 8.2). The code consists, as hypothesized, of three-letter words called **codons** or *triplets*. There are many more codons than amino acids, and most amino acids are specified by more than one codon. Leucine and arginine are each specified by six different codons. The code is said to be **degenerate** because of this redundancy of message. The biological functions, if any, of degeneracy in the code are a matter of debate. Only methionine and tryptophan are specified by a single codon.

Four codons serve as punctuation for the genetic message. AUG always specifies the amino acid methionine, but it is also known as the *initiator codon,* because most messages specifying a protein begin with AUG. Three triplets are **nonsense** or **terminator codons** because they specify no amino acids at all. *Nonsense* is probably a misnomer, however, because these codons (UAA, UAG, and UGA) all terminate the reading of the genetic message. They signify the end of a sequence of codons specifying a particular polypeptide.

The genetic message is thus encoded in a sequence of codons along the length of a DNA molecule. The sequence of amino acids in a polypeptide is specified one at a time by the sequence of codons in DNA. Production of a particular protein or polypeptide stops when a terminator codon is encountered.

The genetic code has been shown to be **nonoverlapping;** that is, for a single gene the sequence of bases is always read in discrete groups of three per codon. Each base participates in only one codon (some rare exceptions are discussed at the end of this chapter). The use of nucleotide bases in sequential triplets has been analogized to the frames of a motion picture. The frames must be read precisely one at a time, or the message is blurred. It also suggests one possible mechanism of mutation, **frameshift.** If at some point in the nucleotide sequence a base is either added or deleted, the entire reading frame is shifted, altering the subsequent message (table 8.3).

Table 8.2
Genetic dictionary of RNA base codons and the amino acids they encode

First base	Second base				Third base
	U	*C*	*A*	*G*	
U	UUU UUC } Phenylalanine UUA UUG } Leucine	UCU UCC UCA UCG } Serine	UAU UAC } Tyrosine UAA† UAG† } Stop	UGU UGC } Cysteine UGA† Stop UGG Tryptophan	U C A G
C	CUU CUC CUA CUG } Leucine	CCU CCC CCA CCG } Proline	CAU CAC } Histidine CAA CAG } Glutamine	CGU CGC CGA CGG } Arginine	U C A G
A	AUU AUC AUA } Isoleucine AUG* Methionine	ACU ACC ACA ACG } Threonine	AAU AAC } Asparagine AAA AAG } Lysine	AGU AGC } Serine AGA AGG } Arginine	U C A G
G	GUU GUC GUA GUG } Valine	GCU GCC GCA GCG } Alanine	GAU GAC } Aspartic Acid GAA GAG } Glutamic Acid	GGU GGC GGA GGG } Glycine	U C A G

*Initiation codons. The methionine codon AUG is the most common starting point for translation of a genetic message.
†Terminator codon.

Yet another type of mutation is **substitution** of one nucleotide base for another (table 8.3). The consequence of base substitution can be far less drastic than frameshift mutation. A simple change in one nucleotide base alters at most one amino acid in a protein. Because of the degeneracy of the code, many substitutions are **same sense;** that is, they do not change the amino acid specified by the affected codon (table 8.3).

A final and most important characteristic of the genetic code is that it is **universal.** With only minor exceptions (such as the DNA of mitochondria of some species, including humans) the code is identical among all species. UUU encodes phenylalanine in humans, snails, cotton, and *E. coli.*

The universality of the code has been shown by experiments in which mRNA from rabbits encoding the structure of hemoglobin was combined with the protein synthesizing apparatus of *E. coli.* The result: rabbit hemoglobin. A more exciting demonstration of the universality of the code was achieved in 1986. The gene coding for the firefly enzyme luciferase (responsible for the firefly's glow) was inserted into tobacco plant DNA. When the tobacco plant was "watered" with a solution of the right compounds, the plant glowed, demonstrating that the firefly gene was expressed by the plant (figure 8.9).

The universality of the code is important in at least two ways. As we have already noted, humans are a difficult species to examine genetically. Because all organisms speak the same genetic language, experimental geneticists have been able to select from the thousands of available species those that have qualities favoring laboratory analysis. Most knowledge of the molecular mechanism of inheritance has been derived from a few well-studied laboratory species: viruses, bacteria, fungi, *Drosophila,* and mice. Because all species share a common genetic apparatus, studying the genes of *Drosophila* tells us much about the genes of humans that could not be learned if scientists studied only humans.

Table 8.3
Types of mutations

Original DNA		CGATCGCAA
Messenger RNA		GCUAGCGUU
Codes for		ala/ser/val/
Frameshift mutation	DNA	* CGGATCGCAA
	mRNA	GCCUAGCGUU
Now codes for		ala/STOP
Substitution mutation	DNA	* AGATCGCAA
	mRNA	UCUAGCGUU
Now codes for		ser/ser/val
Samesense mutation	DNA	* CGGTCGCAA
	mRNA	GCCAGCGUU
Still codes for		ala/ser/val/

* = Mutation

(a)

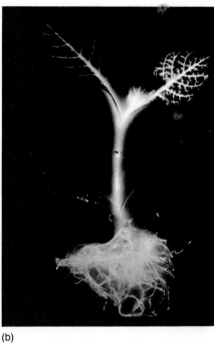

(b)

Figure 8.9 Example of the universality of the genetic code. The tobacco plant below has had the gene for firefly luciferase incorporated into its genome by means of genetic engineering techniques. On the left the plant is seen in ordinary light, on the right it has been "watered" with a solution demonstrating the plant expressing the firefly luciferase gene. (See colorplate 12.)
Source: Keith V. Wood, UCSD.

Additionally, the universality of the code is some of the strongest evidence for a common evolutionary origin of all living species. Such an intricate mechanism, using the same code for all species, argues strongly that the genetic apparatus originated in the progenitors of modern life very early in the history of the earth. Although several billion years have passed for the development and modification of the genetic mechanism, the remarkable similarities shared by all species are clearly the result of a common evolutionary origin.

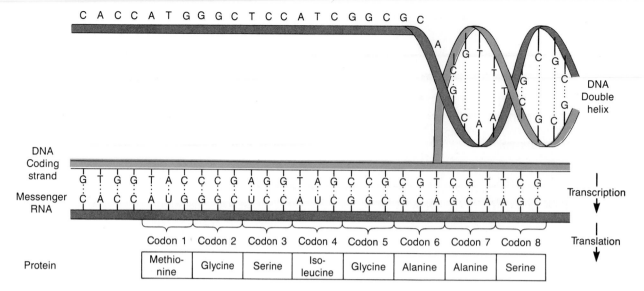

Figure 8.10 Transcription. A portion of the DNA molecule unwinds, exposing the base pairs for use as a template for the formation of mRNA or HnRNA from the *sense strand* only. Following transcription, the DNA assumes its normal conformation.

The structures and functions of RNA

Although DNA is the hereditary material for most species, RNA is fundamental to the function of DNA. DNA is a master file of information, but processing that information to a usable form, making proteins, requires RNA as an intermediary. The structural similarity of RNA and DNA allows DNA to serve as a template for RNA synthesis, just as it does for its own replication. An understanding of protein synthesis requires that we first examine the structure and function of the major types of RNA.

Messenger RNA

The DNA template resides in the nucleus of eukaryotic cells, but proteins are synthesized in the cytoplasm. DNA does not directly participate in protein formation. Instead, a copy of the information encoding the sequence of amino acids for a protein is *transcribed* from nuclear DNA in the form of a complementary molecule, *messenger RNA (mRNA)*.

Synthesis of mRNA occurs primarily during the G1 and G2 phases of the cell cycle in similar fashion to the replication of DNA itself. (During the S phase, DNA is busy replicating itself; during cell division, the chromosomes are too coiled for transcription to occur.) The double helix is believed to unwind at the site of mRNA synthesis (figure 8.10). However, in contrast to DNA replication, mRNA is transcribed from only one of the two strands of DNA, which is often called the **sense strand.** Remember also that uracil (U) rather than thymine (T) pairs with adenine on the sense strand in transcription of RNA. This process is appropriately called *transcription* because it is analogous to a scribe copying a message. The information transcribed into mRNA is a precise, complementary copy of the hereditary message of DNA.

Proteins vary considerably in length. Consequently, mRNAs also vary in length. Most are five hundred to one thousand nucleotides long. However, initial transcription of mRNA in eukaryotic cells usually involves the production of a precursor called HnRNA, *heterogeneous nuclear RNA.* Found only in the nucleus, HnRNA is many times longer than mRNA. For most eukaryotic genes the final formation of mRNA involves removal of segments of nucleotides scattered along the length of the HnRNA molecule.

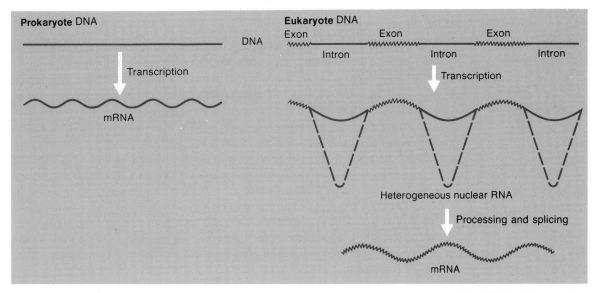

Figure 8.11 Difference in formation of mRNA in prokaryotic and eukaryotic cells.

The existence of HnRNA indicates that many segments of DNA are transcribed, but, because they are excised from HnRNA, they are not later translated into proteins. Thus, a segment of eukaryotic DNA encoding a polypeptide contains two types of nucleotide sequences: **exons,** portions of the template for protein synthesis, and **introns,** which are never translated into amino acid sequences (figure 8.11). Introns vary in length from one hundred base pairs to over one thousand base pairs. The number of introns also varies from one or two in some genes to sixteen in a chicken egg white protein. In some genes, the total length of the introns may even exceed that of the exons. DNA sequences controlling production of most polypeptides consist of alternating sequences of introns and exons.

The existence of non-translated DNA sequences was first recognized in 1977. Introns are extraordinary because they suggest a new level of complexity in the function of DNA. Little is known regarding the function of introns, but present hypotheses argue that introns may regulate the complex functioning of the DNA of eukaryotes.

Bacterial mRNAs seem to have a very short lifetime, usually only a few minutes. Apparently, the various types of mRNA are transcribed as they are required by the metabolic processes of the organism. In eukaryotes, some mRNAs are known to exist for relatively long periods. It is possible that HnRNA serves as a relatively stable form in which mRNA can be stored until needed.

The initial sequence of all mRNAs, the *leader sequence,* provides a binding site for the **ribosomes,** the actual site of protein synthesis. Following the leader is the *initiation codon* AUG, one of four words in the genetic code used as punctuation for the genetic message. Consequently, polypeptides always begin with the amino acid methionine, although this amino acid is often removed upon completion of polypeptide synthesis.

All mRNAs terminate with one of the three nonsense codons, UAA, UAG, or UGA, which stop protein synthesis. Many eukaryotic mRNAs also have long sequences of as many as two hundred adenines at their terminal ends, called *poly-A sections.* The function of poly-A sections is not known, although it is possible that they may help stabilize some of the longer lived eukaryotic mRNAs.

Ribosomal RNA

Messenger RNA carries the information specifying the sequence of amino acids in a polypeptide, but actual synthesis of a protein requires two additional types of RNA. About 80% of the RNA in a cell is *ribosomal RNA* (rRNA), which is transcribed directly from nuclear DNA and moves to the cytoplasm after formation. In the cytoplasm rRNA associates with proteins to form one of the major types of intracellular organelles, the *ribosomes*. In prokaryotes, ribosomes occur scattered freely throughout the cytoplasm, but in eukaryotes many ribosomes are found along the complex network of the *endoplasmic reticulum* (figures 2.4 and 2.6). The proportions vary somewhat between eukaryotes and prokaryotes, but in general ribosomes are about half rRNA and half protein.

Ribosomes are the site of protein synthesis. Each eukaryotic ribosome is composed of two subunits of different sizes. During protein assembly, a ribosome attaches to an mRNA molecule, then moves along it, reading the codons as it goes and attaching amino acids to the growing protein in the proper sequence—the process of **translation.**

Transfer RNA

Amino acids are brought to the ribosomes for assembly into proteins by the smallest type of RNA, *transfer RNA (tRNA)*. Like all RNAs, tRNA is single stranded; but some portions of the molecule appear double stranded, because it doubles back on itself. The resulting structure has been analogized to a cloverleaf with one small loop and three large loops, but it is certainly far more complex than that (figure 8.12). The tRNAs are about eighty nucleotides long, and like the other RNAs they are synthesized in the nucleus, directly from DNA.

The two free ends of a tRNA lie adjacent to each other. One end attaches to a particular amino acid, which the tRNA then carries, or transfers, to the ribosome for incorporation into a growing polypeptide.

The loop of the cloverleaf opposite the free ends has a triplet **anticodon,** complementary in sequence to a specific mRNA codon and identical in sequence to the template DNA codon, except that all thymines in DNA are replaced by uracils in tRNA (figure 8.12). A tRNA with a particular anticodon can attach to only one type of amino acid, and the anticodon sequence of the tRNA for a particular amino acid is complementary to the mRNA codon specifying that amino acid. Thus, when a tRNA bearing an amino acid moves to a ribosome, the match between the mRNA codon and tRNA anticodon insures that the amino acid is inserted in the proper order to manufacture a particular protein. In other words, the ribosome facilitates insertion of an amino acid into a developing protein only when the tRNA anticodon matches the mRNA codon specifying the next amino acid.

Over fifty types of tRNAs have now been discovered, each specifying only one amino acid. Most of the twenty major amino acids can be transported by more than one type of tRNA.

Protein synthesis

Let us now review protein synthesis in its entirety (figure 8.13). Assembly of proteins requires three types of RNA, each performing its own specialized job. The sequence of nucleotide bases in nuclear DNA encoding the *colinear* sequence of amino acids in a protein is *transcribed* to *messenger RNA,* which moves to the cytoplasm. In the cytoplasm, several ribosomes attach to the mRNA, forming a structure known as a **polyribosome** (figure 8.14). A ribosome binds to the leader sequence of the mRNA, and beginning at the initiator codon (AUG) *translation* of the mRNA nucleotide sequence into a sequence of amino acids (a polypeptide) begins.

Amino acid

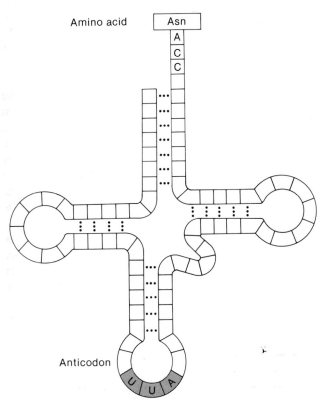

Figure 8.12 A molecule of transfer RNA. Although the "cloverleaf" conformation is characteristic, its three-dimensional structure is actually more complex. Dotted lines indicate hydrogen bonding between complementary bases. The anticodon is the portion of the molecule that interacts with the mRNA. The amino acid is asparagine. Source: Stuart Ira Fox, *Human Physiology*, 2d ed. Copyright © 1987 Wm. C. Brown Publishers, Dubuque, Iowa. All Rights Reserved. Reprinted by permission.

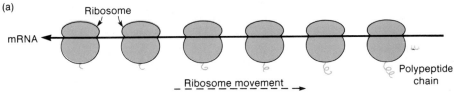

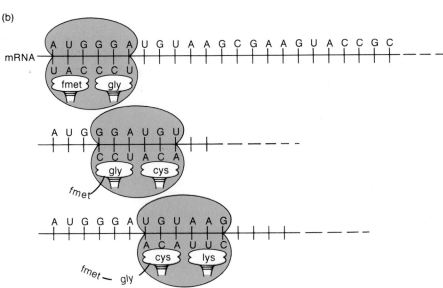

Figure 8.13 Protein assembly: (a) as a series of ribosomes move along an mRNA molecule, each generates an increasingly long polypeptide chain; (b) a closer look at an individual ribosome shows how the chain grows.

Figure 8.14 Electron micrograph of polyribosomes: a sequence of ribosomes attached to a single mRNA.
Source: Alexander Rich.

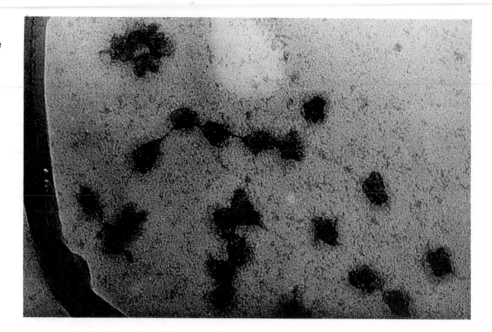

A tRNA with the appropriate anticodon and amino acid attaches to each triplet codon on the mRNA as the ribosome moves along its length. Each tRNA carries a specific amino acid, which is joined by a *peptide bond* to the growing polypeptide chain. The tRNA is then released back into the cytoplasm, where it may attach to and transport another amino acid. When a ribosome encounters one of the three terminator codons, it releases the polypeptide chain into the cytoplasm. Translation is then completed. No known molecular apparatus is associated with formation of the three-dimensional structures of a protein. The sequence of amino acids, the primary structure alone, determines the secondary and tertiary structures of the protein.

Proteins and messenger RNAs are long molecules. Many, perhaps most, mRNAs survive only a few minutes, indicating a rapid rate of transcription and translation. The speed of both activities varies according to a number of factors, but can be truly exceptional. Ten or more amino acids may be added to a polypeptide chain in one second. Translation of the beta chain of hemoglobin, consisting of 146 amino acids, occurs in about ten seconds. Even with a lifetime of only two minutes, a single mRNA molecule could serve as the template for 30 or more proteins that are each 300 amino acids in length. The transcription of mRNA occurs at a rate of about 28 nucleotides per second in *E. coli*.

A unified view of the gene

We have seen three models of the hereditary mechanism: Mendelian, chromosomal, and molecular. Each describes differing aspects of the genetic process. Together, they offer a comprehensive basis for understanding the mechanism of inheritance.

Mendelian principles postulate the existence of genes, discrete particles transmitted between generations that control phenotypic traits. These laws were deduced directly from observations of the phenotype itself. Mendel knew nothing of the actual physical nature of genes.

Analysis of meiotic behavior of chromosomes revealed that chromosomes were the likely location of the genes. Meiosis is a cellular process that explains Mendel's Laws of Segregation and Independent Assortment. Knowledge that chromosomes were the physical location of the genetic material did not, however, explain the molecular mechanism of inheritance.

Over a period of several decades in the twentieth century, molecular geneticists first identified DNA as the hereditary material and then described its chemical structure. The central dogma summarizes contemporary understanding of control of the phenotype by the molecular apparatus of inheritance. As the meiotic behavior of chromosomes suggests, the genetic material, DNA, resides in the chromosomes. The information encoded in DNA is transcribed into RNA, which in turn directs assembly of proteins in the cytoplasm.

The contemporary view of a gene incorporates aspects of all three lines of genetic investigation. Chromosomal and molecular processes are mechanisms explaining the hereditary principles discovered by Mendel. Proteins provide the crucial link between Mendelian genes and the phenotype. Proteins are a primary phenotypic product of the genes, which in turn direct further aspects of phenotypic expression.

This principle is illustrated clearly when mutations in DNA cause formation of a defective protein, as in the case of sickle-cell anemia. Recall that normal hemoglobin consists of two alpha polypeptide chains and two beta polypeptide chains, plus iron-containing heme groups (figure 8.2d). Each alpha chain is 141 amino acids long; each beta chain is 146 amino acids long. Sickle-cell anemia results from a single nucleotide substitution in the DNA encoding the beta chain. The sixth codon in the beta chain is altered so that, during synthesis of the chain, valine is added rather than the normal glutamic acid (figure 8.15). Homozygotes for this substitution have the sickle-cell syndrome of more than a dozen pathologies (figure 4.7).

The one gene—one polypeptide hypothesis

The fundamental importance of proteins has led to the **one gene–one polypeptide hypothesis** of genetic expression. This hypothesis states that each gene codes for the production of one polypeptide chain, which is often an entire protein molecule. This polypeptide, or the protein of which it is a constituent, in turn participates in a series of biochemical reactions that are ultimately expressed as some aspect of the phenotype. For example, in one form of albinism the abnormal DNA sequence leads to the production of a defective form of the enzyme tyrosinase. Tyrosinase normally catalyzes formation of the pigment melanin (figure 8.16). Numerous other genetic diseases, the *inborn errors of metabolism,* are now also understood to result from mutations affecting the production of a key protein. Phenylketonurics, for example, do not have the enzyme necessary to metabolize the amino acid phenylalanine, which accumulates in the nervous tissues and results in mental retardation. The immediate cause of such diseases is a defective protein, but the ultimate defect lies in the sequence of the DNA itself.

Present knowledge of the structure and function of DNA supports many aspects of the one gene–one polypeptide hypothesis: a gene is a sequence of nucleotide bases on a chromosome that determines a colinear sequence of amino acids in a protein. It is clear, however, that some genes have functions other than producing proteins. One of those functions is controlling the function of genes that do produce proteins, the subject of the next chapter.

Figure 8.15 The sickle-cell anemia mutation is an example of the colinearity of the gene and the protein. A single base substitution in the DNA causes a codon change that results in the wrong amino acid being incorporated into the beta chain of the hemoglobin molecule.

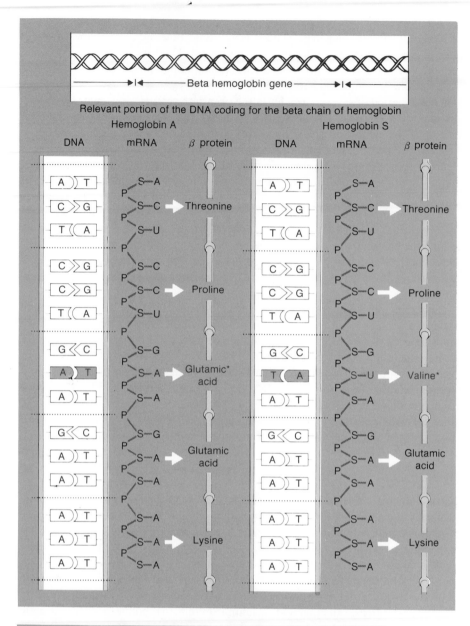

Figure 8.16 Metabolic pathways for some compounds involved in genetic inborn errors of metabolism. Absence of the enzyme converting phenylalanine to tyrosine results in phenylketonuria (PKU). Absence of the enzyme converting tyrosine to DOPA results in classical albinism. Absence of the enzyme converting tyrosine to homogentisic acid results in alkaptonuria.

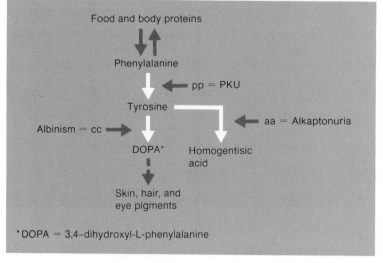

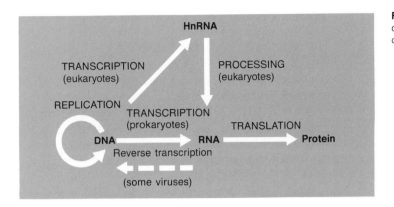

Figure 8.17 A contemporary model of the central dogma.

The central dogma revisited

Since the central dogma was first stated, research has shown clearly that DNA meets all the requirements of the hereditary material stated earlier in this chapter:

1. The sequence of bases in DNA has an immense capacity for encoding information, certainly more than sufficient information to direct the development and life of an organism.

2. A complete set of hereditary information is available in every cell of the body as the result of mitosis. The information is translated into a usable form through the production of necessary proteins, mediated by the action of RNA.

3. As a result of the replication of DNA and the subsequent reduction division of the chromosomes during meiosis, a diploid, sexually reproducing individual transmits a complete haploid set of chromosomes to each offspring. For both mitotic and meiotic division, complementary base pairing of DNA provides an extremely accurate mechanism for the transmission of hereditary information.

4. Addition, loss, or change of nucleotide bases in a DNA molecule provides a mechanism for mutation of a gene, insuring long-term renewal of genetic variability in a population.

Verification and understanding of DNA's role in the hereditary process is one of the great achievements of modern science. The research that allowed verification of the central dogma has also led to some surprises that require modifications of the model. For example, the first step in protein synthesis involves activation of the gene to be transcribed, the process of *regulation.* As we have seen, production of mRNA in eukaryotes involves the additional *processing* of HnRNA. Protein production in eukaryotes, therefore, requires four basic steps—regulation, transcription, processing, and translation—rather than two (figure 8.17).

A few exceptions to general principles have also been found. Some viruses use RNA rather than DNA as their genetic material, and they also have the associated capacity to construct DNA from their RNA as part of their life cycle—*reverse transcription.* Thus the information flow in the central dogma is occasionally reversed.

Even more surprisingly, some viruses have been shown to have **overlapping genes.** That is, a single segment of DNA may encode two or three different proteins. Initiation of protein production begins at different locations in the genetic message, and different frames are read for the different proteins. Yet even in this case, the code for each protein is still nonoverlapping.

Despite these and many other variations, the central dogma will continue to develop as one of the fundamental concepts of modern biology.

Summary

1. Protein fulfills many important functions for the living organism. In particular, proteins that are enzymes catalyze the biochemistry of life.

2. Proteins consist of amino acids joined together by peptide bonds into polypeptides. Either one polypeptide chain or several chains bound together may form a protein. The structure of a protein is described not only by its sequence of amino acids, but also by the twisting and folding of the chains.

3. Proteins were thought to be the hereditary molecules because they are long molecules of variable amino acid sequence and therefore capable of encoding immense amounts of information.

4. The nucleic acid DNA was discovered in 1869 and recognized as a major constituent of chromosomes, although its role as the genetic substance was not immediately recognized.

5. A series of important experiments eventually demonstrated the hereditary role of DNA. Griffith, Avery, MacLeod, and McCarty showed that DNA was the transforming principle of bacteria. Hershey and Chase showed that nucleic acid was the hereditary material of viruses.

6. Watson and Crick described the structure of DNA in 1953. They proposed that DNA is a long molecule in the shape of a double helix, and that the two sugar-phosphate backbones of the double helix are bound together by bonds formed between paired nitrogen bases. Adenine always binds to thymine; cytosine always binds to guanine. Information is encoded in the sequence of bases. Accurate replication results from specific complementary base pairing.

7. RNA is similar in structure to DNA. It is constructed from a sugar with one more oxygen atom than the sugar in DNA; it is single stranded; and it uses uracil rather than thymine. RNA is the hereditary material in some viruses, and three forms of RNA are required for protein synthesis in all organisms.

8. The central dogma describes the flow of hereditary information. Between generations, the accurate replication of nuclear DNA results in precise transmission of genetic instructions. Within the cells of an individual, nuclear DNA is transcribed to messenger RNA, which in turn is translated into proteins. Both mRNA and proteins may be considered phenotypic characters that exert a primary influence over other phenotypic characteristics.

9. Complementary base pairing insures accurate replication of DNA. Each parent strand of DNA serves as a template for the production of a new double-stranded DNA molecule. Replication is semi-conservative, because each parent strand forms one-half of a daughter molecule.

10. The genetic code is contained in the sequence of nucleotide bases in DNA. It is a triplet code, consisting of three-letter words, or codons, each specifying either a particular amino acid in a polypeptide chain or punctuation of the genetic message. Because some amino acids are encoded by more than one codon, the code is degenerate. Essentially the same code is used by all species on earth: it is universal.

11. A gene is a sequence of nucleotides encoding information for the production of a colinear sequence of amino acids in a polypeptide. In most species, genes are nonoverlapping. Some viruses, however, have overlapping genes.

12. Frameshift mutations result from the addition or deletion of a nucleotide base in a DNA sequence, which alters the three-letter reading frame of subsequent codons. Mutations involving the substitution of one nucleotide base for another will affect only the amino acid produced by that codon. A substitutional mutation may have major phenotypic effects, as in the case of sickle-cell anemia. Alternatively, the degeneracy of the code results in many same sense mutations, which have no apparent phenotypic effect.

13. Messenger RNA (mRNA) is transcribed only from the sense strand of DNA. In eukaryotes, heterogeneous nuclear RNA (HnRNA) is first transcribed, from which segments are excised to form mRNA. Segments of eukaryotic DNA that are transcribed but not translated are called introns; segments that are translated are called exons.

14. Ribosomes, formed of ribosomal RNA (rRNA) and proteins, are the sites of translation. Translation often involves the attachment of many ribosomes to mRNA, forming a polyribosome.

15. More than fifty types of transfer RNAs (tRNAs) are known. Each carries a specific amino acid to the ribosome for attachment to the growing polypeptide chain. Insertion of an amino acid into the correct location in a polypeptide depends upon the match between the mRNA codon and the anticodon on a tRNA.

16. The modern view of a gene and the hereditary mechanism combines discoveries from three models of genetic study: Mendelian, chromosomal, and molecular. The one gene–one polypeptide hypothesis relates the molecular definition of a gene to its immediate phenotypic product (the protein) that in turn mediates further phenotypic development.

Key Words and Phrases

amino acid	intron	ribosomal RNA
anticodon	messenger RNA	ribosome
central dogma	nitrogenous base	RNA
codon	nonoverlapping genes	same sense mutation
colinear	nonsense codon	semi-conservative replication
complementary base pairing	nucleic acid	sense strand
degenerate	one gene–one polypeptide hypothesis	substitution
deoxyribose	overlapping genes	terminator codon
DNA	peptide bond	transcription
double helix	polypeptide	transfer RNA
enzyme	polyribosomes	transformation
exon	primary structure	translation
frameshift	protein	universal code
genetic code	ribose	Watson-Crick model
informational macromolecules		

Questions

1. What is the minimum number of nucleotide bases needed to code for the 135 amino acids that comprise the enzyme cytochrome c of some species?

2. If a segment of DNA has the sequence ACACGATTCCCGACG, what is the sequence of bases in the complementary strand? What would be the sequence of a molecule of mRNA transcribed from the first strand? What polypeptide or polypeptides would be encoded by this sequence?

3. What would be the anticodons on the tRNA molecules that would service the message transcribed in the previous question?

4. Genetic engineering techniques now permit scientists to transfer genes from one species to another—often distantly related species. What will the universality of the genetic code mean, in practical terms, to the success of such procedures?

5. Why is it unlikely that a segment of DNA from an individual's father or mother still exists in some cell of the individual's body?

6. Compare and contrast the roles of mRNA, rRNA, and tRNA in protein synthesis.

7. If normal human cells have 46 chromosomes and roughly 6.4 picograms of DNA (table 8.1), could a male produce sperm containing 6.4 picograms of DNA? How?

8. RNA is a single-stranded molecule. How could RNA form double-stranded regions like DNA? What would hold these regions together?

9. Specify the quantity of DNA that you would find in a single human cell at the following stages of cell division: a) prophase of mitosis; b) late anaphase of mitosis; c) prophase of meiosis I; d) metaphase of meiosis II; e) secondary oocyte.

10. The number of unique messages that can be encoded by a segment of DNA is 4^n, where n is the number of bases in the DNA segment. What does the number four signify?

For More Information

Chambon, P. 1981. "Split Genes." *Scientific American,* 244:60–71.

Choi, Y. D., P. J. Grabowski, P. A. Sharp, and G. Dreyfuss. 1986. "Heterogeneous Nuclear Ribonucleoproteins: Role in RNA Splicing." *Science,* 231:1534–39.

Cornish-Bowden, A. 1985. "Are Introns Structural Elements or Evolutionary Debris?" *Nature,* 313:434–35.

Crick, F. H. C. 1966. "The Genetic Code: III." *Scientific American,* 215:55–64.

Gilbert, W. 1985. "Genes-in-Pieces Revisited." *Science,* 228:823.

Kornberg, R. D., and A. Klug. 1981. "The Nucleosome." *Scientific American,* 244:52–80.

Lake, J. A. 1981. "The Ribosome." *Scientific American,* 245:84–97.

Lewin, R. 1985. "Reverse Transcriptase in Introns." *Science,* 229:1083.

Ptashne, M. 1986. "Gene Regulation by Proteins Acting Nearby and at a Distance." *Nature,* 322:697–701.

Racker, E. 1987. "Structure, Function, and Assembly of Membrane Proteins." *Science,* 235:959–61.

Rich, A., and S. H. Kim. 1978. "The Three-Dimensional Structure of Transfer RNA." *Scientific American,* 238:52–62.

Rosenfeld, I., E. Ziff, and B. Van Loon. 1983. *DNA for Beginners.* Writers and Readers Publishing, Inc., W. W. Norton & Co., New York.

Sayre, A. 1975. *Rosalind Franklin and DNA.* W. W. Norton & Co., New York.

Watson, J. D. 1968. *The Double Helix.* Atheneum, New York.

Yanofsky, C. 1967. "Gene Structure and Protein Structure." *Scientific American,* 216:80–96.

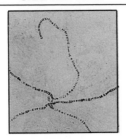

Control of gene expression

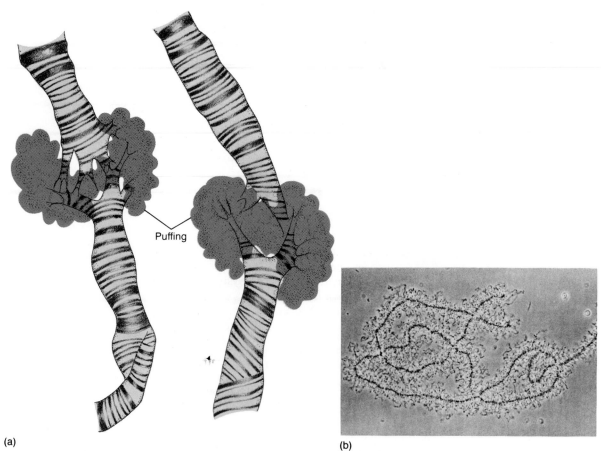

(a)

(b)

Figure 9.1 (a) Puffing associated with gene transcription in a polytene chromosome; (b) phase contrast photograph of a portion of a pair of lampbrush chromosomes isolated from a newt oocyte. Source: (a) Leland G. Johnson, *Biology*, 2d ed. Copyright © 1987 Wm. C. Brown Publishers, Dubuque, Iowa. All Rights Reserved. Reprinted by permission. (b) J. S. Gall.

Each cell in an organism possesses a complete set of chromosomes containing all the information necessary for the metabolic and physiological functions of that organism. Not all of that information is used all the time. Most cells share a small, common set of *background* genes that function more or less continuously to meet basic needs—genes controlling carbohydrate metabolism, for example, necessary to provide a regular energy supply for the cell.

Other genes are turned on for only a short period of one's lifetime. Genes controlling the umbilical cord and the fetal portion of the placenta function only during the first nine months of life. Embryonic and fetal hemoglobin genes (see figure 9.3) function exclusively or primarily during gestation. Other genes function at irregular intervals, often in response to environmental stimulation. A woman may express the genes controlling the maternal portion of the placenta several times over the course of a few decades. Body-building exercises induce a genetically controlled developmental response in affected muscles. The onset of many genetic diseases occurs only at predictable times of life.

The branch of genetics dealing with the control of gene expression is known as *regulatory genetics*. Although the importance of the regulation of gene function has long been recognized, the techniques for studying regulatory processes have developed slowly over the past several decades. The Watson-Crick model of DNA, for example, immediately suggests mechanisms by which DNA can both contain and transmit immense amounts of genetic information. It does not, however, directly suggest factors controlling the expression of that information.

Geneticists now commonly distinguish two categories of genes. **Structural genes** encode the amino-acid sequences for polypeptides destined to become normal proteins, i.e., catalysts, hormones, and supportive, protective, and transport molecules. **Regulatory genes,** on the other hand, function to control the expression of either structural or other regulatory genes. Regulatory genes may produce a polypeptide product or an RNA molecule with a regulatory function.

Secondary and tertiary structures of DNA may also be involved in control of gene expression, and doubtless most phenotypic traits are the result of interactions of many structural loci. Although it is clear that gene regulation is a fundamental aspect of life, the many forms of regulation and their complex interactions are just beginning to be understood.

Gene regulation in prokaryotes

Some of the best-defined regulatory systems are in the bacterium *E. coli*. Recall that prokaryotes have only one chromosome, a single ring of DNA, with no associated proteins. In 1961 French scientists François Jacob and Jacques Monod proposed the **operon** hypothesis to explain gene regulation in *E. coli*. The operon hypothesis, for which Jacob and Monod received the 1965 Nobel Prize for Physiology or Medicine, was the first complete model of regulation of gene function in any organism. It provides the basis for understanding regulation in all prokaryotes. With the advent of genetic engineering techniques in the late 1970s, the operon model has also been put to practical use, providing a technique for controlling the artificial production of synthetic gene products (see chapter 13).

The lac operon

E. coli can derive energy from the metabolic decomposition (digestion) of the milk sugar lactose. Three enzymes produced by *E. coli* are involved in the digestion of lactose, but the enzymes are produced only when lactose is present in the growth medium. Jacob and Monod proposed that the production of these three enzymes is controlled by a structural and functional unit of the DNA of *E. coli* called an *operon* (figure 9.2).

The operon consists of a sequence of contiguous genes in one segment of the bacterial DNA. Three structural genes, designated Z, Y, and A, encode the amino-acid sequence of the three enzymes involved in the breakdown of lactose. Immediately adjacent to gene Z are two regulatory genes—gene O, the **operator,** and gene P, the **promoter.** The entire sequence of promoter, operator, and three structural genes comprises the lac operon, so called because of its role in lactose metabolism.

For transcription of Z, Y, and A to occur, the enzyme **RNA polymerase** (the enzyme that catalyzes the formation or *polymerization* of mRNA from a DNA template) attaches to the promoter region of the operon. The RNA polymerase then moves along the operon, passing the operator. Upon reaching gene Z, it begins transcription of a single, long mRNA molecule, which encodes the amino-acid sequence of the three enzymes required for lactose metabolism. Subsequent translation of this single mRNA produces the three necessary enzymes, which are not attached to each other, though they were produced from a single mRNA.

Transcription of the Z, Y, and A genes does not occur continuously. Transcription is controlled by a *regulator gene,* designated i, located at some distance from the operon and not considered to be a part of the operon itself. The regulator gene, like structural genes, encodes an mRNA that is translated into a polypeptide chain. When four polypeptide chains are produced by the regulator, they bind together loosely, forming a single active protein called the **repressor.** The repressor protein

Figure 9.2 The lac operon. (a) In the absence of inducer, the active repressor protein binds to the DNA at the operator region, preventing mRNA transcription. No enzymes are produced. (b) When an inducer, lactose, is present, it binds to the repressor protein, removing it from the DNA. Now mRNA can be transcribed from the entire operon, and the enzymes used in lactose metabolism are synthesized. When all of the lactose is metabolized, the repressor protein again binds to the DNA, shutting off mRNA production, leading to a cessation of enzyme production. (See colorplate 13.)

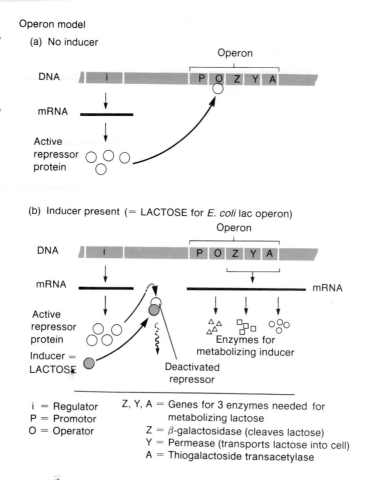

Operon model

(a) No inducer

(b) Inducer present (= LACTOSE for *E. coli* lac operon)

i = Regulator
P = Promotor
O = Operator

Z, Y, A = Genes for 3 enzymes needed for metabolizing lactose
Z = β-galactosidase (cleaves lactose)
Y = Permease (transports lactose into cell)
A = Thiogalactoside transacetylase

binds to the operator site of the operon, physically interfering with the movement of the RNA polymerase and inhibiting transcription. *Transcription inhibition,* then, is the major mode of regulation of the lac operon.

Most of the time, a repressor protein is bound to the operator gene, and transcription does not occur. This is presumably an adaptive trait of *E. coli.* After all, lactose is not continuously present in its surroundings, so continual synthesis of lactose metabolizing enzymes would be inefficient. However, when lactose is present, enzyme synthesis begins. How does this occur?

Lactose in the growth medium of *E. coli* crosses the cell membrane, entering the cytoplasm. There, it combines with the repressor protein, causing a physical deformation of the repressor that releases it from the operator site (figure 9.2). Once the operator site is free, movement of RNA polymerase on the promoter is no longer blocked, and transcription of Z, Y, and A commences.

Transcription continues only as long as the cytoplasm has sufficient lactose to bind repressor proteins, leaving the operator free. Note that lac operon activity (transcription and subsequent translation) reduces the amount of the controlling molecule. In other words, the presence of lactose *induces* transcription, which then leads to production of lactose metabolizing enzymes. In turn, these enzymes break down cytoplasmic lactose, freeing the repressor proteins that then bind again with the operator and turn off transcription, at least until more lactose enters the cell.

Regulatory characteristics of the lac system

Many regulatory mechanisms, particularly in eukaryotes, differ substantially from the lac operon, but the elegance and simplicity of the lac operon illustrates principles of regulatory design that may be common in many organisms:

1. The lac operon is an **inducible** system. Production of enzymes occurs only when necessary and is induced by the presence of the substrate upon which those enzymes are active, in this case, lactose. Such a system is efficient for metabolic pathways that are used only periodically.

2. Although the lac operon has many components, a single regulatory system controls all genes induced in one metabolic pathway. Thus, for example, the lac operon produces only one mRNA, encoding all three enzymes needed for lactose metabolism. Under normal circumstances, either all three enzymes involved in lactose metabolism are present, or none is present.

3. In the lac operon, control occurs at the transcriptional level. In general, however, gene function may be regulated at any point between the gene and its ultimate product. Gene regulation occurs at the level of transcription, translation, or post-translational modification of protein structure. Control of gene function also includes the control of replication of DNA, as well as protein synthesis.

4. The lac operon exemplifies **negative feedback control.** A feedback system is one in which a product of the system exerts control over its future level of function. In a negative feedback system, high levels of the product result in a decrease in future production levels. Low levels of the product cause increased future levels of production. Just as a thermostat maintains temperature at a set level, so the lac operon maintains a relatively constant amount of lactose-metabolizing enzymes in the cytoplasm. Negative feedback is probably a crucial component of many systems of genetic regulation.

5. Regulation of the lac operon involves both the primary structure (nucleotide sequence) and the secondary and tertiary structures (physical conformation) of the DNA molecule. Transcription is inhibited by the repressor molecule physically altering the shape of the DNA molecule. We have previously stressed the overriding importance of the primary structure of DNA, but higher level structures also play a role in gene function.

Gene regulation in eukaryotes

Control of gene function in eukaryotes is presently an area of intense scientific and medical research. The structure of DNA is now reasonably well known, but precise knowledge of its function is developing only slowly.

There is, however, no question that gene regulation is the essence of life. Nothing demonstrates the astonishing precision and complexity of gene regulation in eukaryotes more dramatically than the orderly **development** of an individual organism from the fertilized egg to maturity to inevitable death. The basis of developmental events is differential gene activity—the regular turning on and off of genes. One expression of the loss of normal regulatory mechanisms is cancer, uncontrolled growth of cells.

A primary goal of developmental genetics is to understand the specific mechanisms controlling developmental sequences. The ability to turn off harmful genes or turn on beneficial genes has applications to medicine, agriculture, and business.

We begin this section with an example of the precision of gene regulation at the biochemical level—production of the various types of hemoglobin in an individual. We then consider mechanisms of gene regulation and allelic interactions.

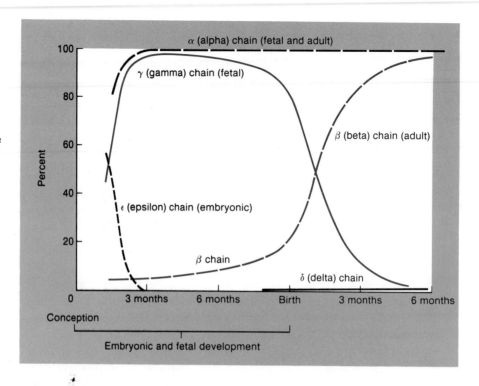

Figure 9.3 Timing of production of different polypeptide chains of hemoglobin during human development. The vertical axis shows the proportion of hemoglobin molecules containing the various chains.
Source: Fig. 14.5 from *Our Uncertain Heritage: Genetics and Human Diversity* by Daniel L. Hartl. Copyright © 1977 by Daniel Hartl. Reprinted by permission of Harper & Row, Publishers, Inc.

Development and regulation at the biochemical level: hemoglobin

Hemoglobin is a physiologically important molecule whose genetics is relatively well known. It combines reversibly with oxygen, which it transports from the lungs to all tissues of the body. The complete hemoglobin molecule consists of four polypeptide chains, each bound to a heme (iron containing) group (figure 8.2d). At least seven different genes in humans encode subunits or chains of hemoglobin, but not all are equally active throughout life. The differential activity of the various hemoglobin genes exemplifies gene regulation at the biochemical level (figure 9.3).

All types of hemoglobin contain two alpha chains (α_2). The gene controlling the alpha chain is turned on early in embryonic life and maintains a high level of activity until death. A gene controlling the epsilon (ϵ) chain also becomes functional very early during gestation, but its function ceases about three months into development. Embryonic hemoglobin consists of two α chains and two ϵ chains ($\alpha_2\epsilon_2$).

At about three months of gestation, the gamma gene (γ) becomes active, maintaining a high level of activity until birth, after which it diminishes in function until about six months of age. Fetal hemoglobin ($\alpha_2\gamma_2$)—designated Hb F—and embryonic hemoglobin both have a very high affinity for oxygen, a characteristic advantageous for a developing fetus who must be able to pull oxygen across the placenta from the red blood cells of the mother.

Also during gestation, the gene for the beta (β) hemoglobin chain becomes active, achieving a high level of function postnatally. About 90% of all adult hemoglobin is designated Hb A ($\alpha_2\beta_2$). A small portion of adult hemoglobin is type Hb A2, consisting of two alpha chains and two delta (δ) chains. The δ gene becomes active at birth and maintains a low level of activity throughout life.

Other types of hemoglobin, containing other chains, have been identified. Additionally, many genetic variants of hemoglobin, each often involving only one amino acid substitution, have been discovered. Over 250 different forms of Hb A alone are known. Some of these types are associated with pathological conditions—sickle-cell anemia, for example.

Thalassemia refers to a group of genetic diseases involving reduction or absence of one type of hemoglobin subunit. Consequently, certain types of hemoglobin are reduced or absent in affected individuals. Beta thalassemia is caused by a gene that inhibits synthesis of the β chain. In homozygotes, the β chain is not produced at all. These individuals are totally deficient in Hb A, causing a serious form of anemia that is usually lethal by the teens. Heterozygotes produce some β chain and thus are less severely affected. Increased amounts of Hb A2 and persistent Hb F compensate somewhat for reduced amounts of Hb A in the heterozygote.

Thalassemia reflects a change in the amount of protein produced, rather than a change in the structure of the protein. Studies have shown that there is a mutation in the gene regulating the synthesis of the β chain of Hb A. The mutation changes one short nucleotide sequence in the regulator gene causing it to resemble a base sequence that terminates transcription. When the transcriptional apparatus reaches this portion of the regulator, it ceases transcription and releases the forming mRNA molecule before the structural gene has even been transcribed. Because the genetic basis of diseases involving defective or deficient hemoglobin is well understood, these diseases are likely to be among the first to be successfully treated by genetic engineering.

Mechanisms of eukaryotic regulation

Although operons have not been identified in eukaryotes, certain aspects of operon structure and function—the regulatory gene and the regulator molecule, for example—are integral components of eukaryotic gene regulation. However, the greater size and complexity of eukaryotic cells probably require more complex types of regulation than are found in prokaryotes.

A thousand times larger in volume than prokaryotic cells, eukaryotic cells have developed specialized organelles to handle particular metabolic functions. Additionally, DNA is isolated in the nucleus. DNA must have means to communicate with organelles. In multicellular organisms, cells located at great distances from one another must have means of communication as well. Thus it is not surprising that eukaryotes have developed a diversity of mechanisms for controlling gene function.

Organization of the nucleus

The physical relationships of genes and chromosomes probably play an important role in gene function. Meiosis demonstrates that chromosomes can identify each other and that intranuclear organization is not random: Homologues pair and remain in close proximity until fission. Assessing physical relationships other than during cell division is difficult because chromosomes are too diffuse to be visible, but functionally related chromosomes may attach to each other or to adjacent locations on the nuclear membrane.

Chromosomal interactions may be mediated by regulatory molecules, as in the lac operon. Even more direct interaction may occur when segments of DNA, called **transposons,** are exchanged between chromosomes. The exact function of transposons is not known, but they appear to occur widely in nature.

Modifications of DNA structure

The physical structure of eukaryotic chromosomes changes predictably during the cell cycle. During mitotic division, chromosomes are tightly coiled and condensed. During the G1, S, and G2 stages of the cell cycle, the chromosomes are relatively diffuse. Replication of chromosomal DNA occurs only during the S phase, while transcription occurs during the G1 and G2 phases. Neither replication nor transcription occurs during mitotic division itself.

Thus, transcription and replication occur only in those phases of the cell cycle when DNA is uncoiled. This suggests that the physical conformation of DNA, particularly the extent of coiling, plays a major role in gene expression. Enzymes that direct transcription and replication may have access to DNA only when it is unwound.

Chromosome puffs   Studies of *Drosophila* have provided some of the best information regarding gene regulation in eukaryotes. One of the great advantages of *Drosophila* and some other insects for chromosomal studies is that in some tissues, particularly the salivary glands, the chromosomes are *polytene*. Recall that in most cells, the chromosomes are unineme, or consist of only one strand of DNA plus associated nuclear proteins. **Polytene chromosomes,** on the other hand, are spectacularly large and easily observed microscopically (figure 9.4). Their large size is the result of the side-by-side arrangement of as many as one thousand identical strands of DNA.

Drosophila salivary chromosomes are not uniform in appearance, but exhibit about five thousand darkly staining *bands*. These bands correspond to *chromomeres*, beadlike bulges along the length of all chromosomes (figure 9.4). The lightly staining segments between bands are called *interbands*. Each *Drosophila* chromomere appears to contain several dozen structural genes.

During development particular chromomeres expand or *puff* in a predictable sequence (figure 9.1a). The puffs result from uncoiling of the chromomeres within each puff. Uncoiling is apparently necessary for transcription of RNA, and **chromosome puffs** have relatively high concentrations of RNA.

Lampbrush chromosomes   Oocytes often exhibit high levels of metabolic activity, including transcription. A special type of chromosome has been discovered in the oocytes of salamanders (figure 9.1b). **Lampbrush chromosomes** have a brushlike appearance due to the presence of many loops of DNA extending out from the main axis of the chromosome. Like chromosome puffs, lampbrush chromosomes are sites of transcription, and high concentrations of RNA are associated with them. Again, physical uncoiling of chromosomes is associated with high levels of transcription.

Chromosome inactivation   One of the most spectacular examples of conformational changes affecting gene regulation is **X inactivation** (chapter 5). Very early in development in cells of all placental and most marsupial female mammals, one of the two X chromosomes is turned off by **heterochromatinization,** an essentially permanent condensation of the chromosome. In its heterochromatic state, the chromosome stains heavily during interphase and is visible as a *Barr body* near the margin of the nucleus.

X inactivation is a major regulatory mechanism that equalizes the expression of genes carried on the X chromosome in the two sexes of mammals, hence its common designation as *dosage compensation*. The extreme condensation of the inactivated chromosome seems to be the mechanism by which transcription is turned off, a finding complementary to the uncoiling and high levels of transcription in chromosome puffs and lampbrush chromosomes.

In most female placental mammals, inactivation of one X chromosome appears to occur at random. In some cells, the paternally derived X is inactivated, while the maternally derived X is inactivated in others. In mules (hybrids between a donkey and a horse) preference has been demonstrated for inactivation of the donkey X chromosome. The importance of non-random inactivation of X chromosomes, as well as the mechanisms controlling it, has yet to be understood.

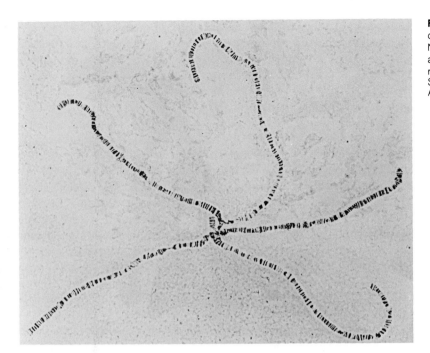

Figure 9.4 Polytene chromosomes of *Drosophila*. Note the dark staining bands and the clear interband regions.
Source: George Lefevre and Academic Press.

A final twist to the X chromosome story lies in the sex cells. In the oogonia of sexually mature female mammals, both X chromosomes are metabolically active, participating fully in meiotic processes. However, X inactivation occurs in *all* cells of the embryo during the first few days of development. Later the inactive X is reactivated in those cells that differentiate to become ovaries. This phenomenon is of considerable importance, because many developmental changes are normally irreversible. Discovery of the mechanism of X chromosome reactivation could eventually lead to control and manipulation of other developmental processes.

Variations in amounts of DNA

Gene function may be controlled by alterations in the amount of DNA in the genome. Unneeded genes may simply be eliminated. Genes whose product is needed in large quantities may be replicated and come to comprise a relatively large proportion of the genome.

Elimination of DNA In a variety of taxa, gene regulation involves elimination of entire chromosomes from the genome. Recall that mitochondria and chloroplasts possess their own DNA, separate from nuclear DNA. In protozoa, organellar DNA inherited from one of the parents is selectively eliminated from the genome. When cells of different species are hybridized in the lab, the chromosomes of one or the other parent are slowly eliminated throughout the life of the cell.

In some marsupials, one X chromosome is eliminated rather than simply inactivated. However, the most extraordinary example of chromosome elimination occurs in some coccid insects (mealy bugs and their relatives) in which the entire paternal genome is destroyed. The resulting individuals are haploids. This system also provides the sex-determining mechanism for these insects: the haploids are male, while diploids—in which no chromosome elimination occurred—are female.

The most interesting point regarding chromosome elimination is its selectivity, which implies that cellular mechanisms exist for precise recognition of specific chromosomes. Chromosome elimination is perhaps the ultimate mechanism of gene regulation. Genes that are destroyed or eliminated can never be expressed.

Increased amounts of DNA   Relatively large amounts of particular RNAs may occasionally be needed during the life of a cell. One way of meeting such a need is to have many copies of the required gene. In the oocyte of many amphibian species, many copies of genes that produce rRNA are present, and the genes are said to be *amplified*. The increased synthesis of rRNA during oogenesis provides for post-fertilization protein synthesis of the zygote.

In fact, the number of copies of particular base sequences varies immensely within the eukaryotic genome. This variation provides clues to the relative organization of the genome into structural and regulatory portions. *Non-repetitive* or *unique sequences* occur only once in each haploid complement. Structural genes encoding proteins probably occur primarily as unique sequences. *Middle repetitive sequences* have ten to one hundred thousand copies of a particular sequence. Middle repetitive sequences encode rRNA, tRNA, and histones, and may also have regulatory functions. The amplified rRNA genes described in amphibian oocytes are an example of middle repetitive sequences. *Highly repetitive sequences* have more than one hundred thousand copies of a particular sequence. Highly repetitive sequences are often associated with heterochromatic regions such as centromeres. They also occur in satellite DNA, where they may play a role in recognition of homologues during meiosis.

The proportion of unique, middle, and highly repetitive DNA in the genome shows considerable variation among species, but in general repetitive DNAs comprise more than half of the entire genome. Unique sequences (i.e., structural genes) occupy 10% or less of the nuclear DNA. This suggests that repetitive sequences play a major role in gene regulation.

Regulatory molecules

Conformational changes in chromosomes play an immediate role in regulating transcription and replication, but as with operons, regulatory chemicals often may initiate or inhibit gene activity. In contrast to operon repressor molecules eukaryotic regulatory molecules often originate outside the cell where their effect is felt.

Hormones are molecules that are a major component of the regulatory system of most eukaryotes. They are a primary mechanism of intercellular communication. *Hormone* is derived from a Greek word meaning "to set in motion," an apt description of its effect in regulating transcriptional activity. In mammals and probably most eukaryotes, hormones are produced by many tissues and organs throughout the body, then released into the systemic circulation, where they are transported to many different sites of activity. Some hormones, for example, interact with histone and nonhistone proteins to initiate transcription in target cells.

Among the hormones that affect transcriptional activity are the steroids, secreted by the adrenal glands, testes, ovaries, and the placenta. Sex hormones such as estrogen and testosterone are steroid hormones. In chapter 6, we discuss the strong influence that testosterone, in particular, exerts over sexual development. Testosterone is produced in the testes, but circulates throughout the entire body. Its effects are not limited to reproductive anatomy, but include aspects of the anatomy and development of many systems. Skin, muscles, and skeleton are particularly affected by testosterone, as are many aspects of physiology and behavior. Clearly, testosterone is a molecular signal controlling gene expression throughout the body.

Other steroids have equally large effects. Most tissues respond to estrogens, the female hormones, as well as to testosterone; both sexes produce male and female hormones, although in different proportions. The balance of male and female steroids is delicate, and many traits respond throughout one's lifetime to alterations in the balance. Sexual function changes regularly, for example, in response to tiny changes in hormonal level. The masculinizing effects of testosterone and the feminizing effects of estrogens demonstrate a fundamental genetic role of these hormones: differentiation of tissues resulting from initiation of transcription in response to a chemical signal.

Pheromones are a type of hormone that has been the subject of intense study in the past decade. Rather than being released into the internal environment, pheromones are released to the external environment, where they play a role in communication between individuals of the same or different species. Part of their effect may lie in mimicking hormones. The active ingredient in catnip resembles one of the sex hormones of cats. Chemicals in truffles, a mushroom greatly favored by gourmets, resemble a sex hormone of pigs; pigs are used by the French to sniff out this luxury growing on the roots of trees.

Pheromones are of particular importance in insect control programs. Sexually mature females of many insect species release pheromones, which attract males. Genetic engineering techniques now allow mass production of synthetic pheromones that can be used in several ways to limit insect reproduction: Traps can be baited with pheromones to attract males; or pheromones can simply be sprayed over wide areas, overwhelming and confusing the sensory systems of males who consequently cannot find mates (see chapter 12).

Interactions of alleles and loci

Thus far, we have examined gene regulation as a molecular and biochemical phenomenon. The most obvious expressions of gene regulation, however, are those affecting gross phenotypic traits, particularly those of medical importance.

In this section, we examine gene regulation at the level of the Mendelian gene and its associated phenotypic traits. Phenotypic variation reflects not only the expression of differing alleles of a gene, but also the way that variation in one gene controls the expression of another. Most major phenotypic traits are the result of interactions of many Mendelian genes. We use an example of a medical condition—PKU—to relate variation in a single Mendelian gene to its many phenotypic expressions. This in turn allows understanding of the key Mendelian concepts, dominance and recessiveness.

Dominance and recessiveness

Dominance and **recessiveness** have thus far been treated primarily as qualitative characteristics of alleles. In fact, the extent to which a particular gene is expressed to the exclusion of another gene often shows quantitative variability. The phenotype of heterozygotes may occur at any location along the continuum of phenotypic variation. If the heterozygote exhibits a phenotypic value identical to that of one of the homozygotes, the controlling allele shows *complete dominance*. If the phenotypic value is intermediate to the two homozygous types, the allele shows **partial** or **incomplete dominance.** Note that *intermediate* can mean any value between the two homozygous values, not just half way between them.

Many, perhaps most, genes exhibit *pleiotropic* or multiple effects (figure 4.7). One cause of pleiotropy is that each allele produces one type of polypeptide, which may subsequently participate in many biochemical pathways, affecting many different traits. One gene affecting different phenotypic traits may show different patterns of dominance relationships with respect to each of the different traits.

The amount of polypeptide produced by the alleles at a locus also explains some aspects of dominance relationships. At the level of protein production, some alleles may produce an inactive or defective form of the protein. Heterozygotes thus may have half as much of a functional protein as do normal homozygotes; that is, they

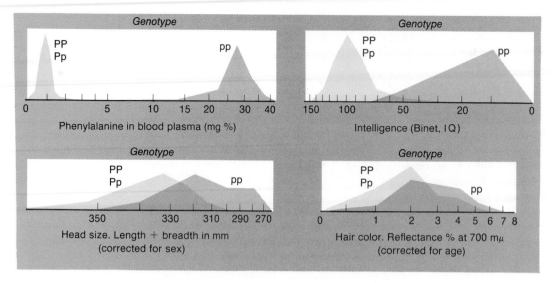

exhibit partial dominance for production of that protein. In some cases, aspects of the phenotype dependent upon those proteins will not be significantly affected by reduced amounts of the protein, as long as *some* protein is present. In such cases, heterozygotes exhibit complete dominance for the traits produced by the protein. In other cases, reduced levels of certain proteins quantitatively alter phenotypic traits; in other words, partial biochemical dominance causes partial dominance for associated traits. Phenylketonuria illustrates both possibilities.

An example: PKU

Phenylketonuria (PKU) is caused by an allele that, when homozygous, results in the inability to produce the enzyme phenylalanine hydroxylase, needed to metabolize the amino acid phenylalanine. Heterozygotes produce half as much enzyme as normal homozygotes, so when considering enzyme production, the normal allele shows partial dominance to the PKU allele.

Only small amounts of enzymes, however, are required for normal phenylalanine metabolism. Remember that enzymes are *not* used up—they are used again and again to catalyze reactions. Even half the normal quantity of phenylalanine hydroxylase is usually sufficient to maintain normal phenylalanine metabolism, so despite reduced levels of phenylalanine hydroxylase, a heterozygote usually has the same level of plasma phenylalanine as a homozygote. The normal allele is dominant to the PKU allele in terms of plasma phenylalanine levels (figure 9.5).

Phenylalanine levels, in turn, affect many other traits, including intelligence, head size, and hair color. Because heterozygotes have plasma phenylalanine levels similar to those of normal homozygotes, the phenotype for the associated traits usually shows the same range for normal and heterozygous individuals (figure 9.5). On the other hand, heterozygotes differ from normal homozygotes in at least one respect (figure 9.6). Phenylalanine tolerance tests show that heterozygotes metabolize phenylalanine more slowly than homozygotes. Heterozygotes can usually be identified on the basis of elevated blood levels of phenylalanine immediately after *loading* (ingesting high levels of phenylalanine). The high levels are only transient, but they permit clinical detection of carriers.

Norm of reaction

The data for PKU illustrate an important principle of phenotypic variation. Note that the genetic types that may occur in a population are discrete; that is, an individual may possess zero, one, or two PKU alleles. Phenotypic values, however,

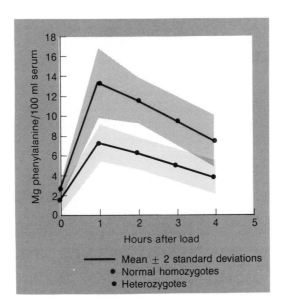

Figure 9.6 Response curves to phenylalanine loading for two genotypes. At time zero, individuals ingest large quantities (0.1 g/kg body weight) of phenylalanine. The subsequent reduction in blood levels of phenylalanine is measured hourly. Heterozygotes for the PKU allele take significantly longer on average to clear high levels of phenylalanine from the blood than do normal individuals.

may be more or less continuous with a relatively broad range. The range of phenotypic responses for a particular gene over *all* environments is called the **norm of reaction.** The norm of reaction comprises all possible phenotypic values that may be assumed by a particular genotype.

The range of variation expressed by a genotype depends upon both the external and the internal, or genetic, environment within which the genotype functions. Even assuming an identical external environment (and two environments almost always differ in some respect), if two individuals have identical alleles at one genetic locus, the modifying influence of other genes will produce variable phenotypic responses. Presumably, the interaction of alleles at a locus with both external and internal environmental factors transforms qualitative genetic variation into quantitative phenotypic variation.

Figure 9.5 illustrates the broad range of phenotypes realized by individuals identical at one genetic locus, but presumably differing for many *background* loci. Untreated phenylketonurics have IQs in the range of zero to seventy-five, while their hair color spans the spectrum of reflectance from 0%-70%. In the first case, IQ distinguishes phenylketonurics with considerable accuracy from the normal population; in the latter case, hair color separates them relatively poorly. In both cases, some overlap in phenotype occurs between normal and affected individuals, even for a relatively serious condition.

The phenotype may be altered dramatically by the external environment, as the norm of reaction of the IQ of phenylketonurics illustrates. The data in figure 9.5 refer only to untreated individuals. Dietary therapy of phenylketonurics results in a range of IQs similar to that for unaffected individuals. Thus, the norm of reaction for IQ for PKU individuals encompasses essentially the same range of IQ values shown by non-PKU individuals.

Until dietary therapy was recognized as an effective preventive measure for mental retardation, the norm of reaction of IQ for phenylketonurics would have been considered to be only seventy-five and below. However, dietary therapy provides the appropriate environment in which the norm of reaction can be greatly extended to include the range of normal IQs. This exemplifies one result of contemporary medicine: understanding the genetic and biochemical bases of diseases often permits affected individuals to be treated (provided a suitable environment) so that they can lead more or less normal lives.

However, because an individual's genes cannot yet be changed, environmental modification has some limitations. For example, women who are homozygous for the PKU allele, but of normal intelligence because of dietary therapy, can marry and reproduce, but they still are unable to provide a normal uterine environment for their children. As described in chapter 4, offspring of such women often are retarded, even though they may only be heterozygous for the PKU allele. The ability to change one phenotypic trait such as intelligence does not necessarily mean a change in all traits affected by one gene.

Expressivity

Yet another term associated with phenotypic variability is **expressivity.** Expressivity measures the extent to which an individual demonstrates various phenotypic effects of a gene. It may refer to one or several characters. For example, expressivity may be high for a phenylketonuric who is severely retarded and very blond, but low for one whose diet is low in phenylalanine. Conversely, most heterozygotes would have low expressivity of the deleterious allele, except where they are subjected to phenylalanine loading.

For many genetically based diseases, the degree of expressivity is extremely variable. Diabetes, for example, may require minimal or major clinical intervention, depending upon the extent of phenotypic manifestations (the symptoms of the disease).

Penetrance

Some alleles have such extreme variation in expression that they may not manifest their presence at all, in which case they are not **penetrant.** With respect to a single individual, a gene is penetrant if it has any detectable phenotypic expression. For an individual, a gene is either penetrant or not. Quantitative variations in the extent of phenotypic effect are discussed in terms of expressivity.

A gene may be common in a family, for example, but not be penetrant in all who have it. Lack of penetrance is often observed when a trait skips generations. Figure 9.7 shows the pedigree of a family with the allele for polydactyly over five generations. Polydactyly (meaning many digits) occurs when an individual has more than five digits on a hand or foot and results from a single dominant gene. If the gene showed complete penetrance, all persons having that gene would be expected to exhibit some degree of polydactyly.

The gene for polydactyly controls the number of bony rays formed in the hand and foot buds of the developing embryo. If the gene is penetrant, it still may exhibit variable expressivity. Individuals III-8 and IV-25 show how variable the expression of this gene can be. III-8 had six fingers on each hand and a normal number of toes. Her son, IV-25, had five fingers on the left hand, six on the right, six toes on the left foot, and seven on the right.

Note individual III-2. Because his mother and three of his children had the gene for polydactyly, it is reasonable to assume that he also had it, but it was not penetrant. Penetrance may depend upon the environmental circumstances at the time the gene is normally expressed.

Delayed onset

Many developmental traits and some genetic diseases are not expressed until relatively late in life. Perhaps the best-known disease with **delayed onset** is **Huntington's disease.** The disease is controlled by a single dominant gene, but onset occurs most commonly between the ages of twenty-five and forty-five (figure 9.8). Huntington's disease affects the central nervous system, and the symptoms include loss of motor control and mental deterioration.

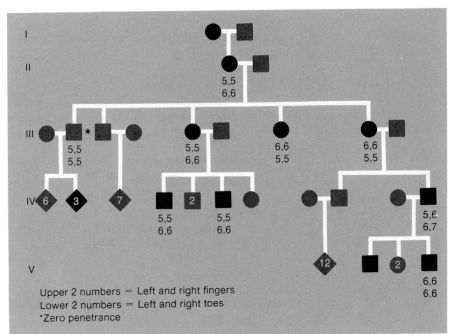

Figure 9.7 Pedigree of five generations of a family with polydactyly, a trait controlled by a dominant allele that is not always penetrant and that has variable expressivity. Where available, the numbers of digits are shown for affected individuals (black boxes or circles). Individual III-2 almost certainly exhibits no penetrance, as his mother and three of his children show polydactyly.

Upper 2 numbers = Left and right fingers
Lower 2 numbers = Left and right toes
*Zero penetrance

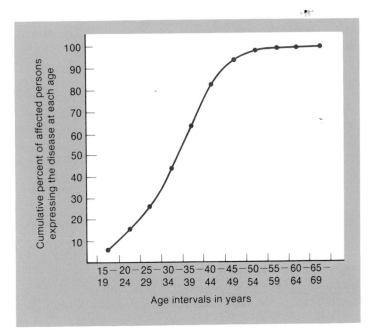

Figure 9.8 Age-specific expression of the dominant gene for Huntington's disease. Of 204 affected individuals, about 90% expressed the disease by age fifty; only two were first diagnosed over the age of sixty-five.

Diseases with delayed onset are particularly insidious. The disease may not develop until after normal ages of reproduction, so potential carriers must face the choice of reproducing without knowing if they can transmit the gene for the disease.

A molecular marker for the gene causing Huntington's disease was identified in 1983, and diagnostic tests for carriers became available in 1986. This technological advance has been fairly widely applied, but its use has merely substituted one set of ethical questions for another because no cure has been found. Many potential carriers prefer not to know the results of such a test. If the Huntington's disease gene is found, serious emotional problems can result, even leading to suicide. Even if an individual is found not to have the gene, he or she may feel guilt for escaping the fate that befalls a parent or siblings.

Codominance

When the alleles in a heterozygote produce different gene products, as opposed to a reduced amount of the normal product, the pattern of expression is **codominant.** The phenotypic results may be normal or pathological. The following examples illustrate both possibilities.

Blood groups   High levels of variation exist for the major blood groups, and in many cases the mode of expression of the gene product is codominance. Genes for blood groups encode complex molecules on the surface of red blood cells, and the presence or absence of particular antigens is not a pathological condition. Individuals who are type AB or type MN express both alleles at the controlling loci. The heterozygote also has a unique phenotype (positive agglutination reactions to each of two types of antibodies) so that both the biochemical and phenotypic expressions are codominant.

At the ABO locus, dominant as well as codominant expression occurs. The O allele encodes no antigenically active blood group molecule. Individuals who are genotype AO are phenotype or blood group A, since they produce only the A antigen, the same as homozygous AA individuals. The cell surface of AO individuals has sufficient type A antigen to produce agglutination with anti-A antibodies. Both the genotypes AA and AO show indistinguishable agglutination reactions, so they exhibit phenotypic dominance of A over O. In a similar fashion BB and BO genotypes are indistinguishable phenotypically.

Sickle-cell anemia   The allele for sickle-cell anemia causes a change in one amino acid of the β chain. Normal individuals produce only normal hemoglobin (Hb A) while victims of SCA produce only Hb S, with two abnormal β chains. Heterozygotes (persons with the sickle-cell trait) produce both Hb A and Hb S. Production of the two types, however, is not equal. About 25%-40% are Hb S, the remainder are Hb A.

Because a heterozygote exhibits both Hb S and Hb A, expression at the biochemical level may be considered codominant. However, in terms of the disease, the sickling allele is recessive. Only homozygotes have sickle-cell anemia, and under normal circumstances heterozygotes seldom express any symptoms.

Epistasis

When several loci affect the same trait, alleles at one locus may mask the effect of alleles at other loci. Such interactions are termed **epistatic.**

Recall from chapter 3 that a monohybrid cross of heterozygotes, where one allele is dominant to the other, yields a 3:1 phenotypic ratio. When more than one locus affects the same trait, many different ratios and more than two phenotypes may occur, depending upon the interactions of the loci.

As an example, consider congenital deafness in humans. Evidence suggests that alleles at one locus affect the development of the auditory nerve, and alleles at a second locus affect the normal development of part of the inner ear, the cochlea. Homozygosity of a recessive allele at either of these two loci causes deafness. If we represent the normal allele at these loci by D and F, then any of the following genotypes produces deafness: DDff, Ddff, ddFF, ddFf, or ddff. Thus, the presence of the recessive homozygous type at either locus is sufficient to override the effect of one or two dominant alleles at the other locus. In a dihybrid cross of double

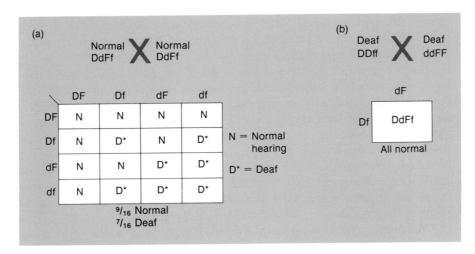

Figure 9.9 Two epistatic loci control deafness in humans: (a) mating of two doubly heterozygous individuals produces children in the ratio 9 normal: 7 deaf; (b) two deaf parents can produce all normal offspring.

heterozygotes, $\frac{9}{16}$ of the offspring will be expected to be normal, $\frac{7}{16}$ deaf (figure 9.9a)—not the 3:1 expected with only one locus or the 9:3:3:1 expected with two non-epistatic loci. Note that a 9:7 ratio is simply a modification of a 9:3:3:1 ratio, in which the epistatic interaction combines the three least-frequent phenotypic categories into only one (3+3+1=7).

Note also that control of a character by more than one locus provides a mechanism by which phenotypic traits do not *breed true*. Mating of two deaf persons whose deafness is a function of different loci produces heterozygous offspring of normal phenotype (figure 9.9b).

Another example illustrates how loci can interact to produce a variety of phenotypes. Coat color in rabbits, as in many other species, is controlled by several loci. Consider the following interaction of two loci, each with two alleles. Allele C at one locus is dominant to c. C __ individuals produce pigment, while cc individuals produce no pigment at all and are therefore albino. At the second locus, allele B for black is dominant to b for brown. However, neither B nor b can be expressed if the individual is genotype cc. Thus, the expression of any color is dependent first upon the possession of pigment. Otherwise, the individual is albino. In this instance, a dihybrid cross of heterozygotes yields the ratio of 9 black:3 brown:4 albino (figure 9.10). Imagine the ratios obtained from epistatic interactions of three, four, or more loci.

The types of interactions that may occur, even considering only two loci, are complex indeed. Only carefully planned breeding tests can sort out the precise patterns of inheritance that involve more than one locus. As the number of loci increase, the complexity also increases. Albinism in humans, for example, can be caused by any one of at least three loci. The allele for albinism is recessive at each locus, and the recessive homozygous condition at any one of the loci interrupts the synthesis of melanin. One of the three loci is X linked, however, so different ratios are obtained for each sex.

Quantitative inheritance

The effect of many genes contributing to a single trait is often additive, analogous to partial dominance when considering the interactions of two alleles at one locus. The phenotypic result, **quantitative inheritance** (introduced in chapter 3), is distinguished from epistasis by the mode of interactions of alleles. Epistasis is characterized by discontinuous patterns of variation due to the dominance of one locus to others. Quantitative inheritance involves the interaction of two or more loci that,

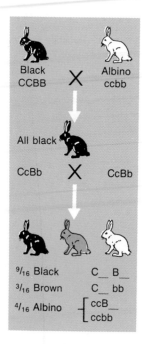

Figure 9.10 Epistatic interactions in the inheritance of coat color in rabbits, assuming only two genetic loci: CC or Cc = color; cc = albino; BB or Bb = black; bb = brown.

Black CCBB × Albino ccbb

All black

CcBb × CcBb

9/16 Black   C__ B__
3/16 Brown   C__ bb
4/16 Albino   { ccB__   ccbb }

individually, exhibit discrete Mendelian patterns of inheritance. However, the pattern of variation shown by these loci is quantitative rather than qualitative. Figure 9.11 illustrates how this might happen.

Consider that in a study population height is controlled, first, by one locus with two alleles (A and a). Individuals of genotype aa are four-feet tall. Each A allele in the genotype contributes one extra foot in height. Thus, Aa individuals are five-feet tall; AA individuals are six-feet tall. Allele A shows partial dominance to a. Figure 9.11a illustrates the frequency distribution of offspring of a mating of Aa × Aa. The frequencies of each genotype and phenotype are derived from a Punnett-square analysis, but displayed in a bar graph to illustrate the relative abundance of each phenotypic class. Note that only three phenotypic categories are observed, and that the intermediate height (five feet) is the most frequent. The mean height of the individuals in this population is also five feet.

Figure 9.11b shows the pattern of variation obtained if height were controlled by two loci rather than one. Again, the height of a completely homozygous recessive individual (aabb) is four feet. Each dominant allele contributes an additional one-half foot in height. The pattern is similar to that in figure 9.11a. The range of heights is still four to six feet, the mean height is five feet, and extreme heights are comparatively uncommon. The most obvious effect of adding one locus is that two additional phenotypic categories, $4\frac{1}{2}$ feet tall and $5\frac{1}{2}$ feet tall, now occur. These two phenotypes partially fill in the gaps that separated the clearly distinct phenotypes in figure 9.11a.

The addition of more loci continues the trend of changes shown between the first two figures. While the range and mean heights remain the same, the number of possible phenotypes steadily increases (figure 9.11c). The result is that clearly discrete or qualitative differences in height become quantitative; one height grades imperceptibly into the next. The dotted line in the three figures connects the peaks of the various frequency classes. Note that as the number of loci increases, the shape of the dotted curve comes to approximate a bell-shaped curve.

A bell-shaped frequency distribution is well known to mathematicians as a **normal distribution.** The frequency distribution of many traits characterizing natural populations shows an approximately normal distribution. When a trait such as height is normally distributed, most members of the population are close to some central value (the *mean* height, five feet in this case). Relatively few individuals show extreme deviation from the mean value—hence, the bell shape of the curve.

Most complex traits of living organisms are governed by the interactions of many genes. Length, weight, color, metabolic rate, and many behaviors, for example, all tend to show a normal distribution of variation.

While such examples illustrate a basic pattern of quantitative inheritance, remember that nature is seldom as simple as the models used by scientists to describe nature. Certainly, almost all continuously varying characters are affected by factors other than those in our simple model. For example, some loci would have more than two alleles, perhaps many more, with each allele potentially making very different contributions. Some loci would likely show epistatic interactions, some of which would produce novel phenotypes beyond the range of normal quantitative variation, such as albinism or dwarfism. The contributions of each of the interacting loci are unlikely to be equal. **Modifying loci,** for example, enhance or suppress the expression of other loci. Thus, a trait may be controlled by a series of primary loci interacting quantitatively and epistatically, plus a number of modifying loci that selectively alter further the expression of the primary loci.

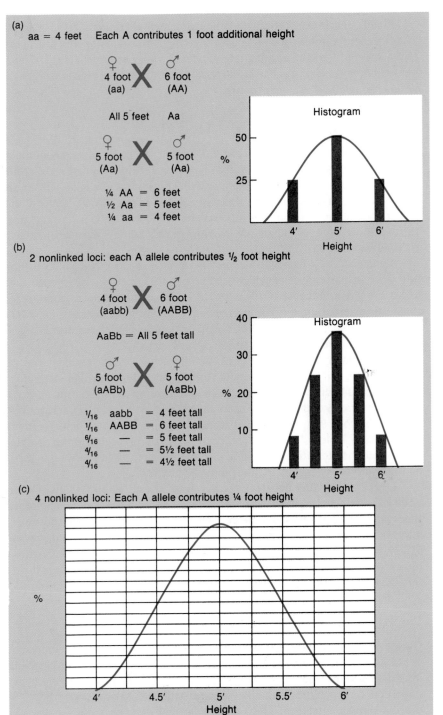

Figure 9.11 Examples of how polygenic inheritance controls quantitative traits: (a) variation due to a single locus with two alleles; (b) the results of two nonlinked loci, each with two alleles; (c) four nonlinked loci, again with two alleles each.

Finally, remember the influence of environmental factors. The norm of reaction of any genetic combination is defined by the environments to which it is exposed. Given the multitude of possible interactions of many environmental factors, many loci, and many alleles, it is hardly surprising that the most obvious variations among natural populations are quantitative.

Control of the phenotype: nature or nurture?

As we have seen throughout this chapter, the phenotype is the final result of a complex process of genetic regulation that begins at fertilization and ends with death. In other words, the phenotype of an individual is always changing. Genes are continually being turned on and off; alleles at a locus are interacting; alleles at different loci are interacting. In countless ways, alleles and loci are influencing and modifying the expression of other alleles and loci, and in turn being influenced and controlled themselves.

Remember, too, that some loci may never be expressed. None of us exhibits the full range of phenotypes of which we are capable. Instead, the genome functions under the direct influence of the external environment. Genes respond to both internal and external stimuli. The environment provides the context within which genes act, and full understanding of the development of phenotype requires knowledge of the influence of environment on specific gene function.

Unfortunately, the complexity of the interactions of environment and genome renders separation of their independent effects virtually impossible. Attempts to extricate their influences have resulted in a longstanding conflict within genetics, psychology, and many of the social sciences called the *"nature-nurture"* debate. Virtually none of the combatants in this debate openly claim that only genes (often called "nature" or "biology") or environment ("nurture" or "culture") entirely control phenotypic manifestations, but individuals differ greatly in the weight they place on the two types of factors in governing phenotype.

The nature-nurture debate is not purely academic. Public policy on social, educational, and medical issues has been decided on the basis of public perceptions of the relative roles of genes and the environment. The heat of public controversy has often reflected both views (prejudices, in some cases) and the virtual impossibility of fully separating the influences of nature and nurture. None of these debates better illustrates the complexity of phenotypic control *and* the social importance of genetic information than the controversy surrounding IQ.

The IQ controversy

During the past century, assessing intelligence and attempting to understand the causes of differences between individuals or various groups have developed into an important field of study. Psychologists, statisticians, and geneticists have all made important contributions to this knowledge. Millions of dollars have been spent to develop suitable intelligence tests and implement them on a wide scale. While intelligence testing has become highly refined, the debate over the causes of intelligence remains unresolved.

Several factors make this problem especially vexing. First, mental capacities clearly differ between persons, but defining, much less measuring, meaningful and objective measures of mental capacity has proven extremely difficult. In the last century, brain size (specifically, volume of the brain case) was used as an index of intelligence. In this century, the most common measure of intelligence has been the

Stanford-Binet Intelligence Scale, a series of standardized questions that most of us have taken to determine **intelligence quotient (IQ).** Many tests other than the Stanford-Binet have been developed to assess intelligence, but all suffer from a set of common problems. In discussing intelligence and mental capacity, we define IQ strictly as the score a person obtains on a standardized test such as the Stanford-Binet. As such, IQ is assumed not to be equivalent to intelligence, although the two may be related.

Perhaps the greatest limitation of all intelligence tests lies in defining intelligence. Most scientists agree that intelligence has something to do with the ability to understand abstract concepts, to solve problems, to express one's self verbally and quantitatively. But those capacities are themselves imprecise and almost certainly embody a variety of mental qualities. A person gifted in mathematics may have limited capacities with languages, and vice versa. Intelligence is a phenotypic trait of many aspects that is poorly represented by a single numerical scale.

Another difficulty in measuring intelligence is that IQ, unlike eye color or blood type, has a wide norm of reaction. Extreme variations in IQ have been shown to occur as a result of environmental factors, particularly during the first decade of life. Just as a person's size as an adult depends upon physical activity and diet during the growth years, so IQ is dependent upon environmental conditions during the psychologically formative years: cultural opportunities, education, diet, and even birth order and number of siblings affect IQ as an adult.

If IQ tests have such severe limitations, why does their use continue? Because, among other reasons, they provide a relatively objective measure of *something.* IQ as defined by the Stanford-Binet and other tests may provide little information about intelligence, but IQ is a good predictor of professional success in our society. The higher one's score on an IQ test, the more likely that person is to succeed in school, and the more likely that person will obtain a professional, well-paid job. Successes in school and on the job, of course, clearly depend upon many factors other than intelligence—creativity, social finesse, and persistence, for example. Nonetheless, IQ as measured on a test is a useful index to the probability of success in particular endeavors, despite the fact that IQ attempts to measure an intrinsically unmeasurable and even unidentifiable quantity.

An important factor confounding the study of intelligence is the persistence of an undercurrent of racism. The goal of some persons involved in the study of intelligence has been to demonstrate the superiority of some races, nationalities, or sexes. No scientist doubts that *differences* exist between racial groups. Various races are characterized by differences in skin color, hair color, eye color, size, proportion, blood group, and countless other phenotypic traits (figure 9.12); so it is reasonable to assume that differences in psychological characteristics may also exist. Such differences reflect only the extraordinary variety found throughout nature, the result of evolutionary processes affecting all living organisms.

The recognition of variety in nature is dangerous when such differences are assigned qualitative values and one trait is claimed to be of higher worth than another. Philosophers call such bad logic the *"naturalistic fallacy"* (the "is-ought problem"), meaning that a fact of nature is used wrongly to establish moral principle. Quite often, an "is-ought" argument is used for racist or sexist purposes. For example, because a woman has a uterus and ovaries (biological fact), some people claim that her destined and appropriate role is to stay home and have babies (moral judgment). Or, because men are on average larger than women (fact), they ought to have the dominant role in society (judgment). Neither statement is logically true,

Figure 9.12 Differences in allelic frequencies at five major blood group loci for three races.

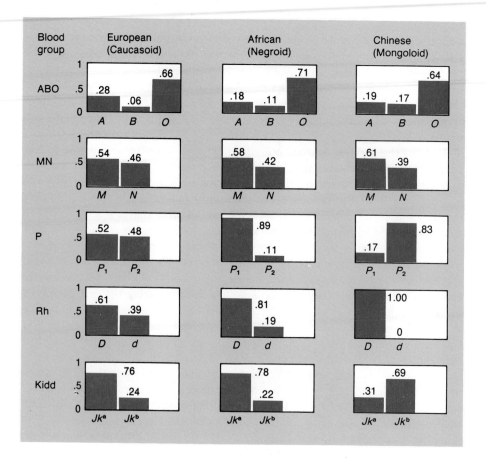

but reflects only the prejudice of those who believe it. Thus, despite the unequivocal fact that some people are more intelligent than others, no relative individual worth can be assigned on that basis.

Some racist applications of presumed intelligence assessments now seem ludicrous. Following Darwin's statement of the theory of natural selection in the late nineteenth century, for example, the public's imagination was captured by the search for the *missing link* between humans and their primate ancestors. For some years, no such link was found, which led some scientists to search in unusual and self-serving directions. The most famous scientists of the time were white and male, just as Western civilization was at the time dominated by white males. Some scientists claimed that the search for the missing link had overlooked the obvious: they claimed that missing links were to be found among the various races of humans still living.

As proof of this notion, evidence on brain size was offered. Ranked on a scale of relative brain size, white males were claimed to have the largest brains. White females and blacks of both sexes exemplified intermediate sizes; chimpanzees, were used as the small end of the spectrum. Thus, missing links were found by searching groups of society other than white males. The comparisons were designed, consciously or otherwise, to indicate a regular progression from chimpanzees through blacks and women to reach a peak—clear superiority—in white males!

How should these data be interpreted? First, you should be aware that data collection was highly biased. Brain volumes for white males were taken from soldiers (young men in their twenties) killed in battle. Brain volumes for blacks were taken mostly from men and women dying at old ages in public hospitals. Brain volume for all persons decreases throughout life, as calcium is deposited within the brain case. Many degenerative diseases also cause a decrease in volume of the brain case. A comparison of young and healthy whites with old and sick blacks is obviously biased, whether measuring speed in the hundred meter dash or brain size.

A second point concerning these data is that they assume that brain size in fact is a measure of intelligence. That clearly is not the case. Within the range of normal human brain sizes, there is no correlation of intelligence or IQ and brain size. In fact, the greatest determinant of brain size is overall body size. Like arm length and waist size, brain volume in humans is larger in large persons than in small persons. Thus, pygmies have smaller brain volume, but not lower intelligence, than most other humans, and women on average have lower brain volume than males.

The assessment of relative brain sizes states more about the culture from which it originated than about any meaningful differences in intelligence between various human groups. Such an example seems both silly and pernicious from our viewpoint at the end of the twentieth century, but it clearly illustrates the energy that has been expended attempting to demonstrate intergroup differences in intelligence. It also demonstrates the mystique that humans impart to intelligence. Perhaps because high intelligence seems to distinguish humans from all other animals, intelligence is one trait persistently used to try to demonstrate purported racial superiority.

The absurd search for the missing link among groups low in the social hierarchy is one of the least directly harmful aspects of the IQ controversy. In the past century, intelligence assessments have been used for purposes ranging from pernicious to vicious. Presumed low intelligence was a major criterion for excluding immigrants to the United States in the early twentieth century. Often, the immigrants could not speak English; naturally their answers to questions asked in English were most often incorrect. Decades later, Hitler and the national socialists (Nazis) claimed that mental superiority was a trait that distinguished Aryans from other races. In fact, Aryan is not even a racial group, but a prehistoric language spoken by peoples of several ethnic groups.

In the United States, differences in intelligence and the causes of those differences are common—although perhaps implicit—areas of contention in determination of amounts of public funding directed at particular disadvantaged groups. Those who believe that differences in IQ result primarily from differences in environment generally favor the expenditure of public aid for programs that increase the opportunities available to the disadvantaged. Project Head Start originated in the 1960s as such an effort.

Another school of thought claims that IQ differences have relatively little to do with environmental differences, but instead reflect genetic differences. If true, then improvements in the environment, they argue, can do little good in increasing IQs and, therefore, public money should not be wasted in the attempt.

Differences in IQ among racial groups in the United States

The theoretical distribution of scores on the Stanford-Binet is shown in figure 9.13. Like many quantitative traits, the distribution is normal. The mean value (the expected average intelligence) is 100, but variation is substantial. About 50% of the population scores between 90 and 110, and about 95% scores between 70 and 130.

The theoretical distribution of IQs is based on the white population of the United States. However, notable if slight deviations from the expected pattern occur. More persons seem to score below 70 or above 130 than expected, and various subdivisions of the white population yield significant differences in IQ scores. Differences have been found between nationalities, ethnic groups, religions, regions of the country, professions, and many other categories.

Figure 9.13 IQ frequency distributions. (Top) Theoretical normal curve, with mean of one hundred; because of rounding errors, total is slightly more than 100%. (Middle) Actual distribution of 2,904 native-born white children (from California, Nevada, Colorado, Texas, Kansas, Minnesota, Indiana, Kentucky, Virginia, New York, and Vermont) taking the 1937 Stanford–Binet. (Bottom) Actual distribution of 1,800 black schoolchildren (from Tennessee, Alabama, Georgia, South Carolina, and Florida) taking the 1960 Stanford–Binet. Source: (top) E. Peter Volpe, *Biology and Human Concerns,* 3d ed. Copyright © 1983 Wm. C. Brown Publishers, Dubuque, Iowa. All Rights Reserved. Reprinted by permission. (Middle) Terman and Marrow, *Measuring Intelligence,* 1937, Houghton-Mifflin, Boston, MA. (Bottom) Wallace A. Kennedy et al., "Unnormative Sample of Intelligence and Achievement of Negro Elementary Children in the Southeastern United States" in *Monographs of the Society for Research in Child Development, 28,* 6, 1963.

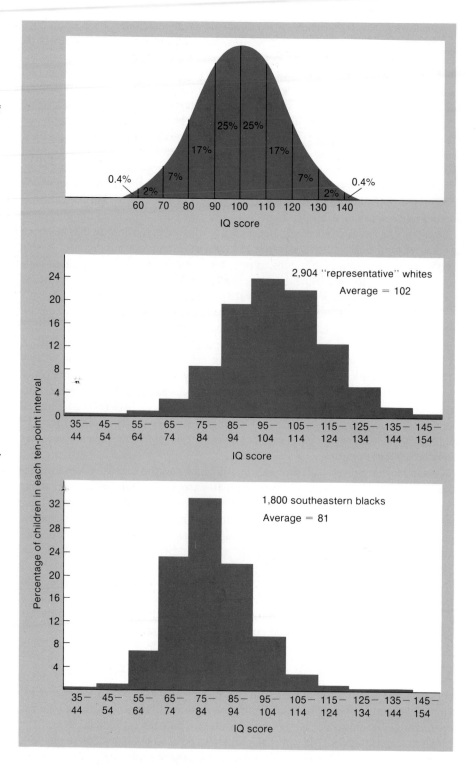

The most controversial differences are those that have been measured between racial groups. The mean score for whites on the Stanford-Binet is 100; for blacks, about 85; for orientals, about 107. The controversy does not surround the quality of the data. The differences between races are repeatable when different tests are given. Furthermore, the IQ differences are clearly correlated with various measures of achievement. Average family income, for example, is highest among orientals in the United States, lowest among blacks, and intermediate among whites. The controversy, rather, surrounds the *causes* of the differences detected. Do differences in IQ scores reflect genetic and therefore unchangeable differences? Or do they reflect serious environmental and therefore changeable differences to which the races have been exposed? No simple answer exists to such questions, but it is worthwhile to review a small sample of the vast but not yet definitive data collected on this subject.

Evidence for the role of genes in intelligence

Specific genes that contribute to normal levels of intelligence have not been identified. Therefore, data supporting the role of genes in IQ have been derived from indirect, often mathematical, analyses of genetic variation among different groups.

Among the most compelling data is the increasing similarity in IQ of persons of increasingly high genetic relatedness (figure 9.14). The average difference in IQ between genetically unrelated persons chosen at random is about 18 points. Therefore, knowledge of the IQ of one person provides little basis for predicting the IQ of any other person chosen at random—a correlation of 0. However, as the genetic relatedness of pairs of individuals increases, the average IQ difference decreases and the correlation increases. Those persons most closely related are monozygotic twins, who are genetically identical. The correlation of IQ between monozygotic twins is almost .90; average IQs of monozygotic twins differ by an average of only about 3–4 points. (Experimental error is 2 points; maximum possible correlation is 1.0.)

These data strongly indicate the genetic influence on intelligence, but influence of the environment is also clear. The correlation of IQ values among persons reared together for a particular degree of genetic relatedness is always higher than for persons reared apart, even for monozygotic twins. Increasing the environmental similarity always increases the similarity in IQ, just as does increasing genetic relatedness.

A second line of support comes from studies of **heritability.** Heritability (h^2) provides a measure of the proportion of variation observed in a population attributable to genetic rather than environmental variation. Heritability is an easily misunderstood concept. It refers to the contribution of genes to *differences* between individuals, rather than to the actual value of the trait itself in one individual. Heritability is a value describing one aspect of the variation within one population; it is *not* the proportion of a particular IQ ascribable to one causative factor or another.

For IQ in white populations, most studies find that $h^2 = .80$. Heritability is a proportion of the total variation, whose range is defined on a scale of 0–1. Thus, about 80% of the average 18 point difference between unrelated individuals is due to genetic differences and about 20% to environmental differences. However, the heritability value says nothing about that portion of the IQ other than the 18 point average difference, which is the major portion of the total IQ. Heritability thus refers to the variation in IQ, not to the total IQ itself: $h^2 = .80$ does *not* mean that 80 points of the normal 100 are determined by the genes.

Furthermore, because heritability can be computed only on a population basis, it is population specific. Heritability refers only to a specific population at a specific time in a specific environment. The same population, measured at either different ages or in different circumstances, will show different heritabilities. The range of heritabilities measured so far is extremely broad, from about 0.30 to 0.90.

The value of h^2 most commonly found (.80) has been interpreted by some persons to mean that 80% of an individual's IQ is due to genetic factors. Clearly, heritability does *not* mean that; it is far more complex. Nonetheless, its misinterpretation has been used forcefully by some to argue against the use of public money to provide improved environments for the disadvantaged in our society.

(a)

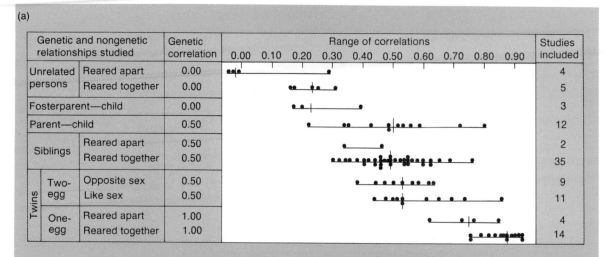

Genetic and nongenetic relationships studied		Genetic correlation	Range of correlations	Studies included
Unrelated persons	Reared apart	0.00		4
	Reared together	0.00		5
Fosterparent—child		0.00		3
Parent—child		0.50		12
Siblings	Reared apart	0.50		2
	Reared together	0.50		35
Twins — Two-egg	Opposite sex	0.50		9
	Like sex	0.50		11
Twins — One-egg	Reared apart	1.00		4
	Reared together	1.00		14

(b)

A comparison of twin pairs, or sib pairs, for three quantitative traits*

	One-egg Twins*		Two-egg Twins*	Sibs*
	Raised Together	Raised Apart		
	50 Pairs	19 Pairs	50 Pairs	50 Pairs
Height	1.7 cm	1.8 cm	4.4 cm	4.5 cm
Weight	4.1 lb	9.9 lb	10.0 lb	10.4 lb
Stanford-Binet IQ	5.9	8.2	9.9	9.8

*The figures in the four columns are simply the average differences within the pairs (rather than the statistical variances)

Figure 9.14 Evidence for a genetic component to IQ from analyses of IQ test results in persons of varying genetic relatedness: (a) data are summarized from fifty-two independent studies, and mean correlation coefficients are indicated by vertical lines; (b) comparisons of height and weight in contrast to IQ from twin studies. Source: (a) "Genetics and Intelligence: A Review," *Science,* by L. Erlenmeyer-Kimling and L. F. Jarvik, Vol. 142, December 13, 1962, pp. 1477–79. Copyright 1963 by the American Association for the Advancement of Science; (b) Newman et al., *Twin: A Study of Heredity and Environment*, 1937, University of Chicago Press.

One final line of evidence directly implicates genetic factors in intelligence. Genes involved directly in determining normal levels of intelligence have not yet been found, but many genetic diseases are associated with reduced intelligence. Many metabolic diseases, such as PKU, affect IQ. Chemical imbalances, whether due to genetic diseases or improper nutrition, can reduce IQ. Chromosomal imbalances, such as aneuploidies, almost invariably produce some degree of mental retardation. IQ is thus the result of the balanced interaction of many components of the genome.

Evidence for the role of environment in intelligence

A long list of environmental factors are associated with differences in IQ between various groups. A brief review of some of these factors illustrates the certainty of an important environmental component to intelligence.

Culture

Differences in IQ have been detected among populations representing different races, religions, nationalities, and regions of the country. Genetic differences also occur between these same groups, so it is almost impossible to determine whether IQ differences should be assigned to genes or to environment. However, the potential role of culture is often ignored. For example, some of the first studies showing black–white differences in IQ compared blacks from the southeastern United States with

Table 9.1
IQs of Israeli children of different genetic backgrounds raised in a home
environment or in a communal environment

Ancestry	Home*	Kibbutz
European	105	115
Middle Eastern	85	115

*Home environment refers also to the region in which the particular ethnic group occurred. Children of European ancestry were raised in Europe; children of Middle Eastern ancestry were raised in the Middle East.

whites from the northeastern United States. The blacks scored lower than the whites. Whites from the southeast, however, also scored lower than whites from the northeast. Differences in IQ among blacks and whites in these studies almost certainly reflect regional cultural and educational differences.

Another relevant line of evidence derives from children raised in a culture different from that of their genetic heritage. In Israeli kibbutzim, or farm communes, children are raised collectively rather than by their parents in a traditional home environment. Comparative IQ studies show that the communal environment has a major effect on IQ (table 9.1).

These data illustrate several points. If only the data from the first column were known, many persons might conclude that European Jews were genetically more intelligent than Middle Eastern Jews. A mean difference of 20 points is highly significant. The second column, however, provides perspective on the data of the first. Children raised in kibbutzim have a relatively uniform environment, whatever the racial background. The identical IQ scores for both groups indicate the primary importance of environmental factors. Sharing a common language, education, and social environment produced similar scores on IQ tests.

It is even more interesting that the IQ scores of both groups increased dramatically in the kibbutz environment. The communal environment may have enriched the learning opportunities of both groups, reflected in the higher IQ scores.

Age and early education

IQ is normally not measurable until about age two. From then until the early teens, IQ tends to increase regularly. The amount of increase is generally related to educational and cultural opportunities available to the child during this period.

Project Head Start was begun during the Johnson administration as a part of the Great Society innovations. Its object was to provide children from disadvantaged segments of society with a culturally and educationally enriched environment early in life. The program also included at least one nutritious meal per day for participating children.

The effectiveness of Head Start is still being assessed, but the first studies indicate favorable results. One study examined children of mothers with IQs less than 70. Children in Head Start scored an average IQ of 125, while those not in Head Start averaged 95. Other studies have shown a higher success rate in high school for Head Start children. They need fewer remedial courses and have a greater probability of graduating than non-Head Start children.

Project Head Start may also influence social behavior. Head Start children have fewer legal problems than non-Head Start children. Children exposed to an enriched social environment early in life may be more receptive to educational opportunities. Education is a cooperative endeavor, an interaction between the student and the environment, as is test taking. Increased facility in social interactions may play a major role in increasing IQ, which depends not only on intelligence and knowledge but equally upon the ability to *display* intelligence in an examination. Altered social behavior could also explain the data from the kibbutzim in table 9.1.

A different interpretation of the success experienced by participants in Project Head Start is that the school officials' knowledge that children had participated in a preschool program kept these children out of the special education classrooms upon entering the public school system. It is well documented that students in special education programs seldom return to the mainstream of public education. They have less expected of them and eventually come to have poor self-images and, consequently, expect less of themselves. In part, the perceived success of Project Head Start may be attributed to the extra attention the program focused on the participants and to their improved self-images.

In spite of the apparent success of Project Head Start, federal funding for it has been greatly reduced. The money saved in the short term must be balanced against long-term costs to society. Reduced levels of education and increased levels of antisocial behavior may result from permanent relegation of a portion of the population to a low-opportunity environment.

Test design and majority effect

When IQ tests are constructed, they are standardized against the test populations to which they are directed. A series of potential test questions is administered to an experimental group. Only those questions are selected for inclusion in the IQ test that (1) are correctly answered by the majority of experimental subjects and that (2) do not distinguish between various subgroups being tested. For the first IQ tests, questions were excluded that gave different results for males and females, and for different adult age groups. Thus, *by design* these tests did not discriminate on the basis of age or sex.

Tests were not standardized on the basis of socioeconomic status or race. It is hardly surprising, then, that significant differences between races have been detected. Recent tests have been designed that attempt to minimize the bias against particular races. Such attempts have been partly successful, but the effort has been criticized on the basis that IQ scores are most useful in predicting success of individuals in the school system and in professions. Because American society is still primarily white in orientation, there is some value in retaining a test with predictive power with regard to success in white society. Recognizing this difficulty, one black leader has noted that the use of IQ tests, even if biased toward white culture, is more fair than the use of no tests at all because "at least [blacks are] being excluded by an objective standard."

Considering the sexist bias of society, it is somewhat surprising that IQ tests were standardized to reduce differences between sexes. The comparative mental capacities of males and females is a topic of serious scientific debate. Females commonly score higher on IQ tests than males until puberty, after which scores decline in comparison to males. This decline has been blamed on social interactions, as the roles of the sexes become more clearly defined at about the time of puberty. Young females, for example, may be reluctant to score higher than males whom they wish to date.

Some scientists claim that the sexes have significant genetic differences in mental capacity. In particular, females have been claimed to be less proficient in mathematical skills than males. Substantial evidence has been raised on both sides of this controversy. No resolution appears in sight.

A final effect relates to majority influence. One experimental study compared the performance of genetically different strains of rats in learning to navigate a maze, a standard measure of intelligence in rats. Six inbred strains were used to establish six different study populations. Each population consisted of a *majority* of a single strain of rat, plus individuals of different *minority* strains. The object of the study was to determine the predictability of performance of minority groups of rats in terms of their abilities to learn to navigate a maze. In each case, the rats belonging to the majority always learned the maze most rapidly. If genes controlled intelligence, one of the strains would have been expected to learn fastest, whether it was in the majority or minority of the population. Instead, a single environmental factor—which strain was most abundant—seemed to govern the measure of intelligence.

The extent to which data from nonhuman species relate to humans is always debatable. Nevertheless, a majority effect may explain certain aspects of IQ comparisons. Immigrant populations, for example, almost always score lower on IQ tests than the resident population. In only a generation or two, most immigrant populations are integrated into American society and show no differences on IQ tests; perhaps they have become one of the majority. Perhaps, also, they have acquired the English language and shared the opportunities of the American educational system. Blacks, however, have not become part of the majority, even after ten or more generations as Americans. Color differences have maintained them as an identifiable and therefore permanent minority group.

Nutrition

The influence of diet upon phenotype is unarguable. Many pathological conditions (rickets, scurvy, diabetes) can be cured or ameliorated by dietary modification. The average height of Americans has increased by over half a foot in the last one hundred years, almost certainly due to improvements in diet rather than to genetic changes. The diet of Americans is so good, in fact, that as a nation we may now suffer more from the effects of overabundance of food and drink than from malnutrition; heart disease, some cancers, obesity, and diabetes have become major American health problems.

Similarly, diet may significantly effect IQ in many ways. During the first few years of development adequate dietary protein is essential for the proper growth and maturation of nervous tissue. Protein-deficient children do not produce sufficient nervous tissue for full mental development. Certain biochemical defects can be treated dietarily. PKU, for example, causes serious mental retardation if untreated; restriction of phenylalanine intake produces normal intelligence.

Recent studies of retarded children indicate that biochemical deficiencies that can be treated dietarily may be one cause of retardation. Groups of mentally retarded children, including victims of Down's syndrome (chapter 10), when given large supplements of vitamins and minerals, have shown increases in IQ of 10 points or more. In the most spectacular example, a child of age seven was still in diapers, had not learned to speak, and had an IQ estimated to be 25–30. After supplements of vitamins and minerals, the child showed major improvements. By age nine he was leading a far more normal life. He could speak, read and write, do arithmetic, play basketball, and ride a bicycle; his IQ was estimated to be about 90.

A related aspect of biochemical defects is their influence on social behavior. Manic-depressive psychosis is now routinely treated by administration of lithium salts. A protein-deficient diet in monkeys induces various forms of antisocial behavior. Some scientists believe that hyperactivity in children may result, in part, from excess dietary sugar. Antisocial behavior may or may not directly influence IQ. It may, however, affect several factors clearly important to IQ, particularly learning. Where behaviors result in inattention at school, or early departure from the educational system due to criminal or psychologically aberrant actions, the IQ almost certainly suffers. Because testing involves cooperation from the test subject, even the determination of IQ of severely antisocial or retarded persons is frequently impossible.

Society's response

The evidence indicates that both genes and environment play crucial roles in determination of intelligence, as they probably do in most complex characters. The relative contributions are extremely difficult to quantify, but present knowledge provides some guidelines for societal behavior regarding IQ.

The range of variation in IQ found in all races is broad, and the extent of overlap is great. Geniuses and mental defectives are found in all races. IQ, much like an individual's height, varies in response to a large number of environmental factors during the formative years. Furthermore, it is uncertain just what IQ measures. It is probably related to aspects of intelligence, but IQ is not precisely equivalent to intelligence. Differences in IQ scores between races reflect only that: differences. Differences in IQ should be treated the same as other phenotypic differences that can be demonstrated between races and, indeed, between any other populations that can be compared. Neither scientific nor moral criteria justify using differences in IQ to measure relative value.

Changing the genetic composition of human populations is neither practical nor desirable. Environmental improvement, on the other hand, is not only humane, but provides tangible results to society through improved social behavior and personal development of its citizens. In either moral or practical terms, an affluent society has little justification for denying its citizenry environmental opportunities for the fullest expression of their psychological capacities.

Summary

1. Regulatory genetics is the study of mechanisms governing the expression of genes.
2. The operon is the basic structural and functional unit unique to prokaryotic DNA. An operon consists of an operator, a promoter, and several structural genes. Transcription of the structural genes into a single message depends upon the interactions of the operator and promoter with a distant regulator gene.
3. Probably due to the greater size and complexity of eukaryotic cells and chromosomes, eukaryotic gene regulation is far more diverse than prokaryotic regulation. Eukaryotic chromosomes differ from prokaryotic chromosomes in being multiple rather than single, consisting of about 60% histones and other nuclear proteins, and probably in possessing a high level of spatial organization within the nucleus, where they are segregated from the remainder of the cell.
4. Physical changes in eukaryotic chromosomes are important to gene function. Uncoiling is associated with transcription, as evidenced by chromosome puffs and lampbrush chromosomes.
5. Hormones control gene function by regulating transcriptional activity.
6. Dominance, recessiveness, and codominance are characteristics of the expression of alleles at a locus. They are meaningful only in reference to particular phenotypic traits. A pleiotropic allele, for example, may be partially dominant at the biochemical level, but recessive at a visible phenotypic level.
7. The norm of reaction is the range of phenotypic responses that a genotype expresses over many different environments.
8. The degree of expression of a gene is called expressivity. Penetrance is an all-or-none measure of phenotypic detectability of a particular allele.

9. Epistasis is a type of interaction between loci in which a gene at one locus controls and may mask the expression of the genes at another locus or loci. Epistasis usually results in departures from Mendelian ratios in hybridization experiments.
10. Most complex traits are governed by alleles at many loci and exhibit continuous or quantitative inheritance. Many quantitatively varying traits exhibit a bell-shaped or normal distribution.
11. Some quantitatively varying traits, such as IQ, are strongly influenced by environmental factors. Data demonstrate that both genetic and environmental factors affect a particular trait, but precisely defining the role of each factor is extraordinarily difficult.
12. Genetic and phenotypic diversity are a fundamental characteristic of all living organisms. No scientific, biological, or moral basis exists for assigning relative values to different individuals on the basis of genotypic differences.

Key Words and Phrases

chromosome puff

codominance

delayed onset

development

dominant

epistasis

expressivity

hemoglobin

heritability

heterochromatinization

hormone

Huntington's disease

inducible

intelligence quotient (IQ)

lampbrush chromosome

modifying loci

negative feedback control

normal distribution

norm of reaction

operator gene

operon

partial or incomplete dominance

penetrance

pheromone

polytene chromosome

promoter gene

quantitative inheritance

recessive

regulatory gene

repressor

RNA polymerase

structural gene

thalassemia

transposons

X inactivation

Questions

1. Why are the physical relationships of chromosomes difficult to study?
2. Of what evolutionary significance is the fact that the lac operon in *E. coli* is inducible?
3. In the absence of genetic engineering technology, why do you suppose that we cannot test a developing fetus for the presence of sickle-cell anemia, when we know that both parents have sickle-cell trait and, therefore, the developing child has a one in four chance of having SCA? (Assume that there is no difficulty in obtaining a sample of fetal blood and tissues.)
4. How were heterozygotes for the PKU allele identified for the study shown in figure 9.6? Why were PKU homozygotes not shown in this figure? What would their phenylalanine response curve look like?
5. In seed corn, two different purebreeding lines of plants with white seeds produce an F1 all with purple seeds. Inbreeding of the F1 progeny produces plants with purple seeds ($\frac{9}{16}$) or white seeds ($\frac{7}{16}$). What are the likely genotypes of the parental lines, the F1 and the F2?

6. Albinism is controlled by the homozygous recessive condition of a gene or genes involved in pigmentation. What are possible genotypes of two inbred and purebreeding albino rats who, upon mating, produce a litter with no albino offspring?

7. Would you expect to find higher heritability for a particular trait in a population that is relatively homozygous for the genes involved, or in a population that is relatively heterozygous for those genes?

8. Some of the strongest arguments for a genetic basis for many behavioral traits come from the study of monozygotic twins raised in different households. What are the assumptions of such arguments? How valid are those assumptions?

9. What is your opinion of the value and validity of IQ tests? On what do you base your opinions? What kind of evidence would make you alter your opinions?

10. Figure 9.7 is a pedigree of polydactyly. What is the genotype of III-3? How do you know? How can you be sure that III-2 is really heterozygous for the allele for polydactyly and that her children are not just the result of mutation?

11. "Stiff little finger" is a condition where the little finger is permanently bent due to defects in the attachments of some of the muscles controlling the joint. This condition is caused by a dominant gene. Some affected individuals have only the right little finger stiff, others have only the left little finger stiff, and some have both bent and stiff. What is the most likely explanation for the observed variation?

12. Early studies of inheritance of combs in chickens showed that if a purebreeding Wyandotte chicken (rose comb) is crossed with a purebreeding Brahma chicken (pea comb), all offspring have a different type of comb, designated walnut (see below). If two of these F1s are crossed, about $\frac{9}{16}$ have a walnut comb, $\frac{3}{16}$ a pea comb, $\frac{3}{16}$ a rose comb, and $\frac{1}{16}$ yet a different type of comb termed single. What is the most likely mode of inheritance?

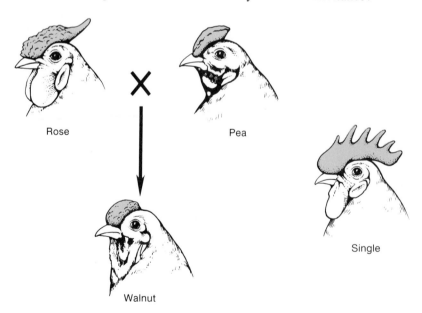

Rose X Pea

Walnut

Single

For More Information

Barlow, H. B. 1983. "Intelligence, Guesswork, Language." *Nature,* 304:207–9.

Beerman, W., and U. Clever. 1964. "Chromosome Puffs." *Scientific American,* 210:50–58.

Benbow, C. P., and J. C. Stanley. 1983. "Sex Differences in Mathematical Reasoning Ability; More Facts." *Science,* 222:1029–31.

Bodmer, W. F., and L. L. Cavalli-Sforza. 1970. "Intelligence and Race." *Scientific American,* 223:19–30.

Bouchard, T. J., Jr., and M. McGue. 1981. "Familial Studies of Intelligence: A Review." *Science,* 212:1055–59.

Bower, B. 1987. "IQ's Generation Gap." *Science News,* 132:108–9.

Britten, R. J., and D. E. Kohne. 1970. "Repeated Segments of DNA." *Scientific American,* 222:15–23.

Brown, D. D. 1981. "Gene Expression in Eukaryotes." *Science,* 211:667–74.

Bryant, P. J., S. V. Bryant, and V. French. 1977. "Biological Regeneration and Pattern Formation." *Scientific American,* 237:66–82.

Davidson, E. H. 1965. "Hormones and Genes." *Scientific American,* 212:36–45.

DeRobertis, E. M., and J. B. Gurdon. 1979. "Gene Transplantation and the Analysis of Development." *Scientific American,* 241:60–80.

Grady, D. 1987. "The Ticking of a Time Bomb in the Genes." *Discover,* (June):26–39.

Kolata, G. 1984. "Closing in on a Killer Gene." *Discover,* (March):83–87.

Kolata, G. 1987. "Early Signs of School Age IQ." *Science,* 236:774–5.

Kolata, G. 1984. "Globin Gene Studies Create a Puzzle." *Science,* 223:470–71.

Lewin, R. 1981. "How Conversational Are Genes?" *Science,* 212:313–15.

Ristow, W. 1978. "IQ Tests on Trial." *New Scientist,* 80:337–39.

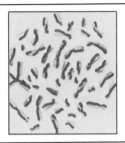

Misinformation

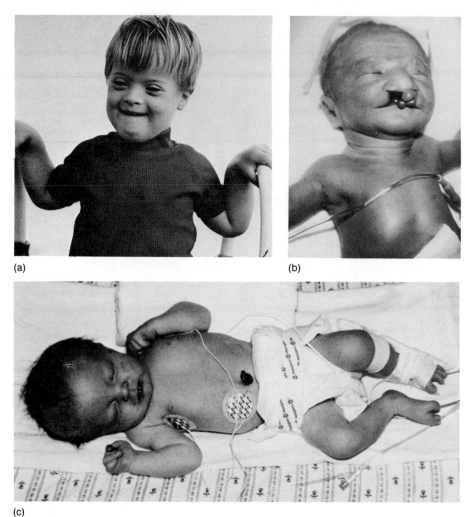

Figure 10.1 Human aneuploid conditions and their associated chromosomal anomalies: (a) Down's syndrome, trisomy-21; (b) Patau's syndrome, trisomy-13; (c) Edwards' syndrome, trisomy-18.
Source: (a) Bob Coyle; (b) Kutay Taysi; (c) James W. Hanson, University of Iowa.

The hereditary mechanisms discussed in the preceding chapters are precise, but they are not perfect. During the S phase of the cell cycle in humans, for example, the entire diploid complement—about three billion base pairs—of each of the ten trillion or more cells in the body is copied (replicated) from preexisting DNA in another cell. Mistakes in copying, which occur at a very low rate during replication, change the base sequence within a DNA molecule.

During meiosis, a second opportunity exists for alterations of DNA. The formation of chiasmata during prophase I of meiosis permits the equal exchange of chromosomal segments between homologues. However, nonequivalent types of chromosomal changes also occur: segments may be totally lost or moved to nonhomologous chromosomes. Division of chromosomes between daughter cells is sometimes not equal, and sometimes the parent cell does not divide at all, ending up with two diploid sets rather than one. Similar mistakes also occur during mitosis. The result is that the genome of any cell is likely to differ in some way from that of its parent cell.

Heritable changes in the DNA and chromosomes are termed **mutations.** Mutation is derived from the Latin word *mutare* meaning to change. A genetic mutation may affect only one nucleotide base, or it may involve long segments of a chromosome. In extreme cases, it may affect one whole chromosome or even the entire diploid complement. Mutations occur at any time during the life of a cell, not just during meiosis or mitosis.

The results of mutation differ greatly depending upon whether they affect germinal or somatic tissues. Germinal mutations affect the gonads or gametogenic tissues. They cause the genetic complement of the gamete to differ from that of the parent, and germinal mutations are passed on to succeeding generations. Somatic mutations occur after fertilization in nonreproductive cells of an organism. The mutation is transmitted via mitosis to all descendants of that cell but, because the gonads are not affected, not to succeeding generations. The relative effect of a somatic mutation depends upon when in development it occurs: the earlier in development, the more cells of the organism will be affected.

Most mutations are harmful. Even the substitution or deletion of a single base may be lethal. However, many mutations appear to be recessive, so their potentially harmful consequences may be masked by a dominant homologue, thus having little effect. On the other hand, mutations also provide the ultimate source of genetic variation for continued evolution of life on earth. All living species result from the genetic modification of preexisting species. The millions of living species represent an accumulation of mutational events whose effects on survival and reproduction were positive.

Chromosome mutation: numerical changes

The most spectacular mutations involve the addition of entire haploid sets, **polyploidy,** or the addition or loss of entire chromosomes, **aneuploidy.**

Polyploidy

Variations in the number of haploid sets are common in many species. Plants such as mosses spend most of their life cycle in a haploid phase, and for all sexually reproducing organisms, the gametic phase of the life cycle, however brief, is haploid. In bees and wasps, differences in ploidy determine sex: females are diploid, males are haploid. Some fish, salamander, and lizard species consist entirely of triploid females; no males at all are found in these species. Furthermore, tissues within an individual sometimes vary in ploidy. Endosperm of plant seeds is usually triploid, and human liver cells exhibit many degrees of ploidy—triploid, tetraploid, and even octoploid.

In plants, polyploidy is relatively common. Within a genus, many species often have a diploid number that is an exact multiple, three or greater, of the smallest haploid number for that genus (table 10.1). Furthermore, many common weeds such as dandelion, crabgrass, and wild oats are polyploid.

Polyploidy in plants originates in at least two ways. In some cases, polyploid cells arise from a failure of the normal mitotic apparatus. A cell duplicates its chromosomes and then completes prophase and metaphase, but the potential daughter cells fail to separate. The parent cell reenters interphase with two entire diploid sets: in effect, it has become tetraploid. Subsequent mitotic divisions produce a clone of tetraploid cells within the plant.

Commonly, polyploid cells and individuals are larger and more vigorous than normal. Tetraploid portions of a plant are often clearly distinguishable from the diploid portions. Many domestic fruits and grains originated as polyploids. They are valued highly because they often produce fruit (including grain) substantially larger than that of their diploid progenitors. In fact, the importance of polyploidy to modern society can hardly be overstated. Food production worldwide has more than kept pace with the exponentially increasing human population, primarily because of the development of new strains of basic grains—rice, wheat, and corn—many of which are polyploid.

Table 10.1
Diploid numbers found in certain genera of plants and animals

Genus	Chromosome number	Ploidy
Chrysanthemum	18	2N
(N = 9)	36	4N
	54	6N
	72	8N
	90	10N
Dahlia	32	2N
(N = 16)	64	4N
Triticum (wheat)	14	2N
(N = 7)	28	4N
	42	6N
Gossypium (cotton)	26	2N
(N = 13)	52	4N
Nicotiana (tobacco)	24	2N
(N = 12)	48	4N
Rubus (blackberries)	14	2N
(N = 7)	21	3N
	28	4N
	35	5N
	42	6N
	49	7N
	56	8N
	65	9N
	84	12N
Hyla (treefrogs)	24	2N
(N = 12)	48	4N
Xenopus (frogs)	36	2N
(N = 18)	72	4N
	108	6N
Cnemidophorus (lizards) (N = 23)	46	2N
(many triploid all-female species)	69	3N

Polyploids also originate from *unreduced gametes*. If bivalents do not separate during anaphase I of meiosis, any gametes subsequently formed are diploid or unreduced. Union of a diploid gamete with a normal haploid gamete produces a triploid zygote; union of two diploid gametes produces a tetraploid zygote.

Sexual reproduction is often problematical for polyploids because of difficulties in *pairing* of homologues. In a tetraploid, for example, all four homologues may synapse during meiosis, or three or two may synapse. Sometimes, balanced gametes are produced, each containing a complete diploid complement, but unequal results are also common. With triploids, pentaploids, and other uneven degrees of ploidy, gamete formation is even more likely to be unbalanced. Triploids are usually not fertile and must be propagated entirely by asexual means. The domestic banana is a triploid derived from a wild diploid species and surviving entirely by human propagation. The small black spots in the center of a domestic banana are remnants of the seeds that did not develop. An even more extraordinary example, the boysenberry, is heptaploid (7N).

Because of the increased productivity of many polyploids, horticulturalists now use techniques to increase the frequency of polyploidization. Application of the chemical *colchicine* to growing tissues of plants stops the mitotic cycle prior to separation of the daughter cells, producing tetraploidy in the dividing cell. Successive colchicine treatments can produce octoploids (8N) and even higher levels of ploidy. The domestic strawberry, for example, is an octoploid.

(a)

(b)

Figure 10.2 The cryptic species pair of grey treefrogs, *Hyla chrysoscelis* (2N) and *Hyla versicolor* (4N). These two species look so similar that they can be distinguished only by karyotype analysis or by differences in the calls of the males.
Source: (a and b) J. Bogart, University of Guelph.

Polyploidy is far less common in animals than in plants, probably because sex chromosomes are more common in animals than in plants. Nonetheless, surveys of chromosome numbers among animal taxa reveal that polyploid events have sometimes occurred. The gray treefrog, *Hyla versicolor,* is a tetraploid, presumed to have arisen from its morphologically indistinguishable diploid relative, *Hyla chrysoscelis* (figure 10.2). The two species were originally distinguished by slight differences in the male's call, and subsequent examination uncovered the chromosomal differences between them.

Polyploidy is apparently always fatal in humans. A few triploids and tetraploids have been born and even survived a few months, but all suffered obvious malformations. The frequency of conceptions involving polyploids, however, is not trivial. About 12% of all spontaneously aborted fetuses are known to be triploid or tetraploid, and the frequency may be considerably higher. In fact, more than 50% of all conceptions are estimated to be inviable due to genetic causes.

Polyploidy in humans results from at least two different causes: unreduced gametes and dispermy. The fertilization of an unreduced diploid egg by a haploid sperm gives rise to a triploid zygote, while the far-rarer union of two diploid gametes produces a tetraploid. **Dispermy,** the fertilization of an egg by two sperm, produces a triploid fetus.

Partially polyploid individuals are also known. Recall from chapter 6 that **mosaic** individuals consist of two or more discrete cell lines. Both 46,XX/69,XXX and 46,XY/69,XXY genotypes have been identified. Such mosaics may originate through the incorporation of a haploid polar body nucleus in a line of cells early in development. The earlier such an event occurred, presumably the larger the proportion of triploid somatic cells. Variation in the proportion of triploid cells might also result from differential growth rates of the two cell lines.

Aneuploidy

A substantial proportion, perhaps as high as 30%, of all human conceptions are aneuploid, differing by one or a few chromosomes from the normal forty-six. The possession of three rather than two homologues of a particular chromosome is referred to as **trisomy** for that chromosome. *Monosomics* have only one, and *nullisomics* have none, of a particular chromosome.

The primary cause of aneuploidy appears to be **nondisjunction** of homologues during meiosis (figure 6.11). Nondisjunction produces gametes with one chromosome too many or one chromosome too few. Subsequent fertilization of such a gamete gives rise to an aneuploid zygote. Nondisjunction also occurs during mitosis, producing individuals mosaic for various aneuploidies. As with polyploid mosaics, the effects of mosaic aneuploidy are more severe when the mosaicism originates early in development.

In most cases, however, aneuploidy has severe phenotypic effects. The genome appears to be in very delicate balance, and any alteration in dosage of chromosomes or chromosome fragments is almost always deleterious. With the single exception of **Down's syndrome,** the only aneuploidies viable to adulthood in humans are those involving the sex chromosomes discussed in chapter 6. The relatively high viability of sex chromosome aneuploids is probably explained by the dosage compensation mechanism. Because only one dose of most genes on the X chromosome is required, the loss of one X chromosome as in Turner's females, XO, is not necessarily lethal. Cytogenetic studies of spontaneously aborted fetuses, however, indicate that over 90% of XO fetuses do not survive to birth, suggesting that not all genes on the inactivated X chromosome are inactivated.

Dosage compensation also explains why XXX, XXXX, XXXXX, and XXY individuals are viable. The additional Xs are simply inactivated, although again phenotypic syndromes associated with these aneuploidies suggest that genes on the extra chromosomes are expressed to a limited extent. Nonetheless, triplo-X females are essentially normal phenotypically, and both triplo-X and tetra-X females may be fertile.

Because the tiny Y chromosome appears to carry only male determining genes, aneuploidies affecting it also are viable. XYY males are relatively normal phenotypically and are fertile. XYYY males are also viable.

Aneuploidies affecting the autosomes are always deleterious. Loss of an autosome is lethal; no monosomies other than Turner's survive past infancy. Trisomies surviving to birth have now been reported for at least chromosomes 8, 9, 13, 18, 21, and 22, but those involving 8, 9, and 22 are extremely rare. Trisomies of the smaller chromosomes (13–22) affect fewer genes and appear to be somewhat more viable. About 25% of all spontaneous abortions involve trisomic fetuses.

Trisomy 13, Patau's syndrome, occurs in a frequency of one in five thousand to one in ten thousand. Trisomy 18, Edwards' syndrome, occurs somewhat more frequently than Patau's. Affected individuals of both syndromes are severely phenotypically affected, and most survive only a few months or less (figures 10.1b,c).

Trisomy 21, Down's syndrome, is relatively frequent (about one in every six hundred births) and also relatively viable. About 10% of all cases of mental retardation in the United States are due to Down's syndrome. Affected individuals suffer numerous phenotypic abnormalities: they are short in stature, and the face, hands, and heart are malformed (figure 10.1a).

The trisomy affecting chromosome 21 is usually caused by meiotic nondisjunction, most commonly in the gametes of the mother. Furthermore, it has been known since 1876 that the incidence of Down's syndrome increases with the age of the mother (figure 10.3). At birth each ovary of a female contains several hundred thousand primary oocytes, in which meiosis has been arrested at prophase I. Meiosis does not resume until the oocyte is about to be ovulated. Apparently, the accuracy of meiotic division decreases with age, and after about age thirty the frequency of nondisjunction associated with chromosome 21 increases substantially. For women over age forty-five, the probability of producing a Down's child is about one in forty.

Recently, higher than normal frequencies of Down's children have also been discovered for mothers under the age of seventeen. The birth rate for girls under age seventeen has increased dramatically in the last decade, permitting statistical studies of birth defects in very young mothers. One theory that may explain the relatively high incidence of Down's syndrome for both older and younger mothers relates the frequency of nondisjunction to estrogen levels. Estrogen helps control the rate of meiosis in primary oocytes. As estrogen levels decrease with age, a slowing of the meiotic process may make nondisjunction more likely. Estrogen levels do not stabilize following puberty until about age twenty. Down's children produced by

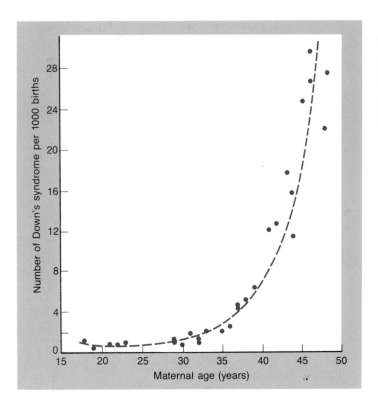

Figure 10.3 The incidence of Down's syndrome as a function of the mother's age. Note the slight elevation in frequency of Down's births in very young mothers and the marked increase in mothers over thirty-five years of age.

very young mothers thus may reflect fluctuations in hormone levels. If age or age-related factors increase the likelihood of nondisjunction, then trisomies involving chromosomes other than chromosome 21 should also be more common in older mothers, and perhaps younger mothers. In fact, the incidence of trisomies 13 and 18 *is* higher in older women. Other trisomies may also be common in these women, but because they are lethal, they are difficult to detect.

In the 1970s improved techniques for chromosomal banding and analysis showed that in as many as 25% of all cases of Down's syndrome, the extra chromosome is derived from the father rather than the mother. The probability of a father having an affected child increases with age, as it does for the mother. The rate of increase is relatively low until age fifty, and steep thereafter.

This finding is of considerable value to genetic counselors. It means that the chromosomes of both parents should always be investigated in trying to pinpoint the source of the genetic problem. Older mothers who have produced Down's children have often borne an unnecessary burden of guilt, assuming they were the "cause" of their child's defect.

Chromosome mutation: structural changes

While DNA is a relatively stable molecule, it may break at any point along its length. Most breaks heal or are repaired with no change in base sequence. Some breaks, however, cause permanent alterations of chromosome structure. The phenotypic effects of such chromosomal mutations range from trivial to lethal.

Deletions

Segments of chromosomes can be lost or **deleted** in several ways. Following a single break, the free ends may not rejoin. The fragment lacking a centromere is subsequently lost during cell division, a *terminal deletion* (figure 10.4a).

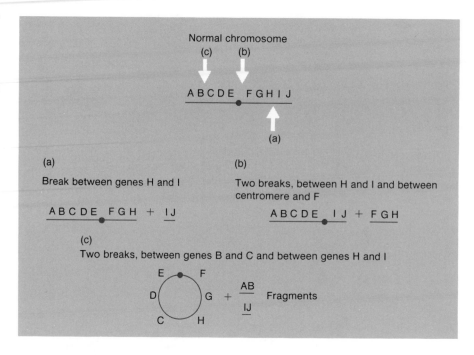

Figure 10.4 Three types of chromosomal deletions: (a) terminal deletions result from one break and cause loss of genetic material, shortening the chromosome; (b) interstitial deletions result from two breaks that do not include the centromere and, again, cause loss of genetic material; (c) two deletions, one on each side of the chromosome, may lead to formation of ring chromosomes if the two "sticky" terminal ends join.

Two breaks in chromosomes may also produce deletions. If the centromere is not located between the breaks, an internal segment may be lost; if the two terminal segments rejoin, an *interstitial deletion* may occur (figure 10.4b). If the centromere is located between the breaks, both terminal fragments may be lost. The free ends then join, forming a **ring chromosome** (figure 10.4c). Chromosome breakage often results from *mutagenic* (meaning change-producing) chemicals and radiation in the environment. Ring chromosomes are often found in the karyotypes of persons exposed to mutagens (figure 10.5).

Another cause of deletions is **unequal crossover** in meiosis. If the exchange of chromatid segments during prophase I is not precisely equal, a segment of DNA is deleted from one chromatid and added to its crossover partner (figure 10.6). Such an event produces four different types of gametes, one representing each parental type, plus one containing the deletion and one containing a duplication.

The phenotypic consequences of deletions are analogous to those of monosomy and presumably increase in severity in proportion to the size of the deletion. As with monosomy, most deletions are probably lethal. The individual is hemizygous for those genes that have been lost, and lethal or deleterious alleles are thus subject to expression. Many viable deletions have now been described for humans, all of which have some detrimental phenotypic effect. Recently, chromosomal deletions have been identified in certain types of cancers (pages 240–41). No deletion has been discovered to be beneficial.

Probably the best-known deletion is one affecting the short arm of chromosome 5, producing the *cri du chat* (cry of the cat) *syndrome*. Affected individuals have microcephaly, wide-set eyes, severe mental retardation, and as infants a distinctive cat-like cry for which the syndrome is named. Survival of these individuals is relatively high, with some living until adolescence.

Duplications

The effect of unequal crossover in meiosis is not only the deletion of chromosome segments, but the addition of those segments to the homologous chromatid—a chromosomal **duplication** in one of the gametes (figure 10.6).

(b)
(a)

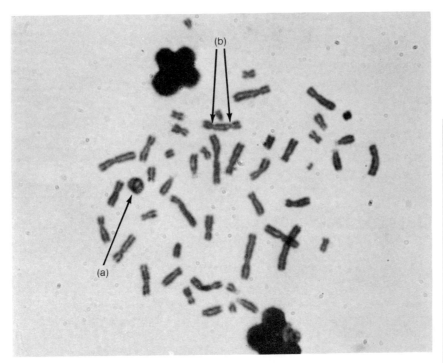

Figure 10.5 Karyotype of a normal person following exposure to X rays. Note ring chromosome (a) and dicentric chromosome (b). Source: Michael Bender.

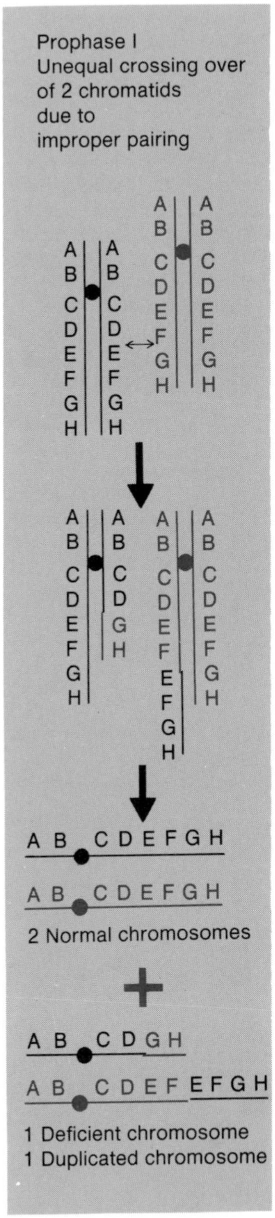

Prophase I
Unequal crossing over of 2 chromatids due to improper pairing
A B C D E F G H
A B C D E F G H
A B C D E F G H
A B C D E F G H
A B C D E F G H
A B C D G H
A B C D E F G H
A B C D E F E F G H
A B C D E F G H
A B C D E F G H
2 Normal chromosomes
+
A B C D G H
A B C D E F E F G H
1 Deficient chromosome
1 Duplicated chromosome

Figure 10.6 Unequal crossing-over produces chromosomes with duplications and deletions. Improper pairing in meiotic prophase I followed by crossing-over of non-sister chromatids results in two normal chromosomes, plus one with a deleted segment and one with a duplicated segment.

Such duplications are unknown in humans, in part due to difficulty in detecting them, but in laboratory species such as *Drosophila,* duplications and their phenotypic effects are reasonably well described. The effects of duplications would be expected to be comparatively less severe than loss of genetic material, although their rarity in humans suggests otherwise. It is generally believed that duplications have a beneficial evolutionary aspect. Duplicated alleles already occur in the disomic condition and should therefore not be required for any phenotypic function. Duplicated genes are thus evolutionarily free to assume new functions, representing an important mechanism of evolutionary change.

Inversions

When two breaks occur in one chromosome, the result need not be a deletion. Instead, the broken ends may rejoin, but with the central segment reversed in orientation—an **inversion** (figure 10.7a). If the inverted segment contains the centromere, the inversion is *pericentric;* if not, it is *paracentric.*

Inversions do not change the amount of DNA in a chromosome; they simply change the order of some genes. In many cases this change has no phenotypic effect. Inversions may substantially alter linkage relationships and can thus be expected to show phenotypic manifestations. Such **position effects** have been demonstrated frequently in *Drosophila.* They have not, however, been clearly associated with inversions in humans, even though a small but significant percentage of humans probably have at least one inversion.

The most obvious effects of inversions are associated with meiosis. Synapsis of homologues heterozygous for an inversion is possible only with the formation of a loop in one of the homologues (figure 10.7b). Such an event is common and produces normal gametes if crossover does not occur. Should a crossover occur, however, some normal and some unbalanced gametes, containing deletions and duplications, result. Reduced fertility may be one consequence of chromosome inversions.

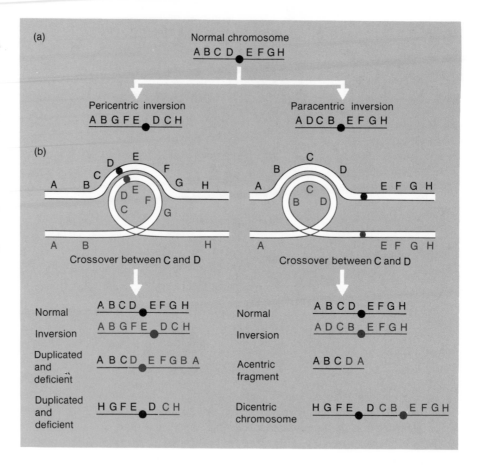

Figure 10.7 Chromosome inversions: (a) production of inversions by two breaks in a single chromosome followed by healing; (b) homologous pairing in meiosis of individuals who are heterozygous for an inversion. Notice that in both inversions only two of the possible four gametes will contain an entire chromosome, one normal and one inverted. The remaining two meiotic products will be missing some genes, have additional genes, be missing a centromere, or even have two centromeres. Both of the latter conditions interfere with normal meiosis. (See colorplate 16.)

Another hypothesized effect of inversions is the protection of well-adapted linkage groups. The meiotic difficulties associated with inversions seem to favor the suppression of crossover in inversion heterozygotes. An inversion associated with well-adapted linkage groups may increase the probability of transmitting such a group intact to future generations.

Furthermore, the meiotic difficulties characterizing inversion heterozygotes could provide a genetic basis for reproductive isolation between populations; that is, inversions may play an important role in the process of species formation. If a particular inversion is highly adaptive, it could become the only type of chromosome in the population, eventually leading to speciation of that population. Such an hypothesis is supported by comparison of banding patterns between closely related species. Species within the same genus often have identical or very similar chromosome numbers, but banding patterns reveal that many inversions have occurred in the course of their evolution. Chromosomal rearrangements, particularly pericentric inversions, appear to have played a significant role in the evolution of humans and our Great Ape relatives (see figure 12.9).

Translocations

A final type of chromosomal rearrangement involves transfer of genetic material between non-homologous chromosomes, **translocation.** In a *simple translocation,* a break occurs in one chromosome, and the acentric fragment then attaches to a member of a different pair (figure 10.8a). If breaks occur in non-homologous chromosomes, the acentric portions may be exchanged, a *reciprocal translocation* (figure 10.8b).

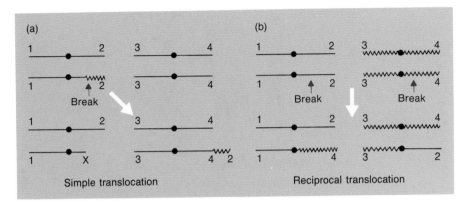

Figure 10.8 Chromosome mutations can involve moving entire segments from one chromosome to another. Homologous chromosome ends are numbered to help keep track of moving segments: (a) a simple translocation occurs when a portion of one chromosome breaks and attaches to a non-homologous chromosome arm; (b) in a reciprocal translocation, breaks occur in two or more non-homologous chromosomes, and the non-homologous chromosomes reciprocally exchange chromosome segments.

Because genetic material is not lost or gained, translocations often have little obvious phenotypic effects. Because they alter linkage relationships, however, the phenotype might be expected to change as a result of position effects. A more significant problem, associated with other chromosome changes as well, occurs during meiosis. When two pairs of chromosomes have undergone reciprocal translocation, synapsis can occur only if all four chromosomes form a single X-shaped configuration (figure 10.9). Subsequent gamete formation produces six different gametic combinations, only two of which are balanced.

Translocations affect many different chromosomes in humans. The best known of these involves a translocation of the long arm of chromosome 21. Most commonly, it is translocated to chromosome 14, but it may also attach to chromosomes 13, 15, 21, or 22. No phenotypic effect is observed in the person in whom the translocation originates. However, if a gamete containing the translocated chromosome 21 unites at fertilization with a normal gamete, the resulting embryo has trisomy 21, Down's syndrome.

Such individuals, referred to as *translocation Down's,* have all the symptoms associated with the standard Down's. They account for about 5% of all Down's cases. Translocation Down's have the normal diploid number (46) with one chromosome having the translocated arm of chromosome 21. One of the parents of a translocation Down's, on the other hand, is aneuploid, having only 45 chromosomes. However, because that parent still has essentially two copies of chromosome 14 and chromosome 21 (one normal 14, one normal 21, and the translocation 14/21), the parent has a normal genetic complement and is thus unaffected. The parent is said to be a *translocation carrier.*

Identification of the type of Down's a child has is important in counseling parents considering further reproduction. Nondisjunction events that cause the standard Down's child are unrelated. Thus, the probability of having a future Down's child is unrelated to the previous occurrence of a Down's child, although it is related to parental age.

Parents of a translocation Down's, on the other hand, have a relatively high probability of giving birth to future Down's children. In effect, translocation Down's is an heritable disease. One of six gametic types produced by the carrier parent bears the extra chromosome 21. Because three of the gametic types are apparently inviable, the theoretical probability of producing a future Down's child is roughly one in three. Further, phenotypically normal children may also be translocation carriers for chromosome 21, thus perpetuating the Down's-producing chromosome in the family. Although parental age appears to have no affect in the origin of translocation Down's, recent studies suggest that the father is more likely than the mother to be the source of the translocation, in contrast to standard Down's.

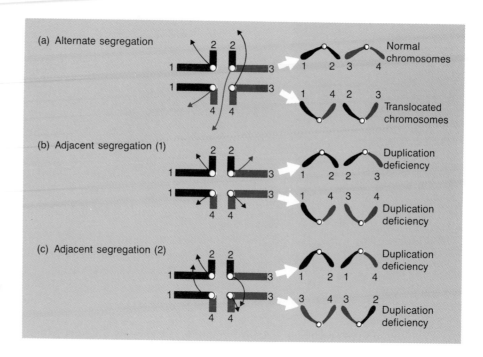

Figure 10.9 Meiosis in a cell that is heterozygous for the reciprocal translocation depicted in figure 10.8b results in different products depending upon whether segregation at anaphase I is alternate (a) or adjacent (b and c). (a) In alternate segregation both gametes have a balanced genome, although half have normal chromosomes and the other half have translocation chromosomes. (b and c) In both types of adjacent segregation, homologous centromeres go to the same (c) or opposite (b) poles, resulting in all gametes being duplicated and deficient for portions of the genome. (See colorplate 17.)

Another translocation is also of considerable medical importance. The **Philadelphia chromosome,** abbreviated Ph1, is associated with 96% of the cases of chronic myelogenous leukemia. The Ph1 chromosome was at first thought to be a shortened chromosome 22, due to a deletion of the long arm, but it has now been shown that the long arm of chromosome 22 has been translocated, most commonly to the long arm of chromosome 9 (figure 10.10). The Philadelphia chromosome was the first karyotypic abnormality to be associated with a specific cancer.

Gene mutation

The smallest type of mutation affects only one or a few DNA bases, a **gene** or **point mutation.** Point mutations cannot be observed by inspection of the chromosomes, and they may have small or large phenotypic effects. The basic types of point mutations are shown in table 8.3. A **frameshift mutation** involves the addition or deletion of one or a few bases, resulting in a shift in the triplet reading frame. A frameshift can render nonsensical all triplets downstream from the site of the mutation. Deletion of even a single base may thus have a major phenotypic effect by causing the production of nonsense mRNA or no mRNA at all.

A **substitution mutation** occurs when one base is replaced by another. If a simple base substitution produces a triplet that encodes a different amino acid than before mutation, the mutation is said to be **missense.** About 30% of all substitutions, however, do not change the amino acid encoded, particularly substitutions affecting the second and third bases in a triplet. These are known as **same sense mutations.**

The effect of point mutations is likely to increase with the number of bases affected. It is generally assumed that most mutations are deleterious, although their expression is often masked because the mutations are also usually recessive. As with chromosomal mutations, the immediate effects are normally not beneficial, but the evolutionary effects may be positive. Point mutations are the source of all new base sequences and provide the ultimate reservoir of genetic variation, continually renewed, upon which natural selection acts. Point mutation is one of the bases for the diversity of life on earth.

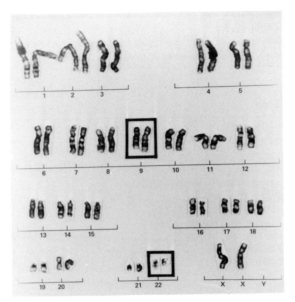

Figure 10.10 Karyotype of a person diagnosed as having chronic myelogenous leukemia. The longer chromosome 9 (upper box) has a piece of chromosome 22 (lower box) translocated to it, and the resultant short chromosome 22 is known as the Philadelphia chromosome.
Source: Victor B. Eichler.

Mutations apparently occur at random. That is, while we can estimate the *rate* at which mutations occur, we cannot predict which base pairs will be affected. No sources of mutation are known to be specific in their action. In that sense, evolution is clearly an undirected phenomenon. The variation available to natural selection cannot be controlled, and mutations do not arise in direct response to environmental changes.

The rate of spontaneous mutation

The background rate of spontaneous mutations is surprisingly constant throughout nature. On average, about one mutation can be expected in a gene in every one hundred thousand to one million gametes per generation. In humans, the rate of spontaneous mutations has been estimated on the basis of the rate of first appearance in a family of genetic diseases that are expressed when the controlling allele is present in only one dose, i.e., diseases caused by autosomal dominant alleles, or those caused by X-linked recessives in males (table 10.2).

Table 10.2 shows that the mutation rate differs among loci. The variation could be due to several factors, such as the number of bases in the gene or the relative mutability of particular base sequences. Estimates giving similar values of the spontaneous mutation rate have been made on many other species, including both plants and bacteria.

While 0.1–1.0 mutations per one million gametes per generation may seem a relatively trivial rate of occurrence, remember two points. First, most mutations are deleterious. If the rate of mutation were much higher, the mortality rate due to lethal genetic combinations might be unbearably high. Recall that more than 50% of all human conceptions are believed to be inviable due to genetic factors, even at this rate of mutation.

Second, humans are estimated to have ten thousand to one hundred thousand structural genes, plus an unknown number of regulatory genes. If the mutation rate is estimated to be one mutation per one hundred thousand gametes per generation and each person has one hundred thousand genes, we can expect that, on the average, each person will have one new point mutation per gamete. Since every person results from the union of two gametes, each would be expected to carry about two new mutations. Other evidence indicates that this estimate is conservative and that,

Table 10.2
Mutation rates of particular alleles expressed as autosomal dominants or X-linked recessives in males

Trait	Mutant genes per one hundred thousand gametes per generation
Epiloia	0.8
Aniridia	0.5
Microphthalmus	0.5
Wardenberg's syndrome	0.4
Facioscapular muscular dystrophy	0.5
Pelger anomaly	0.9
Myotonia dystrophica	1.6
Myotonia congenita	0.4
Huntington's chorea	0.2
Retinoblastoma	0.4
Neurofibromatosis	13–25
Hemophilia	2.7
Sex-linked muscular dystrophy	5.5

Source Table 11.2 from *Human Genetics* by Daniel L. Hartl. Copyright © 1983 by Harper & Row, Publishers, Inc. Reprinted by permission of Harper & Row, Publishers, Inc.

on average, each person carries about three to five new, potentially harmful recessive mutations. By comparison, the rate of chromosomal mutation is considerably lower than the rate of point mutation: only about one gamete in one hundred is estimated to carry a chromosomal aberration.

Causes of mutation

Many factors causing mutation are now known. As noted in chapter 8, mutability is a key requirement for the genetic material. One characteristic of the chemical structure of DNA is a certain amount of instability. Tautomeric shifts occur when hydrogen atoms in the nucleotide bases undergo slight alterations in position. When this happens, the base is unable to pair with its normal complementary partner, but instead pairs with another base; for example, guanine may pair with thymine, or cytosine with adenine. Normally, tautomeric shifts are only transitory states of a base and have little effect. If, however, a tautomeric shift occurs at the time of DNA replication, the mispairing causes a change in the base sequence in the complementary chain that can be transmitted to future copies of the DNA—a substitution mutation.

Tautomeric shifts occur as a result of natural chemical properties of the nucleotide bases and help explain the normal spontaneous mutation rate. Many environmental agents also affect the rate of mutation. Additionally, variations in the mutation rate have been associated with biological factors such as age and sex. Achondroplasia, for example, occurs more frequently in matings involving older fathers.

Chemicals and radiation are the primary physical agents affecting mutation rates. Mustard gas, used widely in World War I, was one of the first chemicals shown to be mutagenic. Mustard gas, a powerful mutagen, reacts directly with DNA bases, causing both genic and chromosomal mutations. Other types of chemicals, **base analogues,** have structures similar to the normal nucleotide bases and are thus subject to accidental incorporation in DNA. Caffeine, for example, is a base analogue of adenine. If incorporated in place of adenine, it may pair with guanine instead of thymine, causing a base substitution during cell division (figure 10.11).

Chapter 10

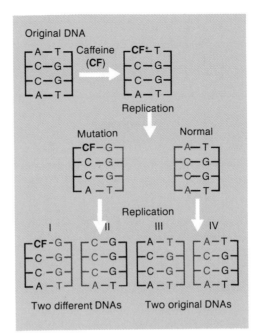

Figure 10.11 How caffeine (CF) acts as a base analogue of adenine and causes mutation. Note that the CG base pair (II) replaces the original AT (III, IV) base pair at the top of the DNA molecule, due to interference of caffeine. The DNA molecule replicates in the normal fashion, except that the caffeine that replaced the A mispairs with G, rather than with T, as A normally does.

The early 1950s marked the beginning of large-scale production of modern chemicals. Now, over sixty-three thousand chemicals are used routinely in industry and commerce, and about one thousand more are added each year. Literally thousands of these are known or strongly suspected of being mutagenic. Some are used only in medical or industrial applications, but many are commonly encountered in everyday life. Among the latter are such chemicals as pesticides and herbicides, caffeine, formaldehyde (used in household insulation), toners (used in photocopy machines), some components of dyes and cosmetics, and sodium nitrites (used as preservatives in meat products such as ham, bacon, and hot dogs). Over twenty-five hundred chemical additives are used in processing food.

A second major category of mutagenic agent is radiation. Two basic types of radiation are of health importance to humans: *electromagnetic* (including **ultraviolet radiation, X rays,** and gamma radiation) and nuclear (including alpha and beta particles, protons, and neutrons). In 1927, H. J. Muller demonstrated that the mutation rate could be increased fifteen hundred times in the progeny of male *Drosophila* subjected to high doses of X rays. Moreover, genetic damage is proportional to the dose administered, and the effect is cumulative. In other words, one large or **acute dose** has roughly the same mutagenic effect as many smaller or **chronic doses** administered over many days, as long as the total dose is the same.

Exposure to some types of radiation, like many chemicals, is unavoidable. Some portions of the world have high levels of background radiation from uranium and other minerals, and most of us are routinely exposed to medical and dental X rays. The ultraviolet (UV) component of sunlight is probably the single most important natural mutagen in our environment. UV light easily penetrates the epidermis and breaks the DNA of underlying tissues. In particular, where two thymine molecules occur adjacent to one another, UV light causes them to bond to one another rather than to the adenines of the complementary strand, forming **thymine dimers** (figure 10.12). If these dimers are not excised and the DNA repaired, mutations occur, as DNA with thymine dimers cannot replicate normally.

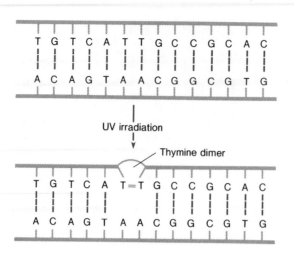

Figure 10.12 Formation of thymine dimers in DNA. Exposure to ultraviolet radiation causes adjacent thymine molecules in a single strand of the DNA to dimerize and break their bonding with opposite adenine molecules. Normally there are enzyme systems that monitor the DNA and excise these dimers and replace them with thymines that bond with the opposite adenine molecules. Source: Leland G. Johnson, *Biology*, 2d ed. Copyright © 1987 Wm. C. Brown Publishers, Dubuque, Iowa. All Rights Reserved. Reprinted by permission.

DNA repair

Given the mutability of DNA and the presence of many mutagens in the environment, it is perhaps not surprising that cells possess mechanisms to repair DNA. One repair process involves removal (excision) of the mutant segment by certain enzymes, followed by its replacement with the proper set of nucleotide bases. Thymine dimers caused by exposure to UV light are normally repaired in this way.

Excessive exposure to UV radiation, however, may produce so much mutation that the repair system is overwhelmed. In such cases, skin tumors eventually form. About four hundred thousand new cases of skin cancer occur annually in the United States due to UV radiation. Fortunately, these growths are normally benign, and reduced exposure to sunlight reduces the risk of recurrence.

The role of DNA repair mechanisms is well illustrated by the genetic disease **xeroderma pigmentosum** (meaning dry, heavily pigmented skin; figure 10.13). Individuals homozygous for this autosomal recessive allele cannot produce enzymes necessary for excision of thymine dimers from DNA and subsequent DNA repair. Affected individuals are thus extremely sensitive to sunlight. Even the slightest exposure causes dry, flaking skin, heavy pigmentation, and eventually numerous disfiguring skin cancers. Death usually ensues by age thirty due to metastases of inoperable skin cancers. This unhappy condition illustrates both the high frequency of natural mutations and the vital importance of the DNA repair processes.

Mutations and human health

It is now known that genetic mutations, both point and chromosomal, cause major human health problems. In some cases, this information allows prevention or reduction in the incidence of certain genetic conditions. In the long term, increasing knowledge should offer treatment for many serious health problems caused by mutations.

In this section we consider aspects of mutation related to aging and developmental abnormalities. In the next section, we consider cancer. We conclude the chapter by briefly considering the politics of dangerous substances.

Aging

Numerous theories have been advanced to explain the process of aging, and several are based largely on genetic factors. The **mutation theory of aging** holds that aging is a degenerative process reflecting the accumulation of deleterious somatic mutations throughout life. The Second Law of Thermodynamics states that, in general,

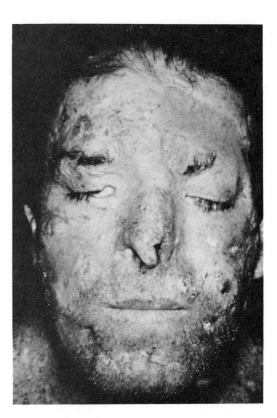

Figure 10.13 A person with xeroderma pigmentosum. Source: Lester V. Bergman and Associates, Inc.

the level of disorganization (entropy or randomness) increases throughout the universe. The high level of organization in living organisms is maintained only by the many homeostatic and repair mechanisms of the body. As somatic mutations accumulate, the genetic repair mechanisms gradually are unable to fix all of them, and the result is a net deterioration of physical and mental function, culminating in death.

Another hypothesis relates the decline in cellular function to the activities of **free radicals** in the cells. Free radicals are groups of atoms that are highly reactive with other molecules in the cell. Formed by ionizing radiation or as by-products of normal cell chemistry, they are usually inactivated by enzymes. Free radicals are so active that some combine with DNA, apparently altering its function permanently. Free radicals could thus provide a mechanism for the mutation theory of aging.

The **commitment theory of cellular aging** suggests that development is a genetically programmed process designed, like all products of natural selection, to maximize the genetic contribution of an organism to future generations. That is, the developmental sequence characterizing a species has evolved because individuals with that sequence produced more offspring. Thus, species differ in their maximum duration of life because they are *genetically programmed* for differing lengths. Bacteria may live only a few minutes, mice only a few years, humans, elephants, whales, and tortoises a hundred years or more. Organisms die at the appropriate time not because of genetic failure, as in the previous hypotheses, but because of their genetically directed developmental program. Scientists have not been able to alter significantly the maximum life span of any species, suggesting strongly the existence of genetically controlled limits that differ between species.

Evidence supporting this view was provided by Leonard Hayflick's studies of the growth of cells in culture. Human fibroblast cells in laboratory culture will not continue to divide indefinitely. The number of doublings that occur in culture decreases with the age of the cell donor. The limit for cells taken from a human embryo is about fifty doublings, but the limit for cells taken from a thirty-year-old human is only about twenty doublings. Species with different maximum life spans also possess different potentials in cell culture. The Galápagos tortoise, one of the longest lived of all animals, may live more than 150 years. In culture, embryo fibroblasts of the Galápagos tortoise undergo 90–125 doublings. A mouse, on the other hand, usually lives only two or three years, and its embryonic fibroblasts are capable of only 14–28 doublings. The **Hayflick Limit** is one of several possible mechanisms controlling the aging process. It describes a phenomenon—the limited growth of cells in culture—and thus provides a clue to the genetic processes related to aging. However, the exact meaning of the Hayflick Limit is still an object of scientific debate.

The **autoimmune theory of aging** is another important hypothesis. This theory holds that decreasing function of one of our primary protective systems, the immune system, leads to deterioration in cellular function with age.

Note, however, that these theories are not mutually exclusive. All are supported by strong but circumstantial evidence, and all may be correct to some extent. Maximum length of life does seem to be a genetically governed trait of a species, but somatic mutation could well be one of the mechanisms enforcing that maximum.

Progeria and *Werner's syndrome* are two autosomal recessive genetic disorders characterized by rapid aging and thus a greatly shortened life span (figure 10.14). Individuals with these conditions age about ten times more rapidly than normal, begin to go bald, have wrinkled skin, and even arthritis and heart disease by the time they are seven or eight years old. Death usually results from a heart attack or stroke before the child is fifteen. Cell cultures taken from individuals possessing these disorders are capable of far fewer doublings (2–10) than normal, suggesting a change in the normal mechanism of control of mitosis.

Developmental abnormalities: Mutagens and teratogens

Aging is a normal and predictable part of the developmental process. As we have seen in chapter 9, development is a highly integrated and complex process of phenotypic changes guided largely by a precisely programmed sequence of genetic events. While medical science will continue to study the causes of aging, we should bear in mind that aging is a natural and generally desirable process. None of us, for example, would wish to retain the phenotype of a five-year-old forever!

The precision of development, however, makes the normal developmental sequence vulnerable to disturbance. If mutation affects a key regulatory gene, the medical consequences can be serious. Similarly, mutations affecting "dosage" of particular genes (in reality, loss of normal controls on expression of those genes) also can cause major developmental anomalies. Down's and Turner's syndromes are examples of the effects of dosage alteration due to mutation.

Environmental agents such as chemicals and radiation can adversely affect human health through *mutagenesis*. **Mutagenic** agents cause genetic mutation. Many environmental mutagens are probably also **teratogenic** (from the Greek *teras,* meaning monster), producing physical defects or deformities in developing embryos.

Mutagens and teratogens are *not* the same thing. Substances that cause mutations need not cause deformities, and vice versa. Genetic mutation probably often has no significant health effects. It is likely, however, that mutagenesis is an important agent of teratogenesis, and as we have seen, developmental abnormalities are often associated with genetic damage.

The first trimester (the first three months) of gestation is a particularly vulnerable time for a developing fetus because it is during this period that the basic tissues and organ systems are established. Chemicals that normally have little or no harmful effect on humans may cause severe damage to a fetus during this period;

Figure 10.14 Fransie Geringer from South Africa, age eight, and Mickey Hays from Texas, age nine. Both are victims of progeria, a rare genetic disease characterized by rapid aging. Source: AP/Wide World Photos.

it is for this reason that pregnant women are urged to stop all use of tobacco, alcohol, and drugs. During pregnancy, even such drugs as aspirin should be taken only on the advice of a doctor.

Perhaps the most notorious teratogen in recent decades has been the drug *thalidomide,* widely used as a sleeping pill and tranquilizer in Europe in the early 1960s. Children of women who used thalidomide during the first trimester of pregnancy were born with abnormalities in the bones of the arms and/or legs (figure 10.15). Phocomelia is a rare genetic disease with the same phenotypic effects. The sudden high frequency of children born with this phenotype led to the discovery that thalidomide produces **phenocopies** of phocomelia. As soon as the teratogenic effects of thalidomide were recognized, the drug was removed from the market.

A drug whose teratogenic effects on the developing fetus have recently been widely recognized is alcohol, the single most widely used dangerous chemical in our society. Although the effects of alcohol addiction or misuse are well known, it is less widely realized that alcohol is a potent teratogen for the developing fetus. The Bible mentions the harmful effects of alcohol on pregnant women, and British doctors at the turn of the century observed that "maternal intoxication" was a major source of fetal damage. The public, however, has only slowly begun to perceive the dangers of alcohol use by pregnant women.

Alcohol is a small molecule that easily crosses the placenta, reaching concentrations in the fetus equal to those in the mother's bloodstream. Its extreme danger to the developing fetus was explicitly recognized medically in 1973 with the description of **fetal alcohol syndrome (FAS).** Alcohol (specifically, ethanol) causes abnormal development of the head and nervous system. The effects are greatest during the first trimester of pregnancy, particularly during the first month. Substantial damage may be done to a fetus even before the mother realizes she is pregnant. Symptoms of FAS include head and facial deformities (figure 10.16), microcephaly, mental retardation, reduced growth and development rates, and ten-times higher probability of perinatal death. About one in 750 births in the United States is severely affected due to FAS. Mothers who drink alcohol have an increased incidence of stillbirth.

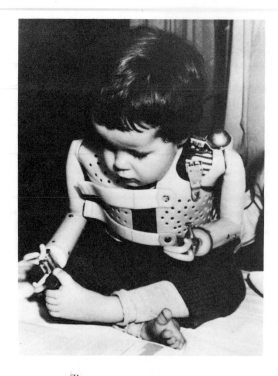

Figure 10.15 A three-year-old girl whose mother took thalidomide during the first trimester of her pregnancy. Absence or shortening of arms are characteristic abnormalities induced by thalidomide.
Source: Omikron/Photo Researchers.

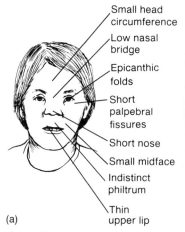

Small head circumference
Low nasal bridge
Epicanthic folds
Short palpebral fissures
Short nose
Small midface
Indistinct philtrum
Thin upper lip

(a)

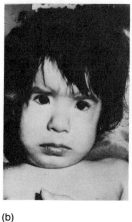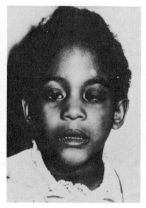

(b)

Figure 10.16 Fetal alcohol syndrome: (a) many facial abnormalities result from exposure to alcohol during gestation; (b) the effects are similar in children of all races.
Source: (b) Dr. Ann Pytkowicz Streissguth, University of Washington: *Science* 209 (18 July 1980), 353–61, © 1980 by the AAAS.

The extent of fetal damage increases with maternal alcohol consumption. Even two drinks per day can cause damage. Recent primate studies have shown that maternal alcohol consumption interrupts placental exchange of oxygen between mother and fetus for up to an hour or more. Such oxygen deprivation may be responsible for the mental retardation characteristic of FAS. Symptoms of FAS have been detected in 2% of the offspring of mothers admitting to two or fewer drinks per day during and just before pregnancy; 11% of infants of mothers having two to four drinks per day; and 19% of infants of mothers having more than four drinks per day. Until FAS is better understood, women should abstain from alcohol during pregnancy, especially during the first trimester.

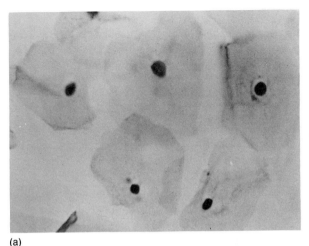

(a)

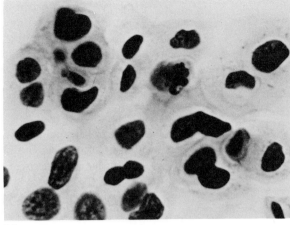

(b)

Figure 10.17 Human cervical cells: (a) normal; (b) cancerous. (Note cells in (b) have much larger nuclei, less cytoplasm, and more of them are in stages of division.)
Source: American Cancer Society.

Cancer

Cancer (figure 10.17) is a disruption of developmental and regulatory systems, arising from the loss of normal controls on cellular growth, differentiation, and migration. Cancer occurs when a cell or group of cells begin uncontrolled growth, forming tumors or **neoplasms** (meaning new forms). Cells within the neoplasm do not differentiate properly, if at all. They are *invasive*, infiltrating tissues and·destroying surrounding cells, and they often spread or **metastasize** from the point of origin to other locations in the body, where they lodge and continue uncontrolled proliferation. Without medical intervention, cancers are normally fatal.

Cancers usually kill by malnutrition and infection. Cancerous cells grow far more rapidly than surrounding cells, robbing the surrounding cells of vital nutrients. In spite of good nutrition, cancer victims become progressively emaciated. Malnutrition makes the body susceptible to disease. Cancers invade the lymph nodes and overwhelm their protective responses. Cancerous cells infiltrate the major organs, blocking circulation, and eventually causing failure of life-sustaining processes.

Three main categories of cancers are identified by the tissues affected. (1) **Carcinomas** affect a variety of organs, accounting for about 85% of all cancers. They include cancers of the skin, glands, gastrointestinal tract, respiratory tract, and urinary and reproductive systems. (2) **Leukemias** and other cancers of the blood-forming tissues account for about 10%–12% of all cancers. They are a major source of mortality among children. (3) **Sarcomas,** the rarest of all cancers, affect the bones and muscles. About 2%–3% of all cancers are sarcomas.

Cancer is a general term, including many separate diseases. Over one hundred different forms of cancer have been described, and the diversity almost certainly reflects many different causes. However, evidence has increasingly indicated that cancer is the result of some type of genetic mutation or damage. Some cancers have long been recognized to be partly hereditary. Chromosomal alterations are commonly found in cancerous cells, although these could be the result rather than the cause of the cancer. Failure of normal mechanisms of DNA repair is linked to cancers, as previously noted for *xeroderma pigmentosum*. Also, substances known to be mutagenic have been shown to be **carcinogenic,** or cancer-causing.

Nonetheless, thus far, only two different, although possibly related, genetic causes for cancer in humans have been conclusively identified: oncogenes and recessive mutations.

Oncogenes and proto-oncogenes

One of the most exciting discoveries in the history of genetics has been the identification of a group of genes that, when damaged, appear to be associated with cancer. About fifty of these genes, called **proto-oncogenes,** are now known. Proto-oncogenes are normal cellular genes whose function is probably related to aspects of cellular growth. In their normal condition, proto-oncogenes are an essential component of the genome of most, probably all, organisms.

Damage to proto-oncogenes may convert them to **oncogenes** (meaning tumor- or cancer-causing genes). The term *proto-oncogene* does not refer to the primary function of the gene in the cell, which is still unknown, but refers instead to the role of the gene as a precursor to a cancer-causing gene.

Exactly how oncogenes cause cancer is also still unknown, but could result in several ways. An oncogene could be continually active, so that its product is always being produced, or produced in excess. Alternatively, the oncogene could produce a protein with a different activity than the normal product. Either of these changes could lead to unlimited cellular proliferation or loss of the ability to differentiate.

Damage to proto-oncogenes that induces oncogenic activity can happen in several ways. The break point in *translocations* may occur in the middle of a proto-oncogene. This appears to explain the cause of chronic myelogenous leukemia, long known to be associated with the Philadelphia chromosome (figure 10.10). The reciprocal translocation involving chromosomes 9 and 22 relocates a portion of the *c-abl* proto-oncogene, in turn producing an enzyme with new functional characteristics. Some cancers are associated with chromosome deletions, which presumably expose recessive oncogenes (i.e., delete a normal dominant allele).

Gene amplification does not normally occur in mammalian cells, but is now known to affect portions of chromosomes bearing certain proto-oncogenes in particular cancers. In some human breast cancers, a proto-oncogene is amplified between two and twenty times. Recall that gene amplification is often associated with high levels of transcriptional activity, which could in turn produce excessive amounts of gene products leading to uncontrolled cell proliferation.

Lastly, *viral activity* is strongly implicated in human cancers via oncogenic action. Viruses are extraordinarily simple entities, consisting only of a small genome—five to several hundred genes—enclosed in a protein coat. Their life cycle can be quite simple (see chapter 11). Viruses inject their own genome into the cytoplasm of normal cells. The viral genes insert themselves into the host genome and then direct the host cell's metabolic machinery to produce more viruses, which are ultimately released from the host cell, continuing the viral life cycle.

Viral activity appears capable of causing cancer in several ways. First, insertion of the viral genome may induce mutations in proto-oncogenes, leading to oncogenic activity. Second, the powerful viral regulatory apparatus may induce hyperactivity in the proto-oncogene. Third, viruses themselves contain oncogenes that induce cancer when inserted into the host genome. In fact, oncogenes were first detected in viruses, and the discovery that human and other animal cells contain comparable and nearly identical genes (now called proto-oncogenes) was a major advance in unravelling the complex story of cancer. It is now generally accepted that viral oncogenes were originally derived from the genome of their hosts, a sort of primeval gene transfer.

Viruses have long been implicated in cancer. Cancers were experimentally induced by injecting viruses into laboratory animals as early as the 1950s. In 1981, **human T-cell leukemia virus (HTLV)** was identified as a cause of adult leukemia. Leukemia, one of the three major types of cancer, is characterized by production of abnormally large numbers of white blood cells that have reduced immune activity. Leukemic white blood cells do not function as normal antibody-producing cells. Instead, they accumulate in the circulatory system, disrupting normal circulation and in general disrupting normal body function. Leukemia is the most common childhood cancer, with about three thousand new cases of childhood leukemia diagnosed annually in the United States.

Acute leukemia is distinquished from chronic leukemia by the rapidity of onset and progress of the disease. Acute leukemia begins with cold symptoms accompanied by bone pain, easy bleeding, and high susceptibility to infections. Health deteriorates rapidly, and death usually occurs within three months of the first appearance of the disease. Chronic leukemia, on the other hand, generally requires several years to run its course. It is usually diagnosed as a result of a routine exam and has symptoms similar to acute leukemia.

A different type of virus, the **Epstein-Barr virus (EBV),** is the cause of **Burkitt's lymphoma,** a common cancer of African children. Burkitt's lymphoma causes tumors of the jaw and facial bones (figure 10.18) and affects the spleen and lymph nodes.

The discovery of oncogenes and their normal counterparts, proto-oncogenes, is extraordinarily important in understanding both the normal mechanisms of cell function and abnormalities such as cancer. It is already clear that the story is complex and that cancer is usually the result of several mutations or of other developmental events, rather than the result of a single mutation.

This new knowledge is already being used to diagnose cancers and to identify new treatments. Victims of cancer can now be assessed for specific karyotypic changes or for the presence of particular viruses. Understanding the function, and especially the gene products, of oncogenes will allow development of new treatments. The discovery of oncogenes thus confirms that cancer has a genetic cause and offers a genuine long-term prospect for successful prevention and treatment of its many varieties.

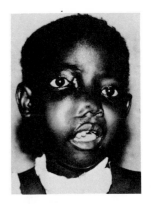

Figure 10.18 Child affected with Burkitt's lymphoma.
Source: National Cancer Institute.

Recessive mutations

Oncogenes appear to be activated by somatic mutations which actively induce cancer—that is, the genetic damage results in a gene product that initiates or participates in a sequence of events leading to cancer. Because oncogenic activity is somatic in origin, it does not explain cancers that are heritable.

Recessive mutations comprise a second general category of cancer-causing genes, which differ significantly from oncogenes in their mode of action. These mutations affect the germ line and thus are heritable. They appear to act by inhibiting gene activity. Individuals homozygous for these mutations are deficient in a gene product, and this loss of function leads to cancer.

Although recessive mutations are considered a separate type of cancer-causing genetic defect, they might be related to oncogenes. Such mutations might, for example, remove normal regulatory controls, producing unlimited oncogenic activity. Wilms' tumor, a cancer of the bladder in children, is associated with a deletion on the short arm of chromosome 11, suggesting loss of an important regulatory gene. Similarly, some cases of retinoblastoma, cancer of the retina in a child under five years of age, have been linked to a deletion affecting a single band on chromosome 13. It is possible that the condition is caused by deletion or point mutation affecting a single gene on this chromosome, but that the deletion is large enough to be visible only occasionally.

Cancer and society

It is often claimed that cancer is a peculiarly modern affliction and that cancer rates have increased throughout the last century because of the increasing exposure of humans to cancer-causing agents as a result of world-wide industrialization. This is a difficult hypothesis to evaluate for many reasons, and it is likely to be only partly true. In particular, claims of increasing rates of cancer cannot be fully verified. Diagnostic techniques are far better now than even a decade ago, and precise rates of particular cancers in the past century cannot be accurately known. Furthermore, cancer is still largely a disease of the old. Average life expectancies have nearly doubled worldwide in the last century, so it is hardly surprising that cancer and other degenerative diseases seem to be more common.

On the other hand, as we saw earlier in this chapter, the rate of genetic mutation can be significantly increased by exposure to many types of chemicals and radiation. We know that these agents may interfere with developmental pathways, resulting in teratogenic effects. Any increase in mutation rate could reasonably be expected to affect proto-oncogenes as well as the rest of the genome, thus having carcinogenic, or cancer-causing effects. Evidence is now clear that some environmental agents, often very common ones, cause cancer.

However, society's response to knowledge about cancer-causing agents is complex, especially when dangerous substances are extremely useful. Furthermore, establishing carcinogenicity, even for extremely toxic agents, is not always easy. For the remainder of this chapter we consider aspects of the identification and use of potentially carcinogenic or mutagenic agents.

Measuring mutagenicity and carcinogenicity

Assays for chromosome damage are difficult and expensive. Because everyone experiences low levels of normal chromosome damage due to age, exposure to sunlight, radiation, and other background mutagenic sources, assays for chromosome damage in a population exposed to a potential mutagen or carcinogen must be carefully conducted. First, the normal level of mutation must be determined in a *control population,* and then the test population must be examined for damage beyond that found in the control group. Such tests should also be *blind:* the geneticist examining the karyotypes should not know to which test group a particular karyotype belongs.

While mutagenicity does not necessarily reflect carcinogenicity, the extent of chromosomal damage is a widely accepted measure of the extent of *exposure* of a test population to potentially harmful environmental agents. Chromosomal damage, however, is meaningful only on a population basis. A particular individual can have many chromosome abnormalities that are totally unrelated to exposure to a particular mutagenic agent. Furthermore, aberrant chromosomes in a test subject cannot be interpreted to mean that the individual will contract a genetic disease.

On the other hand, a high level of genetic abnormalities in a test population may mean that the population is at risk for subsequent development of particular cancers. Because most chromosome abnormalities cannot be directly linked to cancer, the use of information on chromosomal anomalies poses serious ethical questions.

Is it appropriate, for example, to inform a person with karyotypic anomalies that he or she has suffered genetic damage, when the anomalies cannot be directly linked to a specific medical problem? Remember that chromosomal abnormalities are found in many people with no obvious phenotypic effects. Remember, also, that many cancers have long **latent periods,** often ten to thirty years, between the time of exposure to a carcinogenic agent and the development of cancer. Many people would die of other causes prior to onset of cancer. For others, the psychological trauma of anticipating a disease, perhaps unspecified, that might or might not develop, would constitute mental torture as debilitating as the disease itself, despite the fact that they were otherwise healthy during the latent period. Imagined harm can often be more severe than actual physical harm.

Thus, the tests for mutagenicity and carcinogenicity must be conducted with extreme experimental rigor, and the results must be interpreted carefully and explained with sensitivity to the needs of the subjects.

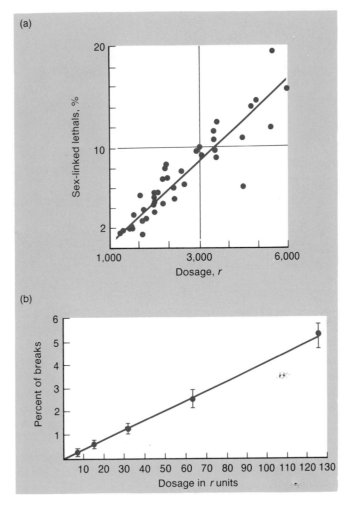

(a)

(b)

Figure 10.19 Increasing doses of X-radiation have been found to increase the number of gene and chromosomal mutations in experimental material:
(a) increased frequency of sex-linked lethal mutations in fruit flies with increasing dosage of X rays (summary of several studies);
(b) increased frequency of chromosome breakage in grasshopper cells exposed to increasing dosage of X rays.

Sister chromatid exchange (SCE) Currently, two types of karyotypic data provide information concerning the extent of exposure to mutagenic agents. The first is the observed level of chromosomal mutations: deletions, translocations, ring chromosomes, and so forth. For example, the number of mutations resulting from chromosome breakage increases in direct proportion to the dose of radiation an individual receives (figure 10.19).

A second indicator of exposure is the frequency of **sister chromatid exchange (SCE).** SCE is essentially a crossing-over between homologous sites on the sister chromatids of a duplicated chromosome (figure 10.20). SCE is not considered to be a mutation, because the exchange of material between identical chromatids is presumably reciprocal, leaving the chromatids genetically equivalent. SCE is not known to have any harmful medical effects, but a high frequency of SCE is associated both with certain pathological conditions and with exposure to certain chemicals. Furthermore, SCE is a more sensitive measure of exposure to mutagens than is the level of chromosome mutations; the level of SCE increases at doses that do not produce chromosome abnormalities. Thus although the genetic significance of SCE is not precisely known, it is a common and easily measured consequence of exposure to mutagens.

The Ames test Both chromosome mutations and SCEs are correlated with past exposure to toxic environmental agents. Obviously, it would be useful to assay for toxicity and, especially, carcinogenicity *prior* to introduction of chemicals into the

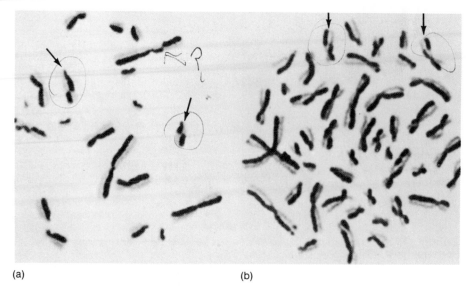

(a)   (b)

environment. This is not a trivial question. The American Chemical Society now lists over six million chemicals in its registry. About sixty-three thousand chemicals are used routinely in medicine and industry, and about one thousand more, never before known on earth, are synthesized in laboratories every year and made available for new uses. Which will cause genetic damage? Which will cause cancer? Deciding which chemicals to test, how to conduct such tests, and how much to spend on testing are major questions for business, government, and society.

Prior to 1974, tests for carcinogenicity in humans used laboratory mammals such as mice—an expensive, time-consuming process. Testing one chemical for carcinogenicity using mice or rats costs about $250,000 and requires three years. Maintaining just one mouse in a laboratory colony for a year costs three hundred to five hundred dollars. Tests with laboratory mice and rats are now used only on chemicals that are seriously suspected of being carcinogens; but these tests suffer basic limitations: chemicals that cause cancer in mice may not cause cancer in humans.

In 1974, Bruce Ames of the University of California–Berkeley developed a test using the bacterium *Salmonella typhimurium* to assay for the mutagenic effects of chemicals. The Ames test is rapid and inexpensive, requiring only two days and a small fraction of the cost of tests using animal models. It is now widely employed as a first estimate of the potential health hazards of new chemicals.

The **Ames test** uses a mutant strain of *Salmonella* that can only grow in a medium containing the amino acid histidine. To perform the test, this histidine-deficient strain of *Salmonella* is added to a medium containing *no* histidine. The chemical to be tested for mutagenicity is then added, along with an extract of enzymes from rat liver. The liver enzymes are added because some metabolic pathways in mammals differ from those of bacteria, producing metabolic by-products unique to mammals that may themselves be carcinogenic. Human metabolism may also produce unique by-products, but human liver extracts are not widely available (for obvious reasons). A set of synthetic human enzymes was developed to make the Ames test more directly applicable to human metabolism.

This concoction is incubated for two days and then examined for bacterial growth (figure 10.21). The only bacteria to grow should be those that have undergone a mutation permitting them to produce their own histidine. The more bacterial colonies present, the greater the number of mutations that have occurred. Because mutations occur at random, most mutations will have affected genes other than the one or ones controlling histidine synthesis. Thus, the observed mutations represent only a small fraction of the total; the Ames test provides a conservative estimate of the total mutational effect of a chemical.

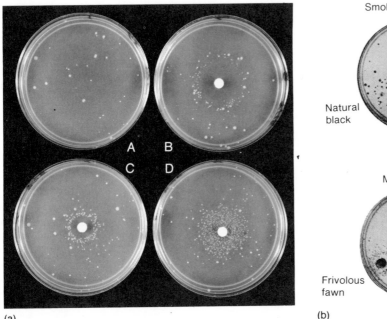

(a)

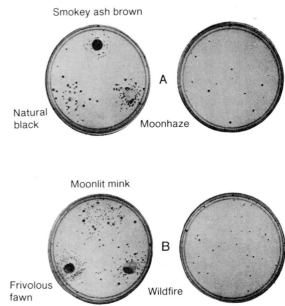

Smokey ash brown

Natural black

Moonhaze

A

Moonlit mink

Frivolous fawn

Wildfire

B

(b)

Figure 10.21 The *Salmonella* assay of the Ames test. (a) Series of plates with (A) control with no mutagen, indicating spontaneous back mutation rate; known mutagens: (B) furylfuramide, (C) aflatoxin, and (D) 2–aminofluorene. (b) Plates exposed to three different hair dyes (A) without liver enzymes and adjacent control plate; (B) with liver enzymes and adjacent control plate. Note increased mutation in B. Source: (a and b) Bruce N. Ames.

The results of one medically and commercially important Ames test are shown in figure 10.22. From 1973–1977, the chemical tris (or, more precisely, tris-2, 3-dibromopropyl-phosphate) was routinely added to children's clothing, especially pajamas, as a fire retardant. To test for mutagenicity, a series of Ames tests using varying amounts of tris was conducted: the larger the dose of tris used, the more mutant colonies of *Salmonella* that were found. Tris is absorbed directly through the skin, so protecting children from fire was also exposing them to doses of a mutagenic and potentially carcinogenic chemical—a truly modern dilemma. Tris metabolites were also detected in the urine of children who had worn tris-impregnated pajamas, even after the pajamas had been washed many times. Needless to say, tris is no longer used in children's clothing.

The Ames test is a direct measure of mutagenicity, not carcinogenicity. However, mutagenicity on the Ames test correlates strongly with carcinogenicity in humans. A positive result for mutagenicity on the Ames test is generally accepted as an indication of carcinogenicity. Over 90% of all chemicals known to be carcinogenic in humans have given positive results on the Ames test.

Are more direct measures of carcinogenicity possible? Perhaps, but as with tests using mammalian models, such tests might be prohibitively expensive. The only direct test would use human subjects, which is ethically unacceptable. A further complication is the long latent period shown by most cancers. Is it more reasonable to use the Ames test to estimate toxicity of a new chemical, or to expose thousands of persons, then wait decades to see if they develop cancer? Present practice uses the Ames test as a first step in evaluation. If a chemical that may be widely used shows mutagenicity on the Ames test, subsequent tests on laboratory mammals can be ordered to confirm the Ames result.

Some examples of common chemical carcinogens and mutagens: recreational drugs

Mutations may be either germinal or somatic. In the long term, germinal mutations are the most troubling because their effects could persist for many generations. However, somatic mutations are increasingly implicated as a cause of cancer. Cancer is now the second leading cause of death in the United States, behind heart disease. In 1981, about 800,000 persons were diagnosed as having cancer, and about 430,000

Figure 10.22 Data from an Ames test on the mutagenicity of the flame retardant, tris, compared to other compounds. All tests used *Salmonella* strain TA100. Note magnitude of revertants (mutations) when exposed to tris.

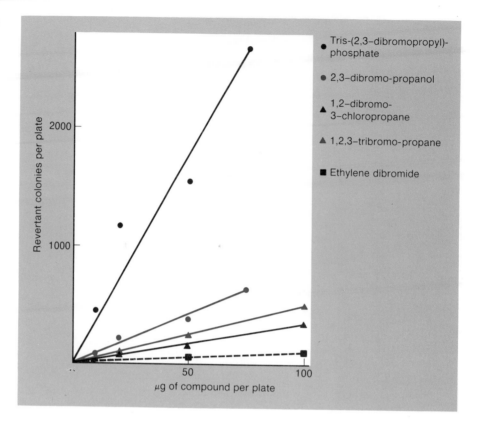

persons died of cancer. Cancers of the lung, breast, and colon account for more than 50% of the cancer-related deaths in the world. While significant progress has been made in diagnosing and treating many cancers, the incidence of some cancers has increased over the past twenty-five years, perhaps due to the increasing levels of toxic chemicals in the environment.

Tobacco   The most dramatic increase in cancer since World War II has involved cancer of the lung, primarily because of the increased rate of cigarette smoking. The scientific evidence concerning the harmful effects of tobacco is abundant and unequivocal. In 1982, the Surgeon General of the United States issued a report saying, "Cigarette smoking is clearly identified as the chief preventable cause of death in our society and the most important public health issue of our time."

There are 350,000 smoking related deaths in the United States each year. This is the equivalent of two fully loaded 747 jumbo jets crashing and killing all aboard every day each year. Of the more than 111,000 deaths from lung cancer, 85% could have been prevented had the victims not smoked. A smoker is one thousand times more likely to contract cancer than a nonsmoker.

Death is not the only cost to society. Medical care for smoking-related illnesses costs about thirteen billion dollars annually, and lost productivity and wages cost twenty-five billion dollars. Society bears a massive economic burden due to the activities of smokers. Even the health problems caused by cigarette smoke are not limited to smokers themselves. Children of mothers who smoke during pregnancy often have low birth weight. Children who live with a smoking parent are at increased risk of respiratory problems. The nonsmoking spouse of a smoker is estimated to have about 30% higher risk of contracting lung cancer than the spouse of a nonsmoker. About 20% of all deaths from lung cancer among nonsmokers are estimated to be caused by environmental tobacco smoke.

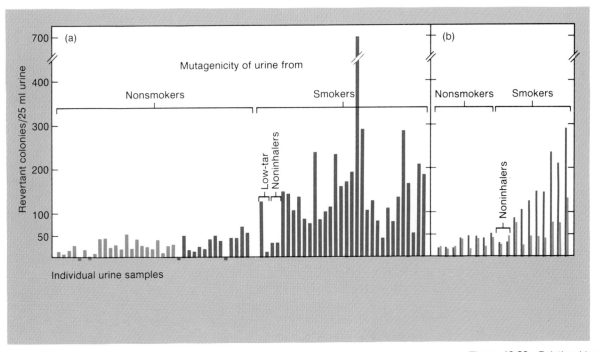

Figure 10.23 Relationships between number of revertant colonies on Ames test (number of mutations) and smoking, as determined by exposure to urine from smokers and nonsmokers. Dark color represents urine collected in the evening; light color represents an early morning specimen: (a) single samples from individuals and (b) paired morning and evening samples from other individuals.

Cigarette smoke contains over sixty-eight hundred different chemical compounds, many of them highly toxic. Included on the list are hydrogen cyanide (which causes emphysema), vinyl chloride, arsenic, the radioactive metal polonium, and nicotine. Urine from smokers is highly mutagenic on the Ames test (figure 10.23), probably explaining the high rate of bladder cancer among smokers.

The substance of most concern in tobacco is nicotine. Nicotine, an alkaloid, is an extremely potent natural poison. A few drops of pure nicotine absorbed through the skin are fatal. A grim fact: before World War II more tobacco was grown in the United States for use in insecticides than for cigarettes. Today, nicotine is used by veterinarians as an external parasiticide.

The evidence relating smoking to cancer is as strong as the evidence for any toxic substance. The tobacco industry, however, is so powerful a political force that the scientific evidence and the immense costs of smoking to society are ignored. Despite the Surgeon General's conclusion that cigarette smoking is the "chief preventable cause of death" in the United States, the government does little to protect the public from the harmful effects of smoking: warning labels are required on cigarette packages and television advertising of cigarettes is restricted. Advertising of cigars, pipe tobacco, and chewing tobacco is not limited. In fact, the government provides a massive financial subsidy to the tobacco-growing industry.

The industry attempts to defend itself by perpetuating the Big Lie, a technique described in George Orwell's *1984*. Tell a lie so big and outrageous that it cannot possibly be true, and tell it often enough, and soon the public will come to believe it. Says the chairman of the Tobacco Institute, "Scientific research has not been able to establish [the causal] link [between smoking and cancer]." In fact, no causal link between carcinogen and cancer is more firmly proven.

Marijuana *Cannabis sativa* (marijuana, pot, grass, or hashish) is the most widely used drug among adolescents. The health effects of marijuana have been seriously debated during the past two decades.

In 1982, the National Academy of Science's Institute of Medicine released a comprehensive review of all data relevant to the health effects of marijuana. The evidence is unequivocal that smoking marijuana leads to respiratory problems similar to those caused by smoking cigarettes. There is not yet evidence, however, to relate marijuana smoking to increased risk of lung cancer. Two reasons may explain the latter finding. First, lung cancer has a long latent period. As marijuana has been widely used in the United States only since the 1960s, sufficient time may not yet have elapsed for the increased incidence of lung cancer among marijuana users to become evident. Second, marijuana users probably smoke fewer marijuana cigarettes than tobacco smokers smoke tobacco cigarettes. Although marijuana smoke has about 50% more carcinogenic material than tobacco smoke, marijuana users may be subjected to smaller doses of carcinogens, thus reducing the relative risk of lung and other cancers. In such a case, a proportionately larger amount of data would be required to establish a carcinogenic role for marijuana.

Marijuana use may also produce temporary sterility, particularly in males. Marijuana suppresses testosterone production, in turn inhibiting sperm production. It interferes with both DNA and protein synthesis. In laboratory mice, marijuana use decreases the size and weight of the prostate and testes. Spermatogenesis is impaired, with many ring chromosomes and translocations visible in gametes. Additionally, among offspring of mice subjected to marijuana treatment, both reduced fertility and meiotic chromosome abnormalities were detected, suggesting that marijuana causes germinal as well as somatic mutations.

Furthermore, the active ingredient in marijuana crosses the placenta and can affect a fetus. It is also secreted by the mammary glands and may thus affect nursing infants. Evidence indicates that marijuana suppresses the activities of the immune system. Marijuana impairs motor response and short-term memory, and it reduces learning capacities. The effects on the nervous system last four to eight hours past the feelings of intoxication, considerably longer than the effects of alcohol.

The evidence is strong that marijuana has many potentially harmful effects on health and may also be mutagenic.

Caffeine Caffeine is a stimulant of the central nervous system. It occurs naturally in coffee, tea, and other plants, and it is an ingredient in soft drinks and aspirin. Although caffeine has no analgesic or pain-relieving effects, it is used in aspirin for its stimulating effect on the central nervous system, increasing alertness and apparently making a person feel better. Caffeine is an habituating drug, but withdrawal is achieved far more easily than with narcotics.

Caffeine is a base analogue (figure 10.11), suggesting that it is likely to be mutagenic. It has been shown to be mutagenic in bacteria. In eukaryotes, it breaks chromosomes only during replication, and its mutagenic effects have yet to be shown as an important health problem. Caffeine crosses the placenta easily, reaching the same concentration in the fetus as in the mother. Recent studies have given conflicting results concerning potentially harmful effects on fetuses. Rats fed the caffeine equivalent of two cups of coffee per day produced offspring with slowed rates of skeletal growth, and a ten-cup-per-day equivalent reduced litter sizes.

Lysergic acid diethylamide (LSD) In 1967 chromosome studies of users of LSD revealed translocations and chromosome breakage. The subjects of these investigations, however, had used the drug illicitly; therefore, these findings must be considered at best circumstantial and not the result of controlled experiments. For example, the amount of chromosome damage prior to use of LSD is unknown. Drugs obtained illicitly often include many impurities, which may also be mutagenic. Furthermore, drug users often mix drugs, and the combined effects of several toxic chemicals are unpredictable and often extremely severe.

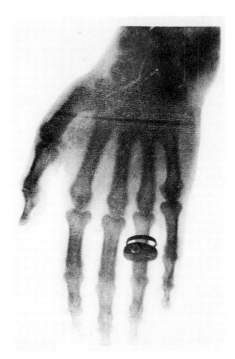

Figure 10.24 One of the first X rays taken, this was done in 1898 by Wilhelm Roentgen, who discovered X rays in 1895.
Source: The Bettmann Archives.

Recent controlled studies still leave the mutagenic capacities of LSD in doubt. Moderate doses of pure LSD have not produced chromosomal mutations in humans. Large doses of LSD are weakly mutagenic in *Drosophila* and fungi; LSD is probably not significantly mutagenic in concentrations that would normally be taken by humans. There is no evidence that LSD is carcinogenic.

Radiation, mutation, and cancer

In 1921, when almost nothing was known of the nature or causes of mutations, **H. J. Muller** said, "Beneath the imposing structure called Heredity there has been a dingy basement called mutation." For the remainder of his professional career, Muller studied mutation. His work on the mutagenic effects of radiation, for which he received the 1946 Nobel Prize for Physiology or Medicine, comprised the first detailed examinations of a mutagenic agent.

X-radiation was discovered in 1895 by Wilhelm Conrad Roentgen. He and his colleagues used X rays indiscriminately to examine the skeletons of patients (figure 10.24). Within a few years some of these scientists and physicians lost their hands and, in some cases, their lives as a result of radiation burns. Despite the early awareness of the potential dangers of X rays, their use was widespread. Until the 1950s, many shoe stores in the United States used fluoroscopes, a form of X ray, to check the fit of new shoes. The devices were so interesting that many children were encouraged by their parents to X ray their feet for entertainment while the rest of the family shopped.

Other forms of radiation were only slowly recognized as dangerous. Radium, a radioactive isotope formed by the decay of the element uranium, was discovered by Marie Curie in 1898. Like X rays, radium was treated more as an innocuous novelty than as a dangerous substance. Between 1915 and 1930, thousands of persons in the United States drank radium as a medicinal substance. In Europe, radium-containing chocolate bars were marketed commercially. As recently as 1953, an American company sold a contraceptive jelly that contained nearly a microgram of radium in a two gram tube. Today, some persons still travel to radon mines in the Rocky Mountains to "take the cure." Radon is a gas formed by the disintegration

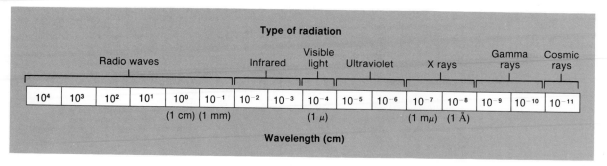

Figure 10.25 Spectrum of electromagnetic waves from long radio waves to very short cosmic rays.

of radium. As humorous as some of these examples may be, the consequences are not always trivial. Cancers of the jaw bone, face, and throat were developed by many women who painted radium on the dials of wrist watches. These women often held the paint brushes in their mouths, receiving large cumulative doses over many years.

Aware of some of these dangers, Muller demonstrated in 1927 that X-radiation increases the mutation rate in *Drosophila*. Within a year, X rays had been shown to increase the mutation rate in plants as well. These observations led Muller to introduce the concept of **genetic load**. Genetic load is a measure of the amount of deleterious mutant genes in a population. Muller believed that increased radiation exposure could significantly increase the genetic load carried by humans. Since most mutations are recessive, their harmful effects would at first be hidden, but at increased frequencies the harmful alleles would eventually be expressed, having serious long-term consequences for our species.

By 1930, radiation genetics was recognized as an important aspect of genetic research. In the half century since Muller began his work, the role of radiation in causing both mutations and cancer has been well established.

Types of radiation

Two types of high energy radiation are genetically important (figure 10.25). *Electromagnetic radiation* is similar to visible light, moving in waves. Examples include microwaves, ultraviolet light, and X rays. Some forms of electromagnetic radiation, such as radio waves, appear to be of little significance genetically, but UV light and X rays contain much higher energy and can cause considerable damage. Because of its ultraviolet component, sunlight is the most potent natural carcinogen, causing about one-third of all tumors in humans. Fortunately, almost all of these are benign skin tumors that can be removed by surgery. This type of tumor is treated with great success, so it is usually omitted from cancer statistics.

Particulate radiation is even higher in energy than electromagnetic radiation. Radioactive elements emit high energy particles (protons, neutrons, and electrons) that can penetrate the human body. Electromagnetic and particulate radiations release energy as they penetrate living tissue, causing reorganization of the tissue's atomic structure. One effect is the alteration of DNA, genetic mutation.

Various units are used to measure amounts of radiation. A **rad** describes the amount of energy absorbed by living tissue. A **rem** (roentgen equivalent man) measures the biological damage done by radiation. For small doses, rads and rems are roughly equivalent. Very small doses are often measured in millirads or millirems, $\frac{1}{1000}$ of a rad or a rem. The exposure and absorption of radiation are seldom distributed equally over the entire body, as illustrated by the women painting radium on the faces of watches. The **genetically significant dose (GSD)** is the amount of radiation absorbed by the gonads, which might produce germinal mutations.

Table 10.3
Radiation doses and units

Rad = amount of radiation absorbed by tissue
Rem = amount of biological damage done by radiation (rem = roentgen equivalent man)
GSD = genetically significant dose; the amount of radiation reaching the gonads

Source of radiation	Average dose in mrem/year (range)
Natural	
Cosmic	45 (30–130)
Terrestrial	60 (30–115)
Internal radionuclides	25
Total	130 (Unavoidable)
Man made	
Global nuclear fallout	4
Nuclear power	0.003
Medical diagnostic	72
Radiopharmaceutical	1
Occupational	0.8
Miscellaneous	2
Total	80
Grand total	210

Sources and effects of radiation

Exposure to radiation may come from natural or man-made sources (table 10.3). Natural or background radiation comes from sunlight and other cosmic radiation, radioactive isotopes in geological formations, and radioactive substances in food. Doses of cosmic radiation vary with altitude. The earth's atmosphere filters cosmic radiation efficiently, but at high elevations the atmosphere is much thinner than at sea level, increasing the exposure to radiation. A person living at sea level is exposed to about 30 mrem per year, but in mountainous regions exposure may be as high as 110 mrem per year.

Ultraviolet radiation is filtered from sunlight by the ozone in the stratosphere, a layer of the atmosphere found between ten and thirty miles above the earth's surface. In the early 1970s, scientists focused public attention on the potential depletion of this ozone layer by the increasing release into the atmosphere of a group of chemicals known as chlorofluorocarbons (CFCs), widely used in the past few decades as coolants in refrigerators and propellants in aerosol sprays. CFCs interact with stratospheric ozone, converting it to molecules of oxygen, which do not filter ultraviolet radiation from sunlight.

These early warnings were greeted with skepticism, but in 1978 the use of CFCs in spray cans was banned in the United States, though not elsewhere. Convincing evidence is now accumulating that a significant reduction in stratospheric ozone has begun and that CFCs are at least partly responsible. In particular, a so-called "ozone hole," or thinning in the concentration of ozone, has been detected in the atmosphere over Antarctica each austral spring during the 1980s. By 1986, the levels of ozone over Antarctica were reduced by about 50%, although they recover to about normal levels during austral summer. Other evidence indicates that smaller reductions in stratospheric ozone are occurring elsewhere in the world.

The consequences of worldwide reductions in ozone could be severe. Ultraviolet radiation causes skin cancer and may contribute to other health problems, such as cataracts. Decrease in the ozone concentration by only a few percent could dramatically increase the incidence of skin cancers in humans. Domestic livestock could be expected to be adversely affected, as well.

An international agreement in 1987 initiated a program for reducing CFC use and production among developed nations, but it is a long-term plan and does not aim for total elimination of CFCs. The program also does not apply to the developing nations of the world, most of which lie in tropical regions and have a great need and demand for refrigeration. If dramatic increases in skin cancer occur, society will be forced to balance the production of these useful chemicals against their harmful effects. Whether the depletion of ozone could be reversed is unknown.

The amount of radiation derived from natural terrestrial sources varies substantially. Some mountainous regions of China have levels as high as 230–440 mrem per year; and in one region of India background radiation approaches three thousand mrem per year.

In general, radiation derived from man-made sources is less than natural radiation (table 10.3). Radiation from global nuclear fallout has decreased since the abolition of above-ground testing of nuclear weapons by the United States and the Soviet Union. France and China, however, continue above-ground tests. In some cases, populations immediately downwind from nuclear test sites have received large doses of radiation, but patterns of global weather circulation insure that eventually the entire earth receives increased radiation from nuclear testing. Radioactive isotopes of the metal strontium are a byproduct of nuclear explosions. These isotopes fall on vegetation, where they are consumed by cattle, then transmitted to humans in cow's milk. Strontium replaces calcium in bones and thus accumulates in human tissue.

Medicine is the primary artificial source of radiation. X rays are an important diagnostic tool, and radiation is used to combat cancers. As the efficiency and precision of X-ray machines have improved, exposure levels have decreased.

The harmful effects of excessive exposure to radiation are well documented. In high doses, radiation causes burns, nausea, loss of hair, anemia, leukemia, sterility, cancer, and genetic mutation. In general, the severity of these effects increases in direct proportion to the dose received. At low doses, the effects are more difficult to document, and considerable controversy exists concerning the existence of a safe lower limit of radiation. Is a certain small amount of radiation exposure tolerable? Or does any exposure, however small, significantly increase the risk to health? While available data are inconclusive, some evidence indicates that, in general, the mutation rate is roughly constant for all organisms, when adjusted for size of the genome (figure 10.26). Thus, because mammals have a comparatively large amount of DNA, the total number of mutations induced by a constant amount of radiation would be greater in humans than in bacteria and other organisms with small amounts of DNA.

At present, the United States government has established 170 mrem per year as the upper limit of acceptable exposure to man-made sources such as X rays and radioactive material. It is estimated that if every person in the United States received the maximum allowable amount (170 mrem) there would be thirty-two thousand additional cancer deaths annually.

For comparative purposes, toxicity is often expressed in terms of an **LD50:** the dose of an agent that is lethal to 50% of a test population or species. For humans, the LD50 of radiation is estimated to be 450 rads. Mammals appear to be among the most susceptible species (table 10.4).

The genetic damage caused by radiation includes chromosome breakage and mutation, sister chromatid exchange, and point mutation. In combination with other types of radiation damage, these genetic changes may lead to cancer, particularly cancer of the blood and blood-forming organs, i.e., leukemia. Leukemia is the most common form of cancer associated with exposure to X-radiation. In the first half of this century, X-raying fetuses was a common medical practice. The children of mothers X-rayed during pregnancy have a 50% higher incidence of leukemia than

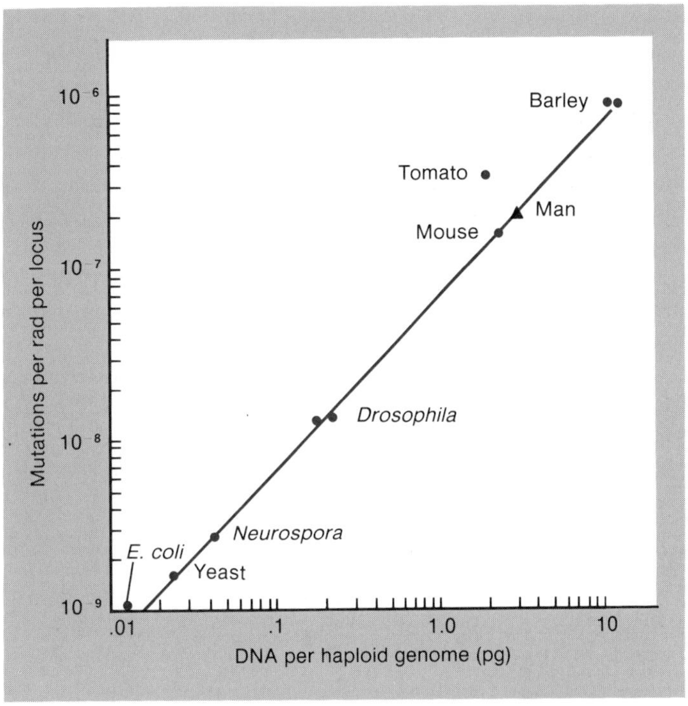

Figure 10.26 Mutation rate in a variety of organisms as a function of genome size and the exposure to radiation.

Table 10.4
Comparison of LD50s among different mammalian species

Organism	LD50 (rads)
Human	450
Monkey	600
Sheep	250
Dog	250
Rabbit	800
Rat	700
Mouse	800
Guinea pig	450

From V. P. Bond, T. M. Fliedner and J. O. Archambeau, *Mammalian Radiation Lethality.* Copyright © 1965 byAcademic Press, Orlando, FL. Reprinted by the permission of Academic Press and the author.

children of mothers not X-rayed. Leukemia has a relatively short latent period, and one of the first effects noted among survivors of the bombings of Hiroshima and Nagasaki was an increased incidence of leukemia.

Other cancers may appear later, of course. Long-term medical tests have been conducted on residents of the Marshall Islands, who received large radiation doses after a test nuclear explosion on the Bikini Islands in 1954. Following the explosion, the wind shifted and unexpectedly carried the fallout to the Marshall Islands. Some of the residents received doses estimated at 175 rads, only slightly less than *half* the LD50. All became ill, but subsequently recovered. Ten years after exposure, over 50% of the islanders still had visible chromosomal mutations in the white blood cells, and five years later a slight increase in incidence of benign tumors of the thyroid was detected.

The Marshall Islanders received a single massive *acute dose,* but because the effects of radiation are cumulative, a series of small or *chronic doses* may be equally damaging. About five hundred pounds of radioactive plutonium are processed annually in Rocky Flats, Colorado. Thus, the residents of Rocky Flats are chronically exposed to radiation. The incidences of both leukemia and lung cancer among the residents of Rocky Flats are about double those found in a nearby control community.

Some of the best data relating to the effects of chronic exposure to low doses comes from regions of high background radiation. In the region in China where the dose of background radiation is as high as 440 mrem annually, studies of large samples of the population detected no significant increase in frequencies of chromosomal damage, incidence of hereditary diseases and congenital malformations, spontaneous abortion rate, or frequencies of various cancers. Only a higher than normal incidence of Down's children suggested the possibility of genetic damage caused by high background radiation.

On the other hand, studies of populations in India living in regions of background radiation about ten times greater than that in China (as high as three thousand mrem annually) detected significantly increased incidences of chromosomal aberrations, Down's syndrome, and severe mental retardation. These problems can reasonably be attributed to the high levels of radiation.

Hiroshima and Nagasaki

The survivors of the atomic bomb explosions at Hiroshima and Nagasaki have been the subjects of the most comprehensive epidemiological studies ever conducted. Since 1945, more than one hundred million dollars have been spent to follow the medical histories of over one hundred thousand persons. Initially, the survivors were examined for the immediate medical effects of the explosion and radiation exposure. During the 1960s, however, karyological techniques for humans were developed that permitted expansion of the studies to genetic effects. (Remember, the correct diploid number for humans was not known until 1956.)

The studies of survivors and their descendants are continuing. Specific doses of radiation have been calculated for each survivor on the basis of their location at the time of the explosion, and the dose estimates have been correlated with the probability of developing certain cancers.

The incidence of cancer is clearly related to the dose of radiation among the survivors of the nuclear explosions. Survivors have sustained rates of cancer far higher than normal, so great in fact that they constitute a cancer epidemic. Furthermore, the data clearly show the variation in latent period associated with different cancers. During the late 1940s and the 1950s, heavily exposed persons developed leukemia at a rate thirty to forty times higher than normal, but the rate decreased almost to normal by the early 1970s. In the late 1950s cancers of the lung and thyroid gland were in high frequency. In the 1960s and 1970s cancers of the breast and salivary glands increased. By 1980, rates of multiple myeloma and cancer of the colon had increased. Exposure to a single high dose of radiation has obviously altered the probability of contracting many types of cancer throughout the entire *lifetime* of the survivors. In 1987 a reassessment of the radiation received by survivors of the bombings was completed. These results indicated that the elevated cancer rates seen in this population were caused by *smaller* doses of radiation than was previously believed. This means cancer risk from radiation is higher than prior estimates and consideration is being given to lowering the maximum allowable radiation exposure by 30–70%.

Genetic studies begun decades after the explosion also reveal residual effects. Karyotypes of survivors show that chromosome fragments and ring chromosomes are common, in spite of the expectation that these abnormalities would not have survived or been transmitted to descendant cells for so many years. Although, again, the genetic changes cannot be proven to have caused the cancers, the associated mutagenic and carcinogenic effects of radiation are abundantly clear.

There are several notable positive aspects of the data from Hiroshima and Nagasaki. First, no latent pathologies other than cancers have yet been shown to occur at higher than normal levels. Second, the harmful effects of the radiation seem limited to the survivors themselves and do not extend to their children. Incidences of death, congenital abnormality, and point and chromosomal mutations are not above normal in the children of the survivors. It was initially feared that genetic mutations resulting from radiation would be passed on for generations, but such does not appear to be the case.

Such findings, however, should not be interpreted to mean that germinal mutations did not happen, or that reproduction of survivors has not been deleteriously affected. The high incidence of somatic mutations almost certainly means that germinal mutations did, in fact, occur. The long-term effects of germinal mutations may have been limited because gametes possessing numerous deleterious mutations were probably selected against. That is, affected gametes either were not viable, or they were so seriously affected that they were relatively unlikely to participate in a successful fertilization. Because most mutations are recessive, however, it is possible that higher than normal levels of germinal mutations have been transmitted but not expressed in offspring of survivors. These mutations would increase the genetic load of the population as a whole. Increased levels of harmful mutations would be expected to be expressed as higher numbers of "genetic deaths" in succeeding generations. Although no evidence yet indicates harmful effects on children of survivors, long-term harmful effects on the Japanese gene pool cannot be ruled out.

Finally, it is worth noting that many survivors (called *hibakusha,* meaning explosion-affected persons) have not married or reproduced. Failure to marry has resulted from disfigurement, injury, or cancer, but in some cases fear of producing genetically damaged children may also have been a reason.

The politics of toxic substances

Of the thousands of chemicals in our daily lives, many are mutagenic, carcinogenic, or teratogenic. Many of these (caffeine, tobacco, and alcohol, for example) are widely used. Other common substances are equally toxic, although their harmful effects are less publicized: formaldehyde (used in household insulation and biological preservatives), toners (used in photocopy machines), hair dyes, and many medical drugs, for example.

Governments throughout the world regulate such toxic substances to varying extents. The decision to regulate depends not only on scientific evidence, but on social and political considerations as well. Many toxic substances have beneficial uses, and their regulation involves balancing the positive and negative aspects of their use. No better example exists than X rays. Radiation can cause cancer, but medical benefits warrant continued use. Not only are X rays a major diagnostic tool, but radiation is an important weapon against cancer. Rapidly dividing cancer cells are more seriously affected by X rays than are normal, slowly dividing cells, so a concentrated dose of well-aimed radiation may selectively kill a growing cancer. Similarly, caffeine is toxic only in amounts far greater than most people ingest in coffee, tea, or soft drinks, so their positive effects seem to outweigh any gain to be derived from restriction of caffeine in beverages.

In reality, of course, political decisions may ignore health considerations. Powerful economic interests underlie the continued tolerance of substances such as tobacco, the single greatest cause of cancer in the United States. Its health and economic cost to society is immense, but tobacco is not restricted. Even worse, the tobacco industry is supported by large government subsidies. Similarly, powerful lobby groups were able to forestall action by the Environmental Protection Agency to restrict use of formaldehyde, a highly carcinogenic substance.

We are not suggesting that use of all toxic substances should be severely limited. That would be unwise and probably also impossible. We do believe, however, that the interests of governments and various political and economic groups do not always coincide with the health interests of the public. The carcinogenic effects of atomic radiation were poorly understood throughout the 1940s and 1950s. As a result, not only the unfortunate survivors of Hiroshima and Nagasaki, but also many American, British, New Zealand, and Australian military personnel and civilians, as well as Pacific Islanders, were exposed to nuclear test radiation. These exposures occurred primarily due to our ignorance of the genetic and medical risks. Even after knowledge advanced, the government was often slow to assume responsibility for its actions or even to admit them.

As genetic knowledge increases, the dangers of many environmental agents will become even clearer. Appropriate regulation of such agents requires a knowledgeable citizenry to insure that decisions are founded on scientific data relating to health risks, and not based on political and economic expediency. The American political system appropriately allows open debate of public issues such as the regulation of toxic or potentially toxic substances. Both sides of arguments may be forcefully expounded in public. In some cases, such as the regulation of sodium nitrite, even professional scientists and toxicologists differ in their opinions concerning the need for regulation. Nitrites indeed increase the risk of cancer in laboratory mammals, but when used as meat preservatives, they reduce the danger of transmitting food poisoning and other bacterial diseases. Which danger should we choose?

In other cases, propagandizing relies not on evidence, but on irrelevant claims that cloud consideration of the issue at hand. For example, the evidence is clear that cigarette smoke has many serious affects on the health of smokers and non-smokers—and no beneficial effects. The Tobacco Institute therefore cannot cite positive aspects of cigarette smoking and is forced to rely on a fraudulent issue: that smoking does not affect others and should therefore be a matter of choice. In addition to concern regarding the health risk created by cigarette smoke, many non-smokers find smoke foul smelling and irritating to their eyes. Some nonsmokers are allergic to cigarette smoke—not surprising, considering the thousands of compounds spewed into the air by burning cigarettes. Smoking clearly affects non-smokers, yet the industry obscures the issue by focusing on the freedom of choice of smokers.

A second example illustrates the same point. In the late 1970s, *dioxin,* one of the most toxic chemicals known, was implicated as a cause of spontaneous abortions, birth defects, and cancer. Dioxin is so toxic that no laboratory experiment has been able to determine a dose so small that it is not toxic—even one part per billion. Because dioxin occurs as a contaminant in several herbicides commonly used by wood product companies, these companies and the chemical manufacturing companies were subjected to considerable public scrutiny and adverse publicity concerning the use of dioxin-containing chemicals in the environment. One major corporation conducted an advertising campaign observing that chemicals are the basis of life, while showing some of the chemicals used by the company to increase the productivity of managed forests.

One such advertisement promoted the virtues of an all-purpose chemical called "dihydrogen oxide"—H_2O, water. The implication was that if this corporation valued water so highly, then all other chemicals it used must be equally benign. Dihydrogen oxide belongs in the same list of good things as baseball, hot dogs, apple pie, and mom. In other words, the corporation was saying, "Trust us—we know what's best." Such commercials are clever and often persuasive. Only an educated and informed public can evaluate how much truth is contained in such propaganda.

Summary

1. Mutations are heritable changes in DNA. They may be somatic, affecting somatic tissues and transmissible only to descendants of a particular cell within the body, or they may be germinal, affecting the gonads and transmissible to succeeding generations. Germinal mutations may produce long-lasting effects on the species, while somatic mutations often have harmful effects on the health of the bearer.

2. Most mutations probably have harmful effects on survival and reproduction. However, mutations also provide the ultimate source of genetic variation and thus play a fundamental role in evolution.

3. Variations in ploidy are common in plants and some animals. The gamete stage of life history is ordinarily haploid for all species, and haploidy may also be a mechanism of sex determination. Polyploids often possess characteristics of great commercial value, and many domestic crops are either artificial or natural polyploids.

4. Polyploidy in animals is relatively infrequent. Triploid and tetraploid human fetuses occur, but never survive.

5. Aneuploidy, the deficiency or excess of one or a few chromosomes, is relatively common and occurs as a result of meiotic or mitotic nondisjunction. In humans, monosomy for the X chromosome, Turner's syndrome, is the only viable monosomy. Trisomies are relatively more viable than monosomies, but they are usually associated with physical and mental abnormalities. Only trisomies involving the sex chromosomes and trisomy 21, Down's syndrome, survive to adulthood.

6. Some mutations affect only portions of one or two chromosomes. Deletions and duplications involve loss or gain of a segment of a chromosome. Inversions occur when a segment of a chromosome is reversed in position within the chromosome. Translocations occur when segments of one chromosome attach to a non-homologous chromosome. Most mutations of these types are deleterious, and a few have been linked to specific diseases.

7. The effects of chromosomal mutation depend upon the genes affected. For example, Down's syndrome may be caused either by an extra chromosome 21 (trisomy) or by an extra long arm of chromosome 21, translocated to another chromosome and inherited from one parent (translocation Down's). Thus, the condition is caused by a gene or genes carried on the long arm of chromosome 21. Similarly, Turner's syndrome may be caused by loss of one X chromosome or by deletion of only the short arm of one X.

8. The Philadelphia chromosome (Ph1) provided the first link of a chromosomal mutation to a cancer.

9. Point or gene mutations affect only a small segment, sometimes only a single base, of a chromosome. Point mutations may be either frameshift or substitutional.

10. Mutations are random. Neither their location nor their result can be predicted. However, the rate with which they occur is relatively constant— about 0.1–1.0 mutations in every one hundred thousand generations of each gene. Each human probably carries about three to five lethal new mutations, which are not expressed because they are recessive.

11. The natural rate of mutation is affected by many different environmental agents. Various forms of radiation and chemicals are mutagenic, causing significant increases in the mutation rate.

12. Cells are adapted to a certain level of background mutation. Most mutations are probably fixed immediately by mechanisms of DNA repair.

13. Nonetheless, mutations are probably involved in many health problems, especially aging and cancer.

14. The mutation theory of aging hypothesizes that aging results from accumulated deleterious mutations, producing a steady degeneration of bodily function. The commitment theory of aging claims that development is under the direct control of genetic programs, which include maximum limits on life span.

15. Teratogens are substances that interfere with normal embryonic development. During the critical first trimester of pregnancy, chemicals that might be otherwise harmless can cause severe abnormalities in development.

16. Alcohol has been shown to cause teratogenic changes in the developing embryo, resulting in a suite of features in the newborn known as fetal alcohol syndrome (FAS). Mental retardation and stunted growth are key features of FAS.

17. Cancer cells are characterized by uncontrolled growth, lack of contact inhibition, and ability to spread throughout the body, metastasis.

18. Based on the tissues involved, three types of cancers are recognized: carcinomas, cancers of the skin, glands, reproductive tract, etc.; leukemias, cancers of the blood-forming tissues; and sarcomas, cancers of the bones and muscles.

19. Proto-oncogenes, normal cellular genes involved in growth, can be converted to oncogenes, cancer-causing genes, in a variety of ways. These include point mutations, chromosome rearrangements, viruses, radiation, and chemicals.

20. Many mutagenic substances are also carcinogenic. Assaying for potential carcinogenicity can be done by scoring cells for sister chromatid exchanges (SCEs) or by performing the bacterial Ames test.

21. Many familiar compounds have been implicated in mutagenicity and carcinogenicity (including tobacco, marijuana, caffeine, and LSD).

22. X rays and ultraviolet radiation are capable of causing genetic damage, including chromosome breakage, SCE, and point mutations.

23. Both acute and chronic doses of radiation cause the same total amount of point mutational damage.

Key Words and Phrases

acute dose

Ames test

aneuploidy

autoimmune theory of aging

base analogue

Burkitt's lymphoma

carcinogenic

carcinoma

chronic dose

commitment theory of cellular aging

deletion

dispermy

Down's syndrome

duplication

Epstein-Barr Virus (EBV)

fetal alcohol syndrome (FAS)

frameshift mutation

free radical

gene mutation

genetically significant dose (GSD)

genetic load

Hayflick Limit

human T-cell leukemia virus (HTLV)

inversion

latent period

LD50

leukemia

metastasis

missense mutation

mosaic

H. J. Muller

mutagenic

mutation

mutation theory of aging

neoplasm

nondisjunction

oncogene

phenocopy

Philadelphia chromosome (Ph1)

point mutation

polyploidy

position effect

progeria

proto-oncogenes

rad

rem

ring chromosome

same sense mutation

sarcoma

sister chromatid exchange (SCE)

substitution mutation

teratogenic

thymine dimer

translocation

trisomy

ultraviolet radiation

unequal crossover

xeroderma pigmentosum

X rays

Questions

1. The calico coat color pattern in cats is the result of the random inactivation of one or the other X chromosome carrying the genes for coat color (a type of mosaicism—see chapter 5). If a male calico cat is produced by a mating of a rust female and a black male cat, can you tell in which of the parents the nondisjunctional event occurred? How?
2. What possible advantage might there be to producing a triploid watermelon, in addition to having a larger and more vigorous plant and fruit?
3. Two parents each have blue eyes. Assume that eye color is controlled by a single pair of alleles, with B for brown eyes dominant to b, which produces blue eyes. Their first child has blue eyes, but their second has one brown and one blue eye. Explain the likely genetic origin of this mosaicism.
4. Why are chromosomal aneuploidies normally a developmental disaster?
5. How can nondisjunction lead to mosaicism?
6. Of what importance are gene mutations; chromosome mutations?
7. Distinguish between nonsense, missense, and same sense mutations.
8. What are the major characteristics of cancer cells?
9. How can mutations cause cancer?
10. Describe the Ames test. Of what value is this test?
11. Describe some of the mutagenic and carcinogenic effects of tobacco; of marijuana.
12. What are the genetic effects of ultraviolet radiation and X-radiation?
13. If a dose of one thousand rems produced sex-linked recessive lethal mutations in 12% of a total of one thousand exposed fruit fly gametes, how many mutations would you predict would occur after exposing one thousand gametes to four thousand rems? How many mutations would you predict if one hundred gametes were exposed to one thousand rems?
14. Why are most mutations harmful?
15. What is the health significance of the Philadelphia chromosome?
16. What are possible mechanisms for converting proto-oncogenes to oncogenes?
17. Which is genetically safer, an acute dose of radiation or a chronic dose? Explain your answer.

For More Information

Ames, B. N.; R. Magraw; and L. S. Gold. 1987. "Ranking Possible Carcinogenic Hazards." *Science,* 236:271–80.

Angier, N. 1987. "Light Cast on a Darkling Gene." *Discover,* 8:85–96.

Arehart-Treichel, J. 1979. "Down's Syndrome: The Father's Role." *Science News,* 116:381–82.

Bearn, A. G., and J. L. German III. 1961. "Chromosomes and Disease." *Scientific American,* 205:66–76.

Benowitz, S. I. 1985. "Retinoblastoma: Unmasking a Cancer." *Science News,* 127:10–12.

Bishop, J. M. 1987. "The Molecular Genetics of Cancer." *Science,* 235:306–11.

Cairns, J. 1978. *Cancer: Science and Society.* W. H. Freeman & Co., San Francisco.

Cairns, J. 1981. "The Origins of Human Cancers." *Nature,* 289:353–57.

"Carcinogen Priorities." 1980. *Sciquest,* 53:26.

Croce, C. M., and G. Klein. 1985. "Chromosome Translocations and Human Cancer." *Scientific American,* 252:54–60.

Devoret, R. 1979. "Bacterial Tests for Potential Carcinogens." *Scientific American,* 241:40–50.

Epstein, S. S. 1974. "Environmental Determinants of Human Cancer." *Cancer Research,* 34:2425–35.

Fung, Y-K. T. *et al.* 1987. "Structural Evidence for the Authenticity of the Human Retinoblastoma Gene." *Science,* 236:1657–61.

Harman, D. 1981. "The Aging Process." *Proceedings of the National Academy of Science,* 78:7124–28.

Harris, H. 1986. "Malignant Tumours Generated by Recessive Mutations." *Nature,* 323:582–3.

Hayflick, L. 1980. "The Cell Biology of Human Aging." *Scientific American,* 242:58–66.

Kolata, G. 1986. "Two Disease-Causing Genes Found." *Science,* 234:669–70.

Kolata, G. 1987. "Oncogenes Give Breast Cancer Prognosis." *Science,* 235:160–61.

Kolata, G. 1987. "Human Cancer Gene Sequenced." *Science,* 235:1323.

Lave, L. B., and G. S. Omenn. 1986. "Cost-Effectiveness of Short-Term Tests For Carcinogenicity." *Nature,* 324:29–34.

Marx, J. L. 1984. "What Do Oncogenes Do?" *Science,* 223:673–76.

Miller, J. A. 1982. "Spelling out a Cancer Gene." *Science News,* 122:316–19.

Muller, H. J. 1955. "Radiation and Human Mutation." *Scientific American,* 193:58–68.

Nicolson, G. L. 1979. "Cancer Metastasis." *Scientific American,* 240:66–94.

Patrusky, B. 1982. "What Causes Aging?" *Science 82,* 3:112.

Reif, A. E. 1981. "The Causes of Cancer." *American Scientist,* 89:437–47.

Roberts, L. 1987. "Atomic Bomb Doses Reassessed." *Science,* 238:1649–51.

Roberts, L. 1987. "Radiation Accident Grips Goiânia." *Science,* 238:1028–31.

Smith, R. J. 1980. "Government Says Cancer Rate Is Increasing." *Science,* 209:998–1002.

"Smoking and Cancer." 1982. *Morbidity and Mortality Weekly Report,* 31:77–80.

Upton, A. C. 1982. "The Biological Effects of Low-Level Ionizing Radiation." *Scientific American,* 246:141–48.

Wallis, C. 1982. "Advances in the War on Cancer." *Time,* November 8:70–71.

Yunis, J. J. 1983. "The Chromosomal Basis of Human Neoplasia." *Science,* 221:227–36.

Yunis, J. J., and O. Prakash. 1982. "The Origin of Man: A Chromosome Pictorial Legacy." *Science,* 215:1525–30.

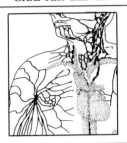

Immune genetics

Nonspecific Defense Mechanisms
Specific Defense: Immunity
Immune Function
Box 11.1 Blood / Specificity / Immune development / Immune response
Immunoglobulins
Types of antibodies / The clonal selection theory of antibody diversity / Box 11.2 Antibody structure
Acquisition of Specific Immunity
Passive immunity / Active immunity / Box 11.3 Behavior and passive immunity
Health, Genetics, and Immunity
Tissue and Blood Typing
HLA type and transplants / The Rh blood group / ABO incompatibility
Malfunctioning Immune Systems
Bubble babies / Allergies and immunity / Autoimmune diseases / Systemic lupus erythematosus
Viruses and Immunity
Life Cycles of Viruses
Reverse transcriptase / Transduction
Viral Diseases
Mononucleosis and EBV / Influenza / Herpes / Multiple sclerosis and measles /
Insulin dependent diabetes and mumps / AIDS

Figure 11.1 Different types of blood cells: (a) red blood cells; (b) platelets; (c) macrophages; (d) scanning electron micrograph of human lymphocyte.
Source: (a) Armed Forces Institute of Pathology; (b) Joseph R. Goodman; (c) American Lung Association; (d) Etienne de Harven and Nina Lampen/ Sloan-Kettering Institute for Cancer Research.

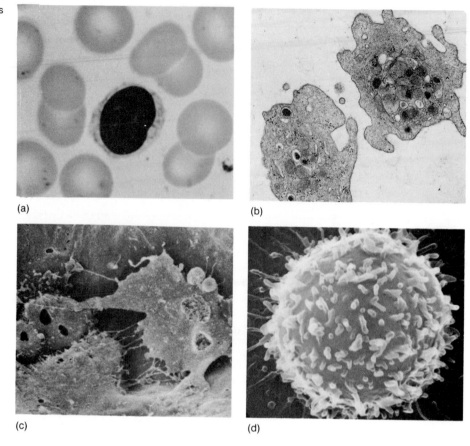

(a) (b) (c) (d)

Living organisms are continually subject to invasion by pathogenic organisms and toxic substances. As a consequence, various defenses against biological and physical assault have evolved. Some defenses are general, or nonspecific, resisting a broad array of potentially harmful agents. Others are specific, defending against only one pathogen or toxin. The function of specific defense mechanisms is closely intertwined with aspects of the hereditary system. In this chapter, we explore the immune system and some of the applications of immune genetics to human health. We conclude with a consideration of viral diseases such as AIDS, whose deadly effects largely depend upon the ability of viruses to avoid and subvert the normal immune defense mechanisms of the body.

Nonspecific defense mechanisms

Nonspecific defense mechanisms protect the body from many different kinds of pathogenic and toxic agents. These nonspecific defenses are often called **natural resistance.** They represent a set of defenses with which all individuals are born, and they function more or less continuously throughout one's entire life.

One important category of natural resistance is the physical protection provided by the external surface of the body. The dry, tough skin is a barrier to almost all pathogens and toxins. The linings of the respiratory, digestive, urinary, and reproductive systems are also considered to be external to the body. All are lined with a warm, moist epithelium; the epithelial surfaces have mechanisms to remove or kill most infective environmental agents. The lungs, for example, are lined with ciliated epithelium. The moist surface traps dust and bacteria, and the action of the cilia sweeps the pollutants out of the lungs to the throat, where they are expelled.

Many fluids of the body surfaces are toxic to bacteria, often because of high acidity or alkalinity. Urine, for example, cleanses the urinary tract. Tears wash particulate matter from the eyes and also contain lysozyme, an antibacterial enzyme. Sweat cleanses and protects the skin. Ingested microorganisms that manage to reach the stomach are almost certain to die in the highly acid gastric juice (pH 2.0–4.0). Lactic acid in the vagina discourages the growth of most microorganisms. The normal flora and fauna of the gut, still technically outside the body, are also effective antipathogens.

If external barriers are penetrated, internal mechanisms of general defense respond. *Phagocytic* (meaning cell-eating) cells attack and destroy infective bacteria. Connective tissue may rapidly grow to surround sites of bacterial and parasitic invasion, isolating the infection from the rest of the body. Many chemicals in the body inhibit bacterial and viral growth. Polypeptides such as histones and spermine act against bacteria, and interferon is an antiviral agent. All of these factors contribute to the general *inflammation* response against infection.

Specific defense: immunity

Specific defense mechanisms are directed against particular pathogens such as bacteria, viruses, parasites, fungi, and the toxins that these organisms may release. The diseases tetanus and botulism, for example, are caused by bacteria of the genus *Clostridium*. Their potentially lethal effects result from potent toxins released by the bacteria, rather than directly from the growth of the bacteria themselves. The resistance of an individual to a particular microorganism or its product is referred to as **specific immunity.** Because specific immunity normally occurs only after recovery from initial infection by that microorganism, specific immunity is also called **acquired immunity.**

Specific immunity results from the interactions within the body of **antigens,** foreign substances that elicit the production of antibodies when introduced into the body, and **antibodies,** the proteins produced by the body in response to foreign antigens. In general, any chemical to which the body has not been exposed early in development is treated as a foreign substance, *non-self,* and is a potential antigen. Occasionally, the *self/non-self* recognition mechanism breaks down. The body may fail to identify certain antigens as foreign, thus not mounting a defensive response; or the body may perceive its own tissues to be foreign and produce antibodies against itself. In either case, the medical consequences are serious.

Following invasion of the body by antigens such as proteins, bacteria, parasites, viruses, and toxins, several types of specific cells respond to neutralize these antigens. **Lymphocytes,** a class of white blood cell, are especially active (box 11.1). **T cells,** one type of lymphocyte, attack foreign antigens directly, a process known as **cellular immunity. B cells,** a second type of lymphocyte, produce *antibodies* that interact with and destroy foreign antigens, a process known as **humoral immunity. Macrophages,** very large phagocytes, not only act directly upon foreign antigens, but also cooperate with T cells (figure 11.1). The total *immune response* reflects the multiple cooperative actions of these three cell types to destroy or neutralize specific antigens.

Immune function

Acquired immunity is an essential protective mechanism of the body, yet there is no anatomically distinct immune system. Instead, immune function depends chiefly upon the actions of B cells, T cells, and macrophages. These cells all originate in bone marrow, but they exist and function within the two major circulatory networks, the **cardiovascular system** and the **lymphatic system** (figure 11.2).

BOX 11.1

Blood

The average person has about five liters of blood. The liquid portion, *plasma,* accounts for roughly 55%; the remaining 45% consists of red blood cells, white blood cells, and platelets (which participate in forming blood clots). In a healthy individual each milliliter of blood contains about five million red cells, ten thousand white cells, and three hundred thousand platelets (figure 11.1).

Red blood cells are produced in the bone marrow. As they mature, they lose their nuclei, becoming bags of hemoglobin functioning in gas exchange. Red cells survive about four months in the circulatory system, eventually deteriorating and being destroyed by the spleen.

White blood cells come in many forms and fulfill many different functions. Most white blood cells, **leukocytes,** have a life span of two weeks or less. Only the antibody producing leukocytes, B cells, survive as long as one hundred days. One major class of white blood cells include the *macrophages* (figure 11.1). Like giant amoebae, macrophages move to a site of infection, engulfing and ingesting foreign matter. Macrophages are present in large numbers early in an infection.

Lymphocytes, a second major class, comprise about 40% of all leukocytes. Depending upon their developmental history, lymphocytes differentiate to one of two basic types. About 65%–75% of lymphocytes are processed by the thymus gland, acquiring a surface antigen that identifies them as T cells. Those lymphocytes that do not develop in the thymus are known as B cells. B cells are the actual site of antibody production, following exposure of the body to foreign antigens.

The cardiovascular system consists of the heart, whose pumping action continually circulates materials throughout the body; arteries, vessels that carry blood away from the heart; veins, vessels that return blood to the heart; and capillaries, vessels that connect the arteries and the veins and whose internal diameter permits passage of only a single red blood cell at a time. Nutrients and other essential materials in the blood diffuse through the walls of the capillaries into the interstitial fluid surrounding all cells in the body; waste products from the tissues diffuse back into the capillaries for ultimate removal from the body.

The lymphatic system, on the other hand, provides only one-way flow of materials from the tissues to the cardiovascular system. Tiny vessels of the lymphatic system drain fluid called *lymph* from all body tissues. Lymph is the liquid portion of the blood, without blood cells and large proteins. Lymph passes through a series of increasingly larger lymphatic vessels until it reenters the cardiovascular system near the heart. Located throughout the lymph system are lymph nodes, which filter bacteria and other foreign particles from the lymph fluid. The tonsils are masses of lymphatic tissue surrounding the throat, forming a protective ring that defends the digestive and respiratory tracts from pathogenic invasion. Decades ago, tonsils were commonly removed during childhood because almost all children experience bouts of tonsillitis, inflammation of the tonsils. Because the protective importance of the tonsils is now better understood, the tonsils are less-frequently removed.

The association of lymphocytes (B cells and T cells) and macrophages with the two circulatory systems is fundamental. The circulatory systems penetrate all portions of the body. Essentially all living cells in the body are no further than the width of one cell from a capillary. Thus, lymphocytes and macrophages can be transported anywhere in the body within seconds to begin their protective action. Furthermore, bacteria, toxins, and other foreign substances are not permitted to circulate freely throughout the body after they penetrate the skin. Instead, they are trapped in nearby lymph nodes, where they are attacked by lymphocytes and macrophages. Concentrations of lymph nodes occur in the neck, armpits, and groin, regions that become swollen and tender during times of illness and infection.

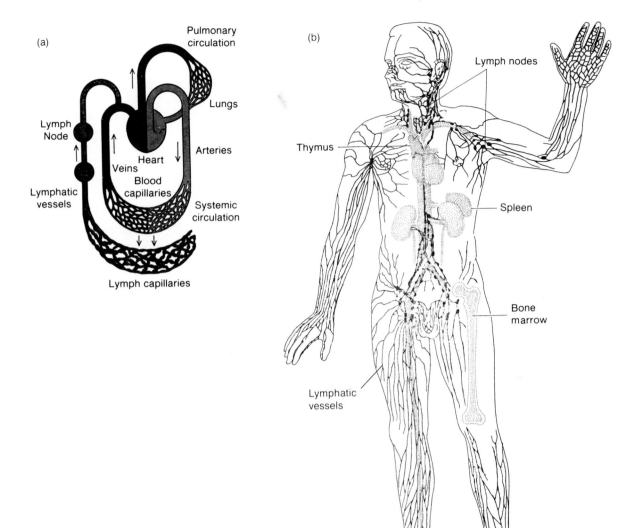

(a)

Pulmonary circulation

Lungs

Lymph Node

Arteries

Heart

Veins

Blood capillaries

Lymphatic vessels

Systemic circulation

Lymph capillaries

(b)

Lymph nodes

Thymus

Spleen

Bone marrow

Lymphatic vessels

Figure 11.2 The organs and vessels of the (a) cardiovascular and (b) lymphatic systems. Source: "The Immune System," by Niels Kaj Jerne. Copyright © July 1973 by Scientific American, Inc. All rights reserved.

Specificity

Immunity depends upon the extraordinary capacity of the body to distinguish self from non-self, i.e., *specificity*. Exactly how foreign substances are identified is still uncertain, but this capacity is essential to normal immune function.

Specific immunity does not develop until after birth. While in the uterus, the fetus is dependent upon its mother's immune mechanisms. During gestation, the fetus learns to identify its own tissues immunologically. The self/non-self distinction is not possible until the individual has catalogued its own tissues. Experiments on laboratory mammals have shown that exposure of fetuses to foreign antigens prior to birth results in those antigens not being recognized as foreign after birth. Skin grafts between mice of different genetic strains are ordinarily rejected. However, if a mouse of one strain is exposed in the uterus to skin of a second strain, postnatal skin grafts are often successful. It appears that during gestation, the body is identifying those tissues present as self.

Immune development

The components of the immune mechanism become functional shortly after birth. A nursing newborn infant receives a large dose of antibodies from its mother in the form of *colostrum,* a thin, yellow substance, rich in antibodies, secreted by the mother's mammary glands immediately before and after birth. Thus, the newborn infant is provided short-term immune protection by the mother's antibodies, while the infant's own immune system matures. Bottle-fed infants miss out on this valuable protection. By about six months of age, the infant's own immune mechanisms are fully functional.

Both lymphocytes and macrophages are produced in the bone marrow and released into the blood (figure 11.3). Lymphocytes originate from *stem cells,* and subsequent differentiation depends upon their fate after entering the bloodstream. Approximately 65% of the lymphocytes enter the **thymus gland,** where they undergo differentiation to become T cells (T refers to the role of the thymus in their development). Following maturation in the thymus, T cells are released back into the blood. The thymus gland, a soft, reddish organ that is part of the lymphatic system, lies directly over the heart and just behind the upper part of the breastbone (figure 11.2). The thymus is relatively large in infancy and childhood. Its decrease in size, beginning at puberty and continuing to death, is believed to be related to a general decline in function of the immune system with age. By age seventy, the immune system is only about 25% as effective as in one's youth. Many deaths of elderly persons are due to insufficient immune response to infections such as pneumonia.

The remaining 35% of lymphocytes do not enter the thymus, but undergo differentiation in the bloodstream to become B cells. (B does not refer to blood in this case, but the association is a good memory aid for distinguishing T cells and B cells). Macrophages also originate in the bone marrow and mature in the blood. Following maturity, T cells, B cells, and macrophages settle throughout the circulatory system. Many reside in the walls of the smaller blood vessels, but a large proportion are concentrated in the major lymphatic organs, the lymph nodes and the *spleen.* The spleen, a large, abdominal organ with a flattened oblong shape, is a reservoir for blood and an important site of defense against infection (figure 11.2).

Immune response

When a foreign substance penetrates the body surface, the lymphocytes and macrophages engage in a complex set of interrelated responses. Some macrophages immediately phagocytize foreign cells, while others become *activated.* They apparently react with a key portion of the antigen, the **antigenic determinant,** which is the portion of the foreign substance that the macrophages identify as non-self. Activated macrophages then interact with *helper T cells.* The particular antigenic determinant received by a helper T cell governs the specificity of its response. Because bacteria and other antigens have many antigenic determinants (blood cells, for example, have several thousand), different macrophages interact differently with the same antigen. Each macrophage interacts with different T cells, but each T cell is programmed to respond to only one antigenic determinant.

Receipt of the antigenic information from a macrophage induces further development and proliferation of T cells. Some T cells attach directly to foreign antigens or to the antigens on the surface of foreign cells. Some T cells release *lymphokines,* chemicals with antiviral and antibacterial properties. Some helper T cells stimulate the multiplication of a few killer T cells, which kill cells that have been invaded by the foreign antigen, releasing the antigen so antibodies produced by B cells can neutralize the invading antigen. A population of T cells is transformed into **memory cells,** and still other helper T cells interact with and activate B cells.

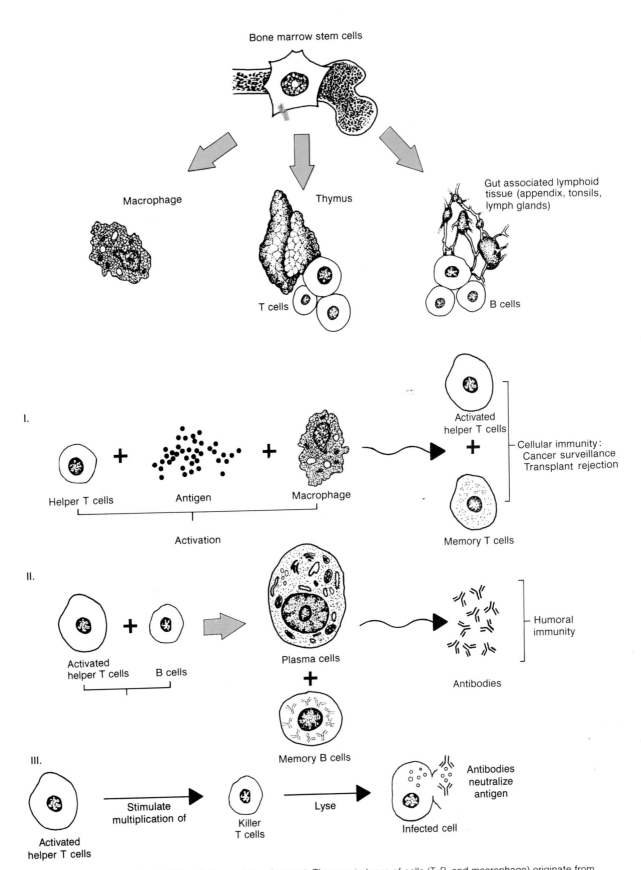

Figure 11.3 White blood cell differentiation and development. Three main types of cells (T, B, and macrophage) originate from stem cells in the bone marrow. T cells and B cells acquire cell specific antigens, identifying them as T cells and B cells. These cells then interact with foreign antigens and macrophages to stimulate further differentiation and proliferation. (See colorplate 15.)

Following an infection, B cells begin to proliferate, their growth enhanced by interactions with activated helper T cells. B cells undergo yet further differentiation to **plasma cells.** The primary activity of plasma cells is the production of specific antibodies or, more properly, **immunoglobulins.** Immunoglobulins attach to the antigens to which they are specific, inactivating toxins or destroying foreign cells. Some T cells and some B cells survive weeks or years and serve as memory cells, preserving the antigenic determinant so that, should a subsequent infection by the same foreign agent occur, the body can respond rapidly. This provides the basis for resistance to specific diseases, once an individual has survived a first infection (figure 11.3).

Immunoglobulins

About five to ten days after an infection begins, immunoglobulins specific for the invading antigen are abundant enough to be detected in the blood. This short period is the time required for B cells to proliferate and mature to plasma cells, following stimulation by helper T cells. During this process, the B cell nucleus is squeezed off to one side as the entire cytoplasm fills with endoplasmic reticulum. This extensive membrane system is the site of massive antibody production, the main function of plasma cells.

An individual B cell produces only one type of antibody in response to the antigenic determinant brought by helper T cells. The maturing B cell reproduces mitotically, generating a clone of identical plasma cells, all producing identical antibodies (box 11.2). Thus, upon exposure to a single bacterial infection, several different antigenic determinants on the surface of the invading bacteria elicit proliferation of plasma cells and subsequent antibody production. Each plasma cell can produce roughly two thousand identical antibody molecules per second. The five to ten day period of immunological response reflects the normal time required for common illnesses such as colds and influenza to run their course.

Types of antibodies
Five types of antibodies (immunoglobulins) are known, each playing a different role in the immune reaction. *Immunoglobulin G (IgG)* comprises about 80% of the antibodies in blood. IgG is active against bacteria, viruses, and toxins. IgG's chemical structure permits it to easily cross the placental barrier and provide the primary specific defense of the fetus.

About 7% of plasma antibodies are *immunoglobulin M (IgM),* a relatively large immunoglobulin. The antibodies to the ABO blood group are IgM. Because IgM molecules are too large to cross the placenta, ABO incompatibility of mother and fetus is not a problem.

About 12% of plasma antibodies are *immunoglobulin A (IgA).* IgA, however, is the most abundant immunoglobulin in the secretions of exocrine glands. Large quantities of IgA are found in saliva, tears, milk, nasal mucus, urine, and fluids of the digestive system. IgA appears to provide an external defense to viral infection.

Two other types of immunoglobulins occur in small quantities, less than 5% of the total plasma antibodies. Neither is known to have protective functions for the body, and they may actually be harmful. *Immunoglobulin D (IgD)* occurs in relatively high concentrations in the serum of some persons with myeloma, a type of bone cancer. Recently IgD has been shown to be involved in differentiation of memory B cells. *Immunoglobulin E (IgE)* is the causative agent of allergies, the bane of millions. IgE also appears to be involved in fighting parasitic infections.

BOX 11.2

Antibody structure

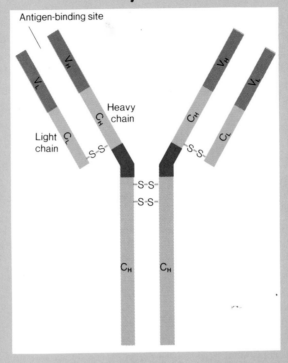

Figure 11.4 An antibody molecule, IgG. Each antibody molecule is composed of four polypeptide chains (two heavy and two light). The heavy chain consists of 446 amino acids; the light chain consists of 214 amino acids. Each chain has a variable region (V) and a constant region (C). The variable regions function in antigen recognition.

Antibodies are large proteins composed of four polypeptide chains, synthesized on the endoplasmic reticulum of plasma cells. R. R. Porter and G. M. Edelman received the 1972 Nobel Prize for Medicine or Physiology for their description of antibody structure.

Two large polypeptides called *heavy chains,* each 446 amino acids long, join for slightly more than half their length to form the Y-shaped framework of an antibody (figure 11.4). Two *light chains,* each 214 amino acids long, are linked to the heavy chains by bonds between the sulfur atoms of sulfur-containing amino acids. Amino acids 1–114 of each heavy chain are a variable region that functions in antigen recognition. The remaining portion of all heavy chains is invariant. Each light chain also has a variable region (amino acids 1–108, functioning in antigen recognition), as well as a constant region. It is estimated that at least four separate genes encode the heavy chain, and at least three separate genes encode the light chain.

The clonal selection theory of antibody diversity

The human immune system has the incredible capacity to produce unique antibodies to nearly every foreign substance it encounters. An individual produces thousands, perhaps millions of different types of antibodies in a lifetime. How is this vast antibody diversity generated and controlled? Evidence now suggests that there are between fifty and three-hundred gene fragments that encode the variable regions of the heavy and light antibody chains, and that somatic mutation and recombination are important genetic events in the origin and maintenance of antibody diversity.

The vast number of antibodies we each produce are the products of cutting and splicing a much smaller set of genes in the maturing plasma cells. Thus each plasma cell is committed to produce a single unique antibody, against one of many determinants on the surface of the invading antigen. This view of the production of antibody diversity is known as the **clonal selection theory.** The apparently limitless capacity of the immune system to respond precisely to a continually changing set of environmental challenges reflects the extraordinary versatility of the genetic mechanism. Antibody production (a highly variable phenotypic character) is a finely tuned survival response resulting, like all phenotypic traits, from a complex interaction of the genotype and environment.

Acquisition of specific immunity

One may acquire immunity actively or passively. An individual's immune defenses are not fully developed until about six months after birth. During the preceding fifteen months of life, immune protection is provided by antibodies produced by the mother. Any infections or toxic agents can gain access to the fetus while *in utero* only by penetrating the mother's defense mechanisms. If that is achieved, the fetus is still protected by maternal IgG that freely crosses the placenta.

Passive immunity
Passive immunity is the immune protection resulting from the transmission of antibodies from one individual to another. The example of antibodies passed from a mother to her nursing child is probably the most common form of passive immunity. As noted earlier, colostrum, the first portion of mother's milk, is exceptionally rich in antibodies. Not surprisingly the incidence of colds is significantly lower among nursing infants than among bottle-fed infants. The fact that the mother has been exposed to a much wider variety of antigens in her lifetime than has the newborn means that her immune system responds faster and more efficiently than the immune system of her infant. By receiving antibodies from its mother, a nursing infant is better protected from infectious diseases (box 11.3).

Artificial passive immunity is now commonly employed in medical practice. Persons at risk for specific diseases are given injections of *gamma globulins,* the antibody-containing fraction in blood. Mumps, for example, is a mild disease in children but a serious disease for adults. Prior to the development of a mumps vaccine, adults exposed to mumps were given injections of gamma globulin from persons who had previously had mumps and thus had anti-mumps antibodies.

Gamma globulin injections provide only temporary immunity. If long-term protection is needed, successive injections are required. In rare cases, bone marrow transplants can be used to transfer the entire antibody-producing machinery from one person to another. The recipient benefits from a lifetime of passive immunity.

Active immunity
Throughout life, an individual's immune mechanism produces antibodies in direct response to substances encountered in the environment, i.e., **active immunity.** Upon first exposure to a foreign antigen, some B cells are stimulated to differentiate into plasma cells, initiating the immune response. Within two weeks, antibody production is well underway. Some T cells and some B cells survive for years as *memory cells.* Upon subsequent exposure to the same antigen, the immune response is immediate. The rapid response of the memory cells is the basis for immunity to measles, mumps, and other childhood diseases *after* one bout of illness.

Behavior and passive immunity

(a)   (b)

Figure 11.5 Passive immunity may protect nursing infants from many diseases.
Source: (a) Ron Monner; (b) Mary Eleanor Browning/Photo Researchers.

Maternal fondling and kissing of babies is a trait that may have been favored by natural selection throughout mammalian (including human) evolution. Mothers of many species continually lick and groom their young, just as human mothers kiss their infants (figure 11.5). This behavior exposes the mother's immune system to the bacteria and other organisms on the infant's face and body. If the child is exposed to any antigens not yet encountered by the mother, kissing behavior insures the mother's exposure to these antigens and insures that her system makes the antibodies most needed by the child. The mother's antibodies will then be passed in her milk to the nursing infant.

Maternal exposure to bacteria and other antigens carried by her newborn has a second beneficial effect. The mother has previously been exposed to many of these antigens and has memory cells that recognize them. In effect, she has been vaccinated against most common environmental agents. Thus, the mother's immune response to "old" antigens is swifter and stronger than the initial response of her infant to "new" antigens. The normal fourteen-day lag in immune response time could be fatal to a fragile infant. Kissing behavior may thus provide a nursing mammalian infant with double immunological protection against its new environment.

Vaccination against disease is a common method for inducing active immunity without having to suffer any disease symptoms. First, the antigen is prepared by either making an extract of the disease-causing organism, or by genetically crippling the disease organism so it cannot reproduce and using this defective strain as the antigen. Subsequent injections of the same vaccine (*boosters*) restimulate the plasma cells producing higher concentrations of antibodies and memory cells. Upon later exposure to the live disease-causing organisms, there is immediate antibody production that destroys the organism *before* it can cause disease.

Polio vaccines illustrate the different methods of preparing vaccines. The Salk vaccine, introduced in 1955, uses killed polio virus as antigens, whereas the Sabin oral vaccine, introduced in 1961, uses an attenuated strain of the live virus. An attenuated virus is genetically disabled so it usually cannot cause disease. There were a few scattered cases of paralytic polio after administration of the Sabin vaccine because the virus had not been sufficiently disabled, due to problems in the initial manufacture of some batches of the vaccine. When live virus is used, there is always a small probability that the virus will revert to its virulent form and cause polio, a probability not associated with the Salk vaccine. Despite the small risk, initial evidence indicated that the Sabin vaccine provided better protection from polio than the Salk vaccine. More recent studies show that both vaccines afford equal protection.

In either case the original exposure to the polio virus triggered an immune response resulting in a proliferation of B cells producing antibodies specific to determinants on the polio virus. Boosters (injections or drinks) of either of these vaccines restimulated the T cells and B cells and produced a higher concentration of antiviral antibodies. Some of these cells function as memory cells, so that if the person is ever exposed to live polio virus, within one to three days of exposure, their immune system will begin producing anti-polio antibodies. This secondary response is quite rapid in contrast to the initial response that takes place over five to fourteen days, depending upon the nature of the antigen and the individual's overall health.

Rubella is normally a childhood disease that has no severe complications. However, the rubella virus can cross the placenta and damage a developing child during the first trimester of pregnancy. Depending upon the stage of pregnancy, rubella virus may cause deafness, cataracts, heart abnormalities, or even spontaneous abortion. Immunization against rubella is now recommended for all girls over one year of age in order to protect their future children against birth defects.

Induced active immunity by vaccination is one of the greatest success stories of modern medicine. Immunization programs have effectively eliminated such diseases as smallpox, tetanus, typhoid, typhus, polio, and diphtheria in the developed world—diseases that have killed thousands or millions throughout history.

Health, genetics, and immunity

Many diseases result from a weakening or breakdown of the immune system. In other cases, normal immune function results in an inappropriate response directed against an individual's own tissues rather than an invading organism or toxin. Increased knowledge has permitted advances in medical treatment of diseases associated with the immune system.

We now review some important genetic aspects of immunological medicine. Particular attention is given to viruses, organisms with a special capacity for subverting immune mechanisms, as illustrated by the extraordinary modern disease AIDS.

Tissue and blood typing

In chapter four we introduced the ABO and MN blood groups associated with red blood cells, and the HLA group associated with white blood cells. All three groups, and many others as well, are the result of proteins called *cell surface antigens* occurring on the blood cells. These groups are of value for forensic purposes, such as paternity testing and identification of criminals from blood and other tissues left at crime scenes. Additionally, patterns of variation among ethnic and cultural groups have often provided important anthropological evidence for understanding the history of human populations.

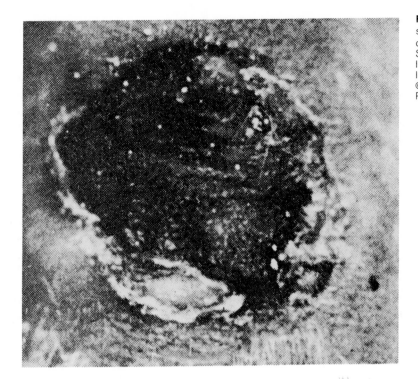

Figure 11.6 Rejection of a skin graft, showing necrosis of the tissue. Source: L. Brent, from Roitt, Ivan: "Essentials of Immunology" 3rd ed. © Blackwell Scientific Publications.

These blood groups are clearly of medical importance, as shown by the necessity for matching ABO types for transfusions, or by the association of specific ABO and HLA types with diseases (see chapter four). In the next sections we consider another important medical aspect of HLA type and introduce the Rh group.

HLA type and transplants

During gestation, the developing immune mechanism identifies all tissues and all cells of the fetus as *self*. Postnatally, those antigens recognized as self are tolerated. All others are attacked and eventually destroyed; this is the immune response. The immune response thus poses a major barrier to *transplantation* of tissues and organs from one person to another. Except in a few special cases, the recipient of a transplant experiences an immediate tissue rejection response. This reaction primarily involves the T cells, which isolate and ultimately kill the transplanted tissue (figure 11.6).

A few sites in the body are not susceptible to transplant rejection. These **immunologically privileged sites** include corneas, bone, cartilage, heart valves, and testes. It is thought that because the cornea has little vascularization and little contact with blood and lymph, circulating T cells that monitor the body for foreign tissue never "see" the cornea. Thus, corneas can be transplanted with little fear of rejection, and such operations save the sight of many persons every year. In spite of the ease of cornea transplants, not enough corneas are available to meet present demand.

Most tissues in the body, however, are subject to continual immune surveillance and to a rejection reaction following tissue or organ transplants. The *human leukocyte antigens* (**HLA**) govern the tissue rejection response. Recall that HLA type depends upon at least 5 loci and over 110 alleles, generating millions of unique genotypes and a very low frequency of any one of them. Although blood transfusions are performed routinely as a result of the limited number of ABO genotypes, such is not the case with tissue and organ transplants.

The extraordinary HLA diversity in humans explains the immunological barrier to tissue and organ transplants. Even parent/child or sibling/sibling transplants are seldom successful, as HLA genotypes within a single family can all be distinct. In figure 4.11, for example, considering only the major loci, the parents share no HLA alleles and produce four children, all of differing HLA genotype. Thus, if a child in the family needs a transplant, *none* of the family members is a suitable donor. If crossover occurs and the minor loci are considered, the possible genetic variation makes an HLA match within any pedigree extremely unlikely.

A notable exception to the rarity of HLA matching is the case of monozygotic twins. These individuals are genetically identical, therefore, all tissues have the same surface antigens. Tissue rejection is no barrier to transplants between identical twins.

Tissue matching is a serious medical problem. Only rarely is a healthy organ with the right genotype available for transplant at the right time. Physicians now accept a partial match in serious cases. If a match at the A and B loci can be achieved, the use of immunosuppressant drugs can limit the rejection response due to other genetic variation. Current research is examining many drugs and other techniques that can inhibit the rejection of transplanted tissues. One method that may prove useful for transplanting tiny amounts of tissue involves enclosing the donor cells in an ultra-fine mesh bag. Diabetics, for example, need only a small number of insulin-producing cells. Experiments suggest that it may be possible to transplant such cells into the pancreas of a diabetic, enclosing donor tissue in a mesh fine enough to keep antibodies out, but large enough to permit the entry of nutrients and the exit of insulin.

Heart transplants   The first heart transplant occurred in 1967 in South Africa, and the first heart transplant in the United States occurred the following year. By 1986, over fifteen hundred transplants had been attempted worldwide. Ten years is about the maximum survival time after a heart transplant. Most recipients died of tissue rejection, usually within two to three years after the operation. The most severe cases of rejection occurred within three months of the operation. In all of these instances, immunosuppressant drugs, radiation, and chemical therapy were employed, as well as attempts to match the HLA-A and HLA-B antigens.

Autografts   Sometimes tissue from a person's own body is transplanted to a new site, a procedure known as an *autograft* ("auto-" means self). An autograft has the special advantage of a perfect tissue match—self to self—and therefore tissue rejection due to HLA mismatch is not encountered. Autografts of skin are a standard procedure for certain types of plastic surgery and for treating severe burns, where areas of skin are so badly damaged that they must be replaced. Autografts of large veins from the leg are now used to replace cardiac veins and arteries damaged or occluded by atherosclerosis. In some cases, an amputated or congenitally damaged thumb or finger can be replaced by autografting a toe to the hand. A spectacular type of autograft to treat Parkinson's disease suggests that autografts may continue to develop as an important type of treatment for many medical conditions.

Parkinson's disease is characterized by gradual and severe loss of voluntary motor function; victims suffer tremors, lose the ability to speak, and eventually become totally incapacitated. The disease is caused by deterioration of the brain's capacity to produce dopamine, a chemical required for communication among brain cells. The only available treatment is administration of the drug L-dopa, which is similar but not identical to dopamine. Victims usually respond to L-dopa for a short time, but gradually its beneficial effects diminish. Some victims suffer serious side effects from the drug.

The new treatment takes advantage of the fact that dopamine is also produced by the adrenal glands, which are located on top of the kidneys. Cells from one of the adrenals are autografted to the patient's brain. First trials of the procedure produced no improvement, but in several subsequent cases, responses to autografts were almost immediate. Men who had been severely incapacitated improved rapidly to the point that they could resume many of their normal activities.

Success of the procedure will depend upon a sustained improvement, so the autograft will not be confirmed as standard treatment for many years. However, the potential advantages of this type of therapy are obvious. A person only needs one adrenal gland, so the second is always available for transplant with no problems due to tissue mismatch. Additionally, the autograft provides the brain with its own source of dopamine identical to that originally produced by its own defective cells.

Fetal-cell surgery A second type of transplant therapy for Parkinson's disease is being developed experimentally using laboratory animals: defective dopamine-producing brain cells are replaced with fetal brain cells. Transplants of fetal cells are not subject to the tissue rejection response of adult tissue and may be accepted as the patient's own tissue. If successful, this procedure would have the same advantages as autografting adrenal cells.

As noted in chapter 7, however, fetal-cell surgery confronts society with serious ethical problems. If transplanting fetal tissues proves widely successful for treating Parkinson's disease and other conditions, the number of spontaneously aborted fetuses that might be available for such transplants will be far exceeded by the potential demand. How will decisions on the use of fetal tissues be made? Who should have access to fetal tissues? Should *in vitro* fertilization be used to generate excess embryos to supply the transplant market? Might mothers be under increased pressure to abort children with only minor defects diagnosed prenatally? Rapidly advancing medical technology will force society to answer these and other questions in the near future.

The Rh blood group

Karl Landsteiner received the 1930 Nobel Prize for Physiology or Medicine for his discovery of the ABO blood group, but this extraordinary scientist was still far from retirement. In 1940, Landsteiner and Philip Levine used RBCs from a rhesus monkey to produce antibodies in rabbits. When tested against blood from humans, the rabbit antibodies to rhesus monkey RBCs caused agglutination of the blood of 85% of the Caucasian test population. These individuals were thus designated *Rh positive*. The remaining 15% whose blood showed no agglutination reaction were designated *Rh negative*. These data indicated that 85% of the test group had an antigen on their RBCs similar to that present on the rhesus monkey RBCs, accounting for the agglutination reaction.

Rh blood type is an heritable character, but it follows a complex pattern that is still not fully understood. We can understand medical aspects of the Rh blood group if we use a normal Mendelian model for one locus with two alleles. Allele R determines the presence of the Rh antigen and is dominant to r. Thus, RR and Rr individuals are Rh positive; rr individuals are Rh negative.

The Rh blood group has considerable medical importance. Human serum contains no naturally occurring antibodies to the Rh antigen, unlike the ABO system. If an Rh negative person is exposed to Rh positive blood (as might occur with a transfusion, for example), he or she develops anti-Rh antibodies. The antibodies are not produced rapidly enough that they can attack the RBCs of the first transfusion, but should subsequent transfusions of Rh positive blood occur, the anti-Rh antibodies will be present in large quantities, causing agglutination and a shock reaction. (This is exactly the principle behind immunization against various diseases.) Donor blood and recipient blood is routinely cross matched to avoid such a reaction.

However, an analogous process sometimes results from Rh incompatibility between mother and developing fetus. If an Rh negative female mates with an Rh positive male, any child they conceive has the possibility of being Rh positive, like the father. If the father is homozygous Rh positive, all their children will be Rh positive; if the father is heterozygous Rh positive, half the children will be Rh positive. (The reciprocal mating, Rh positive female with Rh negative male, has no important medical consequences.)

Normally, large molecules or cells cannot cross the placenta between mother and fetus. However, at the time of birth the placenta may tear, allowing entry of some of the child's blood into the mother's system. If the mother is Rh negative and the child Rh positive, the mother is essentially being immunized against the Rh antigen, just like the rabbits that Landsteiner injected with blood from rhesus monkeys. She will develop large quantities of anti-Rh antibody.

Rh antibodies are chemically different from anti-A antibodies and anti-B antibodies and are small enough that they *do* pass through the placenta. Should the Rh negative mother conceive subsequent Rh positive fetuses, they will be at severe risk from the mother's anti-Rh antibodies. The antibodies attach to the surface of the RBCs of the fetus and destroy them, causing the condition **erythroblastosis fetalis** (literally, "destruction of the red blood cells of the fetus"), a type of hemolytic anemia. These infants are called *blue babies* at birth. The bluish tint of their skin is due to the destruction of the RBCs, reducing the blood's capacity to carry oxygen. Destruction of the RBCs also causes *jaundice,* a condition resulting from abnormally high levels of *bilirubin,* a pigment released into the blood from the breakdown of hemoglobin. High levels of bilirubin may cause palsy and mental retardation in the newborn, and if RBC destruction is severe enough, the infant may be stillborn.

Fortunately, understanding this condition has led to a variety of effective therapies. Newborn infants with erythroblastosis fetalis can be given a complete transfusion, replacing all their blood (which contains the mother's anti-Rh antibodies) with blood of their own type *without* those antibodies. In rare cases, such transfusions are given prior to birth. Careful exposure of the infant to fluorescent light degrades the excess bilirubin rapidly enough that it causes no damage.

Preventive therapy is also given an Rh negative woman having a child with an Rh positive man (and thus at risk of producing a blue baby). Immediately after the birth of her first and all subsequent Rh positive children, the mother is given massive injections of anti-Rh antibody. These antibodies destroy the child's RBCs *before* they trigger an immune response in the mother. Thus, the Rh negative mother does not develop anti-Rh antibodies in her system that could harm future children.

ABO incompatibility

Because antibodies to the ABO blood group occur naturally in the serum, and because maternal-fetal incompatibilities are more common for the ABO group (85% are Rh positive), you might expect hemolytic anemia to occur frequently for the ABO and other blood groups. Although some infant anemias are due to incompatibility of blood groups, hemolytic anemia occurs far less frequently than would be expected. Anti-A antibodies and anti-B antibodies are so large that they normally do not cross the placenta from mother to fetus, in contrast to the smaller anti-Rh antibodies.

An unexpected occurrence is the reduction in hemolytic anemia due to Rh incompatibility as a result of simultaneous ABO incompatibility. For example, suppose a mother of blood type O, Rh negative carries a fetus of type A, Rh positive. As fetal RBCs leak across the placenta, the mother's natural anti-A antibodies attach to and destroy them *before* the mother's system can begin production of anti-Rh antibodies. Thus, that child and future children are protected from hemolytic anemia due to Rh incompatibility.

Malfunctioning immune systems

We have learned most about immunity through studies of individuals with defective immune systems. About one person in fifty thousand is born without a thymus gland. These children cannot process T cells and thus have no tissue-mediated immunity. They are extremely susceptible to infectious disease and have a high mortality rate. They do produce B cells and can mount a weak humoral response, even without helper T cells. Some athymic children can be cured by transplanting donor thymus tissue into their abdominal region (the tissue may come from anyone, since without T cells, tissues are not rejected). In some cases the transplanted thymus grows and functions normally, eventually processing leukocytes into mature T cells and bestowing normal immunity on these children.

Other children have a thymus but suffer from a sex-linked recessive condition known as **agammaglobulinemia** (meaning without gamma globulins). These children do not produce B cells and thus cannot produce antibodies. Simple bacterial infections are life threatening to these children because antibodies are the primary defense against bacterial disease. They can, however, mount an immune reaction to viruses, indicating that viral immunity is a function of the T cells.

Agammaglobulinemia can be treated by a bone marrow transplant, which provides a source of normal B cells. This surgery is successful only with precise HLA matching. The T cells of the recipient are fully functional and will reject unmatched tissue. In addition, the donor bone marrow is also immunologically functional and can mount an immune reaction to the recipient, called *graft-versus-host disease*. Incompatible grafts can be fatal.

Bubble babies

Rarely, a baby is born with no immune capability at all, the result of a sex-linked recessive gene. Victims of *severe combined immune deficiency (SCID)* have no thymus gland, no lymph nodes, and no plasma cells. With no immune protection, the boy (why not girl?) succumbs to the first infection contracted.

In 1971, a boy with SCID was born to a Texas couple who had already lost one son to this disease. At birth, David was isolated in a sterile plastic bubble. Initially doctors had expected to find a suitable bone marrow transplant for David, but his HLA type occurs in the population with a frequency of only one in thirty-two thousand. After twelve years of searching, a suitable donor was not identified. In 1984, successes with a new experimental technique for preventing graft-versus-host disease in bone marrow transplants led David and his family to opt for a transplant using bone marrow from his sister.

His sister's HLA type was not a perfect match with David's, so her marrow was pretreated to destroy all mature T cells, which play a major role in graft-versus-host disease. The marrow was subsequently transplanted, but after several weeks, David died from complications of the transplant. A few successes with this experimental treatment, however, make the potential for curing HLA-mismatched SCID patients appear promising.

Allergies and immunity

An *allergy* is immunological hypersensitivity to a foreign antigen or *allergen*. Much evidence is accumulating on the genetic basis of allergy, and some allergies (for example, ragweed, table 4.5) have been associated with particular HLA genotypes. Over thirty-five million Americans have some type of allergy. The effects may be innocuous or annoying, as with seasonal hay fever or contact dermatitis. Alternatively, some allergies are life threatening, as in the case of allergies to insect stings or hypersensitivity to penicillin or other drugs. Allergies to common foods, such as milk and wheat, may require major alterations in life-style and habits. Virtually any substance is capable of inducing an allergic response: some women have been found to be unable to conceive because they are allergic to their husband's semen.

IgE antibodies are the culprit in an allergic reaction. These antibodies are highly specific, combining first with the offending allergen and then attaching to special cells in the body known as **mast cells.** Mast cells are found in many tissues in the body, but are most abundant in the dermis (the layer just under the dry outer skin) and the tissues surrounding the gut, respiratory tract, and blood vessels. IgE triggers mast cells to release copious amounts of **histamine.** Histamine causes blood vessels to dilate and fluid to leak from the circulatory system into surrounding tissues. It also stimulates the copious production of mucus. Food allergies and hives are other results of histamine release, caused by an outpouring of fluid into the dermis, which results in fluid-filled, itchy pustules.

Antihistamine drugs relieve allergy symptoms by blocking histamine receptors on nerve endings. They do not stop histamine release by the mast cells, thus accounting for the limited effectiveness of the drugs. Present research is seeking a class of drugs that would occupy the binding sites on the mast cells to prevent histamine release altogether.

The life-threatening allergic response of some persons to bee and wasp stings results from a massive histamine release. So much plasma pours out of the bloodstream that the victim's blood pressure drops rapidly. Without medication, the victim may enter shock and die.

Autoimmune diseases

In some instances, the normal mechanism for distinguishing self from non-self fails, causing one of a group of **autoimmune** diseases. Autoimmune disease occurs when antibodies against one's own tissues, *autoantibodies,* are produced.

In normal individuals a very low level of autoantibodies is always present, but with no harmful effects. As an individual gets older, the production of autoantibodies tends to increase, perhaps contributing to the aging process. Although the normal function of autoantibodies is unclear, they may provide a mechanism for destroying cells in the body that spontaneously transform into cancerous cells, before these cells can cause disease.

Autoimmunity has been implicated in many diseases. The symptoms vary with the tissues affected. In multiple sclerosis, autoantibodies are produced to the myelin sheath of the brain and spinal cord. Autoantibodies to muscle protein occur in myasthenia gravis; in rheumatoid arthritis, autoantibodies are produced to IgG itself. Rheumatic fever is a complication that sometimes follows a childhood streptococcal infection. Antigens on the strep bacterium resemble those on heart tissue. The antibodies developed to defend against the strep bacterium may function as autoantibodies, causing permanent heart damage.

Systemic lupus erythematosus

Systemic lupus erythematosus (SLE) is an autoimmune disease affecting about one-half million Americans, mostly women. One symptom of SLE, a red rash across the nose and cheeks, accounts for its name, lupus erythematosus (meaning red wolf). Victims of SLE produce antibodies to their own DNA. These autoantibodies attack DNA, destroying cells in the kidneys, lungs, heart, and brain. About one-fourth of SLE victims suffer severe kidney damage and another one-third suffer severe brain damage. SLE is a chronic condition with bouts of remission and relapse. Nine of ten patients are women. It has been observed that, in general, women have a more active immune system than men, but also have a higher frequency of autoimmune diseases.

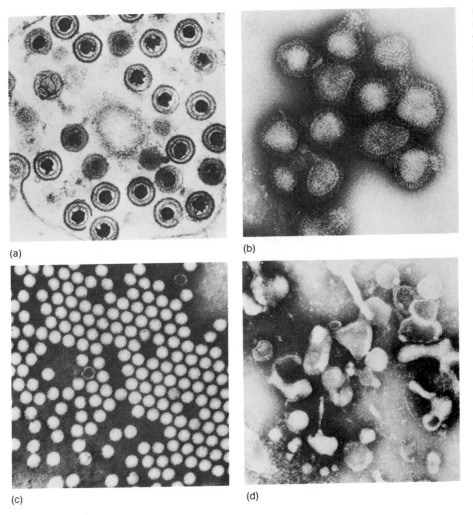

(a)

(b)

(c)

(d)

Figure 11.7 Electron micrographs of several different viruses: (a) herpes virus; (b) influenza virus; (c) polio virus; (d) feline leukemia virus.
Source: (a–d) Centers for Disease Control, Atlanta, GA.

SLE is treated with immunosuppressant drugs and *plasmapheresis* therapy. In plasmapheresis, a patient's blood is slowly removed. All antibodies, including the damaging autoantibodies, are filtered from the blood, which is then returned to the patient. Temporarily lacking antibodies, the patient is susceptible to infection, but plasmapheresis often induces remission of the disease.

Twin studies indicate a genetic predisposition for SLE. If one identical twin gets SLE, the chance of the second twin contracting SLE is 50%–60%. In fraternal twins, the concordance of the disease, however, is only 2%–3%.

Viruses and immunity

Viruses (figure 11.7) are remarkable yet simple entities, which are composed of a protein coat enclosing a core of nucleic acid, either DNA or RNA. They may have as few as five genes and never more than several hundred. Viruses probably infect all living things and are generally species specific. That is, a tulip virus infects only tulips, not roses or humans or bacteria. Additionally, many viruses are organ specific, growing only in one type of tissue. Leukemia viruses, for example, grow only in white blood cells.

Viruses reproduce only inside a living cell. The virus subverts the regulatory apparatus of an infected cell, forcing it to make viral protein (for the coat of the

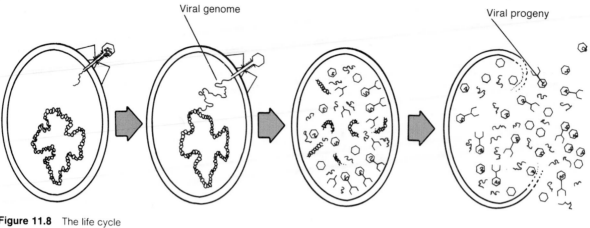

Viral genome

Viral progeny

Cell lysis

Figure 11.8 The life cycle of a lytic virus. The virus infects the cell by injecting its genome into the cell. The viral genes use the cell's nucleic acid and protein-synthesizing apparatus to make viral protein and nucleic acid. The viral particles are assembled and then the cell ruptures, releasing all the viral progeny.

virus) and nucleic acid, at the expense of the host cell's normal metabolism. Such infections generally end with the death of the cell and the release of hundreds of new viruses.

Many familiar diseases are caused by viral infections. DNA viruses cause smallpox, warts, mononucleosis, fever blisters or cold sores, shingles, and genital herpes. Among the common RNA viruses are those causing polio, rabies, mumps, measles, encephalitis, and influenza. As described in chapter 10, viruses cause certain types of cancers. For the remainder of this chapter, we consider aspects of the life history of viruses and focus on the ways that viruses interact with the immune system to produce certain diseases, including AIDS.

Life cycles of viruses

Viral infections have different outcomes. The virus may immediately use the host's metabolic machinery to replicate itself many times. Following replication, the progeny viruses are released, killing the host cell. This is known as a **lytic infection** (figure 11.8).

Alternatively, the virus can enter the host cell in a **latent** or relatively inactive form. Viral DNA attaches to and becomes an integral part of the host's DNA. When normal cell division occurs, the viral DNA is replicated along with the host DNA. Thus, each descendant cell bears a copy of the latent virus genome within its own genome. This type of infection, known as **lysogenic infection** (figure 11.9), may persist for years within a host cell and its descendants without producing pathogenic effects. While in its latent phase, a lysogenic virus sometimes appears to confer benefits upon the host cell. For example, the host may be protected from infection by other viruses. Infected cells may have growth advantages over other cells.

Following infection, the host cell grows normally until subjected to an environmental stress such as radiation, heat, or toxic chemicals. Such an assault can cause the lysogenic virus to enter a lytic cycle. The viral DNA then excises itself from the host chromosome and causes the host to produce hundreds of new viruses. After a short period, the host cell lyses, releasing infective viral progeny. Thus, the life cycle of a lysogenic virus terminates in a lytic phase.

Finally, some animal viruses are able to infect and remain latent within cells without integrating their genomes into that of the host cell. These viruses may cause little harm to the host cells and replicate with the host until some stimulus triggers them to become virulent or cause cellular transformation.

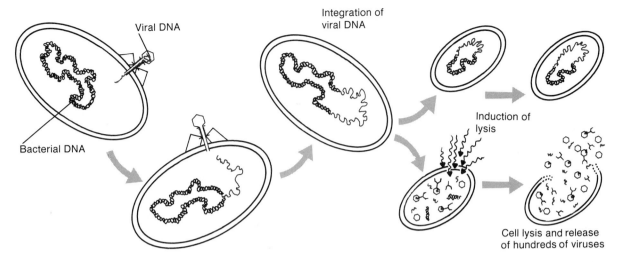

Viral DNA

Integration of viral DNA

Bacterial DNA

Induction of lysis

Cell lysis and release of hundreds of viruses

Figure 11.9 The life cycle of a lysogenic virus in a bacterium. Following injection of the viral genome into the cell, the viral DNA integrates into the bacterial DNA. As the bacterium replicates, the viral genome is copied also, thus spreading throughout the bacterial population. Later, often in response to shock or stress, the viral DNA is excised from the bacterial genome. The virus then enters a lytic phase, culminating in the death of the bacterium and the release of viral progeny.

Reverse transcriptase

RNA viruses cannot insert their RNA genome into the DNA of the host cell. For an RNA virus to be integrated into the genome of an host cell, it must first make a DNA transcript of itself. In addition to their RNA genome, some RNA viruses (**retroviruses**) carry an enzyme known as **reverse transcriptase.** Upon entering the cell, this enzyme builds viral DNA from the viral RNA genome, in direct contrast to the information flow predicted by the central dogma (chapter 8). Following the action of reverse transcriptase, the DNA transcript of the viral RNA genome is incorporated into the DNA of the host cell. The remainder of the life cycle of the RNA virus is similar to that of the DNA virus (figure 11.10).

A large group of retroviruses have now been identified. These viruses have a normal lysogenic life cycle, inserting a DNA copy of their genomes into infected host cells. Additionally, as described in chapter 10, they have been found to cause cancer in many species.

Transduction

Viruses sometimes move segments of DNA from one cell to another, a process known as **transduction.** Transduction can be considered a form of bacterial genetic recombination. When a bacterium lyses due to a viral infection, some of the viral progeny may include segments of the DNA of the bacterial host. When the virus subsequently infects another bacterium, the DNA from the first host can be transferred to the new host.

Transduction is one mechanism responsible for the widespread drug resistance of many bacteria. In bacteria, genes controlling drug resistance are often carried on pieces of extrachromosomal DNA known as **plasmids.** Transducing viruses are responsible for spreading these drug resistant genes among a wide variety of bacteria, particularly in hospitals. While drug resistant strains of bacteria pose an obvious medical problem, genetic engineers hope to use transduction for spreading beneficial genes between valuable domestic species. Genes for resistance to disease or insects could be transmitted among plants by such a technique.

Transduction, however, is only one way in which genes can spread through a population. More often, a gene for drug resistance is disseminated because its carrier survives and transmits the gene to its descendants, i.e., through the action of

Figure 11.10 The life cycle of an RNA tumor virus involves an extra step over that of a lysogenic DNA virus. The viral RNA and reverse transcriptase are injected into a cell. A DNA copy of the entire viral genome is then made and this DNA is incorporated into the host's genome. Then by normal transcription, viral RNA, more enzyme, and viral protein are made, producing mature viral progeny.

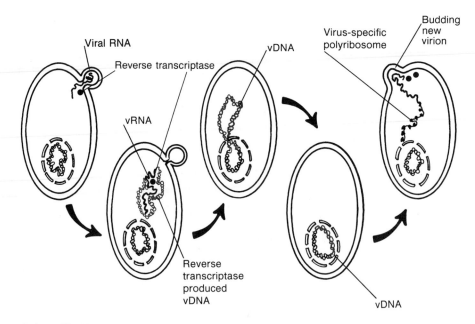

Color = Viral DNA made from vRNA using reverse transcriptase●

natural selection. Transduction differs from natural selection in spreading drug resistance because (1) only one, or a few, genes are transmitted at one time and (2) genes may be transmitted across species boundaries. It is this latter trait that makes transduction of particular importance to genetic engineers.

Viral diseases

As noted earlier, viruses are the cause of many serious diseases. Viral diseases are particularly pernicious for two reasons. First, unlike bacteria, viruses are not usually susceptible to treatment with antibiotic drugs. Viruses must simply be allowed to run their course, treating the symptoms but relying upon the body's own immune mechanisms to fend off the disease.

Second, after the first infection, **latent viruses** can retreat to tissues of the nervous system and remain hidden there for years. They may cause no further symptoms or they may recur with varying severity from time to time, as with oral herpes (which produces cold sores in the mouth) and genital herpes. The presence of latent viruses can be detected by the presence of viral antigens in the host's tissues or by viral DNA sequences in some of the host's cells.

The confirmation in 1981 that the **human T-cell leukemia virus (HTLV)** causes a rare form of human leukemia was a major advance in understanding viral disease, but the ability of viruses to attack cells, using their own metabolic apparatus, means that the fight against viral disease will be especially difficult. We now briefly examine a few viral diseases whose pathology appears to depend upon their ability to disrupt or evade the immune mechanism.

Mononucleosis and EBV

Burkitt's lymphoma (figure 10.18) is a relatively common cancer in African children but is rare elsewhere in the world. Tumor cells from virtually all persons with Burkitt's lymphoma are infected with the **Epstein-Barr virus (EBV),** strongly implicating the EBV as the cause of the cancer. Surprisingly, about 80% of the world's population are infected with EBV, although the cancer is limited primarily to Africa.

Commonly an EBV infection produces only a mild cold, and the affected individual develops antibodies to the virus, a response similar to that of any viral influenza. In the United States, a small fraction of persons exposed to EBV develop **infectious mononucleosis.** First described in 1920, mononucleosis is characterized by fever, tonsillitis, swollen lymph glands, and an enlarged liver and spleen. It is *not* highly infectious and usually runs its course in one to four weeks, depending upon the patient's health and genotype. Only about 1% of all patients develop complications such as pneumonia or kidney failure. Mononucleosis is also characterized by an increase in the white blood cells known as mononucleocytes, causing some scientists to consider mononucleosis a form of self-limiting leukemia.

EBV enters the body through the mouth or respiratory tract. Mononucleosis develops in four to seven weeks. An EBV infection normally produces permanent immunity, but the virus persists in a lysogenic form in the B cells. It is sporadically released and thus spreads through the population. Approximately 35%–80% of students entering college have antibodies to EBV. Each year about 10%–15% of those students without antibodies to EBV catch mononucleosis from their peers, who probably had no more than a cold from their initial infection with EBV. Mononucleosis may be a disease whose incidence has actually increased with improved hygiene. Prior to the twentieth century, essentially all persons may have been exposed to EBV (and numerous other viruses) at an early age. Such early exposure, at a time when the immune system is most active, probably resulted in a mild disease, in contrast to the potentially serious illness experienced by college students, who may not be exposed to the virus until well past puberty.

Influenza

Three major types of **influenza** viruses (figure 11.7b), designated A, B, and C, are recognized by unique characteristics of their surface antigens. Types A and B are responsible for flu epidemics, while type C usually produces only minor symptoms. Influenza viruses, unlike other viruses, regularly change the antigenic determinants on their protein coat. These slight alterations, known as **antigenic drift,** permit the virus to escape recognition by the immune system, making humans susceptible to repeated bouts of flu.

Major alterations in the antigens of influenza viruses, **antigenic shifts,** occur less frequently, about every ten years. These changes can produce widespread, potentially severe flu epidemics. In only six weeks in 1918, a new flu virus was responsible for the deaths of at least twenty million persons and affected perhaps one hundred times that number. After major antigenic shifts, the Asian flu (1957) and Hong Kong flu (1968) epidemics also affected large portions of the world's population.

Influenza viruses are so insidious and ubiquitous because of their ability to undergo antigenic drift and antigenic shifts. These viruses accomplish antigenic drift by rearranging their genes, as well as by deletions, insertions, and point mutations in their genomes. These genetic changes can alter the antigenic determinants of a specific strain of virus, making the strain immunologically unrecognizable to antibodies produced in response to earlier infections. Such genetic changes are highly adaptive for viral survival. In effect, the virus is engaged in a continual arms race with the immune system of the host, the virus changing just enough to elude the host's antibodies.

Antigenic shifts and antigenic drift require that new flu vaccines be produced annually. Samples of influenza virus from around the world are monitored regularly in an attempt to identify altered strains that might be particularly virulent. Once such strains are identified, new vaccines are prepared.

Herpes

Herpes viruses (figure 11.7a) are DNA viruses that are found in all animal populations. Five distinct types of herpes infect humans: (1) type 1 herpes, causing facial cold sores; (2) type 2 herpes, responsible for genital herpes; (3) EBV, causing infectious mononucleosis; (4) varicella-zoster virus, producing chicken pox and shingles; and (5) cytomegalovirus, which results in serious mental retardation in an estimated 1% of all newborn infants in the United States. The effects of herpes viruses have been known to humans for at least two thousand years (the Roman Emperor Tiberius banned kissing at public ceremonies in an attempt to stop an epidemic of oral herpes).

Oral herpes infects at least eighty million Americans. Genital herpes, a far more serious problem, affects an estimated twenty million Americans. About six hundred thousand new cases are reported annually. Recent evidence shows that both oral and genital herpes viruses can be transmitted to either part of the body during oral sex. The initial manifestation of genital herpes is fluid-filled blisters on the genitalia. The extremely painful symptoms persist for several weeks.

Neither oral nor genital herpes is curable. As the blisters heal, the virus retreats to cells in the nervous system, where it remains in a latent condition. Some victims never experience another outbreak of the sores. However, many herpes victims have a recurrence of symptoms once or twice a year, while in others recurrence is far more frequent. Furthermore, *all victims* harbor the virus and are subject to an outbreak at times of stress.

Herpes viruses may have an even more serious effect on human health. Herpes viruses are known to cause cancer in frogs and chickens. Such a relationship has yet to be proven in humans, but women with genital herpes are eight to ten times more likely to develop cervical cancer than noninfected women.

Multiple sclerosis and measles

First described about 130 years ago, **multiple sclerosis (MS)** afflicts about two million people worldwide. MS is an autoimmune disease in which an individual's own antibodies attack the myelin tissue lining the central nervous system. Myelin is essential for transmission of information throughout the nervous system. Onset of the disease most commonly occurs between the ages of twenty and forty; twice as many women as men develop the disease. As the disease progresses, scar tissue forms in the myelin, preventing the flow of messages throughout the nervous system. Gradually the victim loses motor control and balance, eventually becoming paralyzed.

Viruses are strongly implicated in MS. Essentially all MS victims have had measles. Measles viral nucleic acid can be identified in one of four MS patients studied. The characteristic measles viral antigen has also been detected in the central nervous systems of MS patients, but not in other tissues of the body. Thus, MS may be initiated by entry of the measles virus into the central nervous system, where it subsequently flourishes. Normally, the central nervous system is strictly protected from infections of any kind. Persons who contract MS may have some genetic defect that permits the measles virus access to their central nervous system. Recent studies report a deficiency of interferon, which normally aids in the fight against viral infection, in patients with MS.

Measles is a **membrane budding virus.** When it replicates, it packages its RNA into the viral coat at the host cell wall (figure 11.11). During packaging, cell protein may be accidentally incorporated into the new virus. Although the nervous system is normally immunologically privileged (i.e., isolated from the body's immune mechanisms), viruses accidentally bearing proteins from the nervous system could

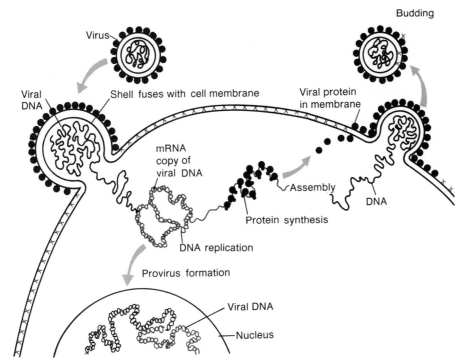

Budding

Virus

Viral DNA

Shell fuses with cell membrane

Viral protein in membrane

mRNA copy of viral DNA

Assembly

DNA

Protein synthesis

DNA replication

Provirus formation

Viral DNA

Nucleus

Figure 11.11 Membrane budding viruses fuse with the host cell membrane and inject the viral genome into the host cell. When new viruses are produced, portions of the host cell membrane are incorporated into the viral protein coat. Such viruses thus serve to move pieces of host membrane to distant parts of the body, perhaps inducing autoimmune diseases.

enter the circulatory system as a result of membrane budding. The immune system does not recognize the protein (myelin, in the case of MS) as self and thus begins producing antimyelin antibodies, initiating the disease.

This scenario, analogous to the postulated relationship of EBV and mononucleosis, is reasonable, but has not been proven conclusively. Evidence from experiments with mice, however, is compelling. If samples of myelin are removed from a mouse, ground up, and then injected into the same mouse, symptoms identical to MS develop. Thus, it seems reasonable that the measles virus or other membrane budding viruses may provide a vehicle for exposing a person to their own myelin, as if the myelin were foreign tissue, which triggers the immune response and results in MS.

Age is another factor involved in the development of MS. Most persons have measles, but do not develop MS. Persons who contract measles after their teens are at higher risk of developing MS than those who had measles before age thirteen. Because the immune system is most efficient in juveniles, persons contracting measles early in life may mount an immune response sufficient to rid the body of the virus entirely. If, however, measles is not contracted until the late teens or twenties, the immune system may not cope with the virus as effectively. A latent reservoir of the virus could thus persist in the body, making the person prone to later development of MS.

Insulin dependent diabetes and mumps

Diabetes is an inability to metabolize carbohydrates properly due to a deficiency in the production of insulin. One form, *maturity onset diabetes,* can normally be controlled by diet, exercise, and oral drugs. This type of diabetes has a strong genetic component. Over 85% of the victims have at least one diabetic parent, and twin studies show high concordance for this condition. If one twin contracted this type of diabetes, the other twin did also.

Insulin dependent diabetes is a severe disease, requiring daily insulin injections. It normally affects children, only 11% of whom have a diabetic parent. In insulin dependent diabetes, the white blood cells appear to participate in an immune reaction that destroys the insulin-producing cells of the pancreas. A relationship of this form of diabetes with mumps (caused by a virus) was first suggested by a Norwegian doctor in 1864. Like the measles virus, the mumps virus is a membrane budding virus. If the mumps virus is able to persist in the pancreas, the viruses could incorporate pancreatic antigens that stimulate an autoimmune response, just as in MS.

Insulin dependent diabetes is also related to HLA types; about 60% of the victims are either type B8 or type BW15 (table 4.5). These particular HLA genotypes may make individuals more susceptible to viral infection or less able to rid their bodies of the offending virus, thus triggering diabetes. Both MS and insulin dependent diabetes clearly illustrate the complex interactions of latent viruses, genotype, and chance in the development of serious disease.

AIDS

The 1980s will be remembered in medical history as the decade that saw the onset of the extraordinary disease AIDS. No disease has ever been studied so intensively and so immediately once it was recognized as an epidemic of worldwide consequence. In less than a decade, AIDS has become one of the major causes of death by infectious disease throughout the world. Its rate of increase is geometric and shows no sign of slowing down (figure 11.12).

AIDS is the acronym for **acquired immune deficiency syndrome.** Acquired means that it is an infectious disease caught from another person. Immune deficiency means that the disease affects the immune system, drastically reducing its ability to fend off other infectious organisms or parasites. Syndrome means that the single condition has many effects.

AIDS was first clinically identified in 1981, when several physicians in the United States noticed an increased incidence of the rare cancer Kaposi's sarcoma in young homosexual males. Kaposi's sarcoma was previously associated primarily with patients who had been treated with drugs that suppressed the immune system, often in conjunction with transplants. Once the condition was clinically recognized, the number of cases increased rapidly, and in 1982 the name AIDS was given to the condition.

The HIV virus Medical researchers throughout the world immediately realized the potential threats posed by AIDS and established research programs on both the medical and social aspects of the disease. In 1983 the cause of AIDS was identified as a virus, and in 1986 this virus was officially designated the human immunodeficiency virus—HIV. The entire genome of HIV has now been sequenced.

HIV is a retrovirus, incorporating itself into the host's genome and then releasing up to fifty thousand progeny at a time, time after time. As part of the host's genome, HIV is well protected from drugs. To destroy the cause of the disease, the drug would have to destroy the host's cells. Several experimental drugs are now undergoing human trials. The most promising drug, AZT, interferes with the replication of the AIDS virus.

Prevention of AIDS by vaccine is difficult because of several traits of the HIV virus. First, the HIV virus is subject to antigenic drift. In fact the AIDS virus mutates five times faster than the influenza virus, formerly thought to be the most variable virus. Thus, a single vaccine is unlikely to prove effective against all variants of the virus, which are continually undergoing minor genetic alteration. Second,

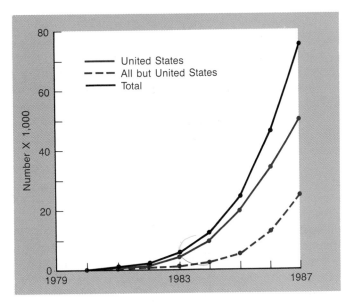

Figure 11.12 Cumulative number of cases of AIDS in the United States and those countries who reported to the World Health Organization by year of diagnosis.
Source: P. Piot et al., "AIDS: An International Perspective," *Science*, Vol. 239, pp. 573–79, February 5, 1988. Copyright 1988 by the AAAS.

several major variants or types of HIV virus have now been identified, differing by over 25% of their genome. Since AIDS may be caused by several types of the HIV virus, a single vaccine would be only partially effective at best.

The HIV virus has a long and variable latent period. About half the people with the HIV virus or antibodies to it develop AIDS disease within five years. The disease appears to have 100% mortality; victims die, on average, about four hundred days after first diagnosis, most of that time being spent in hospitals. Roughly 80% of AIDS victims die within two years of diagnosis.

AIDS is lethal because the HIV virus attacks the body's primary immune defenses. The main target is helper T-cells, but macrophages and B-cells are also affected. This weakens the entire immune mechanism, and AIDS victims generally die as a result of secondary infection from one or more of over thirty types of disease organisms—other viruses, bacteria, protozoa, or parasites, none of which would normally be fatal.

Additionally, the HIV virus appears to be able to penetrate the central nervous system, and progressive dementia is a common symptom of the disease. Psychological impairment is particularly dangerous because of its intrinsically distressing nature to the victim, family, and friends and because it may reduce the judgment of the victim. Thus the victim may behave irresponsibly, engaging in behaviors dangerous to others and increasing the likelihood of further spread of the disease.

Epidemiology The HIV virus can live in a wide variety of body fluids, but the disease is contracted primarily by contact with the semen or blood of an infected person. The primary modes of infection have been sexual contact, sharing of dirty needles by drug addicts, and blood transfusions. Some health workers have contracted the disease by accidental needle sticks. Additionally, mothers may transmit the virus to their children either *in utero* or via maternal milk.

Contrary to early claims, AIDS is not a "gay disease"; AIDS is not restricted to any particular societal group. All segments of society have been exposed to the HIV virus, and anyone exposed to it may develop AIDS. Certain behaviors are especially risky. Sharing needles while using drugs greatly increases the risk of contracting AIDS, as does promiscuous sexual behavior, particularly without the use of a condom. Hemophiliacs as a group are at risk because of their need for frequent blood transfusions, although screening of donated blood has reduced the risk.

Male homosexuals have comprised a major "at risk" group partly because of high levels of promiscuity. The onset of the AIDS epidemic in North America has convincingly been linked to a homosexual man called Patient Zero, an Air Canada steward who died of the disease in 1982. He or his sexual partners are known to have had sexual contact with a majority of the first cases in the United States and Canada. Patient Zero had as many as 250 sexual partners per year, even after his condition had been diagnosed. The AIDS epidemic is considered to have begun in 1981, although antibodies to the AIDS virus have now been retrospectively diagnosed in samples of blood stored in blood banks since the late 1950s.

Data are limited, but it seems clear that anal intercourse is an effective mode of transmission, probably because the cells lining the rectum do not form an efficient nonspecific defense against the virus. In contrast, levels of transmission of the HIV virus from males to females during vaginal intercourse are lower, perhaps because the vaginal lining is a better barrier. For unknown reasons, the transmission rate from infected females to males is even lower. The importance of an efficient barrier is demonstrated by the relatively high rate of transmission via sexual intercourse to individuals who have an open lesion such as a herpes lesion.

Since 1981 when the Centers for Disease Control in Atlanta (CDC) began gathering statistics, the rate of increase of AIDS has been tremendous. In mid-1985, about 14,000 cases had been diagnosed in the United States. By late 1987, 45,000 persons were diagnosed and 25,000 had died from AIDS. Because of the latent period of the disease, the situation is far worse than these statistics indicate. Over 1,000,000 people in the United States are believed to be infected, and many will certainly develop the disease. By 1991, about 270,000 people in the United States are predicted to have developed the disease, and 54,000 will be dying annually. About 74,000 new cases are predicted in 1991 alone.

Worldwide, other developed countries are lagging behind the United States in numbers of AIDS cases, but in parts of the Caribbean and especially in Central Africa, infection rates are believed to be as high as 10%. In some Central African cities, over 80% of the prostitutes have AIDS antibodies. The potential of AIDS to cause death on a worldwide scale cannot be underestimated.

Ethical issues   The massive and deadly impact of AIDS raises many ethical questions for society. It also poses a logistical nightmare to the public healthcare system. Medical costs for AIDS victims average about fifty thousand dollars per person per year. There is no cure, so treatment is aimed at reducing pain and lengthening survival. Drugs such as AZT are now being used to treat patients, but costs are high, about ten thousand dollars per year and initially supplies were limited. Can society afford these costs as the number of cases increases in coming years? Who should, and who should not, receive the expensive drugs? Remember that the effect of present drugs may complicate the situation by prolonging life and increasing the cost that must be absorbed by an already stressed healthcare system.

Part of the difficulty in developing treatments and vaccines arises from the ethics of experimentation on humans. In first trials of drugs on humans, some victims are used as a *control group* in the study and given a *placebo* treatment; that is, persons in the control group are administered a treatment that has no drug in it as a standard for comparison against the group given the test drug. But in the case of a fatal disease, if the group given the test drug shows significant improvement, it is ethically unjustifiable to deny treatment to those individuals in the placebo group. In the case of AZT, for example, the initial positive results immediately led to termination of the experiment by administering AZT to the control group. This may be beneficial for individuals in the control group but it means that proper scientific tests of the new drugs are never completed. Serious long-term side effects, for example, may easily be missed.

Similarly, in the case of test vaccines, if the vaccine is faulty, it may actually induce the disease in individuals who are experimentally vaccinated. Or, if the vaccine is ineffective, people who are vaccinated and then exposed to the disease may also develop the disease. The first experimental AIDS vaccine is now being tested, and the French scientist who developed it insisted on being the first person to be tested.

The development of tests for both AIDS antibodies and the HIV virus have led to calls for mass screening of the public and of particular groups. In the United States, testing is required for members of the military and foreign service. Some insurance companies require AIDS testing prior to granting a life insurance or a medical policy, as do some companies prior to hiring. Some persons suggest that all prisoners and immigrants should be tested for AIDS.

Civil rights groups, on the other hand, have objected to mass screening. Screening immigrants, is reminiscent of the racist immigration policies of the early twentieth century. More significantly, having AIDS antibodies is not the same as having AIDS. The latent period of AIDS is still poorly defined. Is it fair to exclude a person with AIDS antibodies from a job when she or he may not develop the disease for ten years or may never develop AIDS? The proportion of people with antibodies who will develop AIDS in unknown, but estimated at only 10%.

There is neither a cure nor a vaccine for AIDS, so prevention is generally accepted as the most suitable way to halt spread of the disease. Prevention is possible only by education that causes people to reduce or eliminate their risky behaviors. Programs have begun to educate drug users about the dangers of sharing needles. Some people advocate allowing drug users to trade in their needles for clean ones. Other people find such programs objectionable on the grounds that these programs encourage or condone drug use.

Worldwide, two major types of sex education programs have been initiated to reduce the spread of AIDS. First, people are encouraged to limit their sexual contacts. Second, the use of condoms is advocated as an effective barrier to the HIV virus. In many countries, public service advertisements for condoms have been extremely explicit, clearly making the point that failure to use condoms during intercourse may lead to the death of sexually active individuals.

Television advertisements for condoms in the United States have met with a mixed response, again raising objections from those who believe they may encourage promiscuity or other objectionable behaviors. Although the federal government has been willing to spend large amounts of money on AIDS research, it has been unwilling to lead an aggressive program of AIDS education, despite the near unanimity of the scientific and medical communities on this issue.

An interesting comparison is with the position of the federal government and public media with respect to tobacco use and advertising. Use of tobacco is a major public health problem affecting hundreds of thousands of persons annually and costing billions of dollars. Recall that an estimated 350,000 persons in the United States die each year from smoking related diseases. Tobacco is perhaps the only product widely marketed that is guaranteed to be harmful to your health if used as recommended. Nonetheless, the federal government has only reluctantly limited tobacco advertising on television and other media. Additionally, the tobacco industry is massively subsidized by public money to support a habit that has been widespread in the United States only since World War II.

Sexual activity, on the other hand, is not merely a common practice in all human societies, it is a fundamental human behavior.

Summary

1. The body is protected from invasion by dangerous foreign substances by two different categories of defenses. Nonspecific defenses, such as the skin and various secretions, are effective against a broad range of harmful agents. Specific defenses or immune mechanisms develop after exposure to a particular foreign antigen.

2. Lymphocytes, a type of white blood cell, differentiate to either T cells or B cells, the primary agents of the immune response. T cells provide cellular immunity by directly destroying foreign antigens. B cells differentiate further to plasma cells, which in turn produce antibodies. Antibodies destroy foreign antigens in a process called humoral immunity.

3. Macrophages interact with helper T cells, assisting in the immune response.

4. Specific immunity does not develop until after birth. Prior to birth, the immune system is active in identifying its own tissues as self.

5. Antibodies are produced against antigenic determinants, portions of foreign molecules that are recognized as non-self.

6. Following an immune response, some T cells and some B cells survive for long periods as memory cells. In the event of further infection by the same agent to which the cells initially responded, memory cells facilitate an extremely rapid second response.

7. Five types of antibodies or immunoglobulins are known. IgG, IgM, IgA, IgD, and IgE.

8. The clonal selection theory of antibody diversity attributes the specificity of the immune system to recombination among multiple genes, as well as somatic mutation in differentiating plasma cells.

9. Specific immunity is acquired after birth. It may be acquired passively, by receiving antibodies from another person; or it may be acquired actively, by developing one's own antibodies. Vaccinations induce actively acquired immunity.

10. Transplant rejection is primarily a function of T cells. It can be prevented by precise matching of the human leukocyte antigen (HLA) type of the donor and recipient. The HLA system is genetically extremely diverse. When HLA types cannot be precisely matched, immunosuppressant drugs are used.

11. Allergies are a form of immune hypersensitivity and appear to have a genetic basis. Particular allergens elicit overproduction of IgE and histamine release, hallmarks of the allergic response.

12. Autoimmune diseases occur when individuals produce autoantibodies to their own tissues. Systemic lupus erythematosus (SLE), multiple sclerosis, rheumatoid arthritis, and rheumatic fever are important autoimmune diseases.

13. Viruses are extremely simple organisms, consisting only of an outer protein coat surrounding an inner core of nucleic acid. Viruses cannot replicate independently of living cells.

14. Lytic viral infections cause immediate death to the host cell accompanied by the release of progeny viruses. In lysogenic viral infections, the viral genome is incorporated into the host cell genome and is replicated each time the cell divides. Upon appropriate stimulation, the virus ends the latent phase and enters a lytic cycle, killing the host as it reproduces.

Chapter 11

15. The life cycle of an RNA virus involves an additional step for a lysogenic infection: the RNA must be copied into a DNA strand that can then be incorporated into the host cell genome. Reverse transcriptase is an enzyme made by RNA viruses that permits the synthesis of a DNA molecule from an RNA template.

16. Transduction is the transfer of DNA from one cell to another by a virus. Resistance to antibiotics has been transferred among medically important bacteria by this process.

17. Viral involvement has been shown in mononucleosis and multiple sclerosis. These diseases appear to develop when the victim is exposed to EBV or measles viruses after their teens. The seriousness of the disease is related to several aspects of the victim's genotype, including their HLA genotype.

18. The herpes viruses are an example of latent viruses that may persist for decades in a relatively inactive form in the victim's nervous system.

19. Influenza is a persistent human disease because the flu viruses are capable of antigenic drift and antigenic shifts. These two genetic abilities permit the virus to escape recognition by the immune system and prevent populations from developing immunity to these potentially deadly diseases.

20. AIDS is an insidious infectious disease caused by HIV, a retrovirus that is so genetically variable that it is difficult to produce an effective vaccine for it. AIDS can be prevented by appropriate behavior, but no cure is presently available. The AIDS virus infects cells of the immune system, leaving its victims unable to defend themselves against normally nonfatal infections.

Key Words and Phrases

acquired immune deficiency syndrome (AIDS)

acquired immunity

active immunity

agammaglobulinemia

antibody

antigen

antigenic determinant

antigenic drift

antigenic shift

autoimmune disease

B cell

Burkitt's lymphoma

cardiovascular system

cellular immunity

clonal selection theory

Epstein–Barr virus (EBV)

erythroblastosis fetalis

herpes virus

histamine

HLA

human T-cell leukemia virus (HTLV)

humoral immunity

immunoglobulin

immunologically privileged site

infectious mononucleosis

influenza

insulin dependent diabetes

latent virus

leukocyte

lymphatic system

lymphocyte

lysogenic infection

lytic infection

macrophage

mast cell

membrane budding virus

memory cell

multiple sclerosis (MS)

natural resistance

passive immunity

plasma cell

plasmid

retrovirus

reverse transcriptase

specific immunity

systemic lupus erythematosus (SLE)

T cell

thymus gland

transduction

Questions

1. Distinguish between active immunity and passive immunity.
2. What are autoimmune diseases and how might they be treated?
3. Why is the HLA system potentially more useful than the red blood cell antigens in paternity suits?
4. Distinguish between T cells, B cells, macrophages, lymphocytes, and leukocytes.
5. Of what importance is the lymphatic circulatory system?
6. Why is Immunoglobulin E (IgE) called the allergy antibody? How does it produce the uncomfortable allergy symptoms?
7. The testes are an immunologically privileged site. If testicular material were surgically removed from a laboratory rat, minced finely, and reinjected under the skin, as with a vaccination, what do you predict would happen?
8. What success would you predict for transplants of testes between unrelated males; between twin brothers?
9. Distinguish between transformation and transduction.
10. Distinguish between lytic viruses and lysogenic viruses.
11. Why are unconventional viruses so difficult to study and the diseases they cause so enigmatic?
12. What is the significance of the enzyme reverse transcriptase?
13. How are viruses implicated in some autoimmune and degenerative diseases?
14. What is the relevance of antigenic drift to the difficulty of producing effective influenza vaccines?
15. What is the best way to protect yourself from AIDS?
16. Why do only some of the persons infected with the AIDS virus develop the actual AIDS disease?

For More Information

Ada, G. L., and G. Nossal. 1987. "The Clonal-Selection Theory." *Scientific American,* 257:62–9.

Arehart-Treichel, J. 1982. "Probing the Causes of MS." *Science News,* 121:76–77.

Barnes, D. M. 1986. "Nervous and Immune System Disorders Linked in a Variety of Diseases." *Science,* 232:160–61.

Barnes, D. M. 1986. "Strategies for an AIDS Vaccine." *Science,* 233:1149–53.

Brockes, J. P. 1982. "Nerve, Myelin, and Multiple Sclerosis." *Engineering and Science,* March:9–14.

Clark, M., and M. Gosnell. 1979. "Battling Diabetes." *Newsweek,* December 10:117–25.

Cromie, W. J. 1981. "Genetic Trade-offs: The Hidden Cost of Survival." *Sciquest,* January:10–14.

Edwards, D. D. 1987. "Still Stalking MS." *Science News,* 132:234–35.

Gallo, R. C. 1986. "The First Human Retrovirus." *Scientific American,* 225:88–98.

Gallo, R. C. 1987. "The AIDS Virus." *Scientific American,* 256:46–56.

Gonda, M. A. 1986. "The Natural History of AIDS." *Natural History,* 95:78–81.

Hamburger, R. N. 1976. "Allergy and the Immune System." *American Scientist,* 84:157–64.

Henle, W.; G. Henle; and E. T. Lennette. 1979. "The Epstein-Barr Virus." *Scientific American,* 241:48–59.

Hood, L. E. 1979. "Immunity, Disease, and Cancer." *Engineering and Science,* Jan.–Feb.:6–12.

Hunkapiller, T., and L. Hood. 1986. "The Growing Immunoglobulin Gene Superfamily." *Nature,* 323:15–16.

Jaret, P. 1986. "Our Immune System: The Wars Within." *National Geographic,* 169:702–34.

Kolata, G. 1982. "Alzheimer's Research Poses Dilemma." *Science,* 215:47–48.

Lewin, R. 1981. "New Reports of a Human Leukemia Virus." *Science,* 214:530–31.

Marx, J. L. 1975. "Cytomegalovirus: A Major Cause of Birth Defects." *Science,* 190:1184–86.

Marx, J. L. 1983. "Human T-Cell Leukemia Virus Linked to AIDS." *Science,* 220:806–9.

Maugh, T. H., II. 1975. "Diabetes: Epidemiology Suggests a Viral Connection." *Science,* 188:347–51.

Miller, J. A. 1982. "Slow Viruses: The Body's Secret Agents." *Science News,* 121:332–33.

Piot, P. et al. 1988. "AIDS: An International Perspective." *Science,* 239:573–79.

Rapp, F. 1978. "Herpes Viruses, Venereal Disease, and Cancer." *American Scientist,* 66:670–74.

Thompson, L. 1988. "AIDS Diary." *Discover,* 9:36–8.

Time. 1982. "The New Scarlet Letter and Battling an Elusive Invader (Herpes)." August 2:62–66, 68–69.

Tonegawa, S. 1985. "The Molecules of the Immune System." *Scientific American,* 253:122–31.

Treichel, J. A. 1982. "Monoclonal Antibodies Correct Severe Immune Deficiency." *Science News,* 122:244.

Varmus, H. 1987. "Reverse Transcription." *Scientific American,* 257:56–65.

Wallis, C. 1987. "Back to Normal: Hope for Parkinson's Victims." *Time,* April 13:39.

Weiss, R. 1987. "Cell Grafts Proceed, Value Uncertain." *Science News,* 132:341 (re: Parkinson's treatment).

Young, J. D., and Z. A. Cohn. 1988. "How Killer Cells Kill." *Scientific American,* 258:38–44.

Genetic diversity

Figure 12.1 The different races of humans: (a) African black (Negroid); (b) European white (Caucasoid); (c) American Indian (Mongoloid); (d) Australian aborigine (Australoid).
Source: (a) Margaret Thompson/Anthro Photo; (b) Rick Smolan; (c) Shostak/Anthro Photo; (d) Neg. #333218, Courtesy Department of Library Services, American Museum of Natural History.

(a)

(b)

(c)

(d)

Genes are instructions for survival. A set of genes whose information matches (is *adapted* to) requirements of an individual's environment allows that individual to survive and reproduce. Maladapted genes lead to early death without reproduction.

Genes do not have an independent existence. They occur as integrated sets of information, analogous to a computer program, called the **genome.** In sexually reproducing species, there is a larger unit of integration, as well—the **gene pool.** The gene pool consists of all the genetic information, all the alleles, found in a group of interbreeding individuals. Thus for a population of N diploid individuals, the gene pool consists of all 2N alleles. All humans have the potential for interbreeding and thus all belong to one gene pool. The human gene pool, however, consists of many smaller gene pools determined by religious, social, and geographic factors. Because of this subdivision we recognize a number of human **races** (figure 12.1). In genetics, races are characterized by the frequency of alleles in a given population, and by any alleles that may be unique to that particular gene pool. In reality, all humans have the same genes. We differ only in the particular alleles found at specific loci.

Each species may comprise a large gene pool which is commonly subdivided into numerous smaller local gene pools. The total genetic variation found within the gene pools of individual species and also found in all species together is referred to as **genetic diversity.** Genetic diversity is the basis of all life on earth. Without genetic differences between individuals and between species, changing environments would quickly exterminate life.

Additionally, life is interdependent. Animals depend on plants and other animals for food. Plants depend on animals for dispersal and fertilization. Without other species, human life would not exist. Thus, humans need to understand the processes behind the present diversity of life on earth. How is genetic diversity generated and maintained? How can we ensure that the genetic basis for our continued existence will not be progressively depleted? How can we use genetic resources to sustain us in the future?

Population genetics is the study of gene pools and genetic diversity. Population geneticists try to understand how populations differentiate and how they will change in the future. In short, they study the processes of **biological evolution.** Population genetics is one of the most elegant and mathematical of the biological sciences because it is based on a central unifying principle, the **Hardy-Weinberg equilibrium equations.** In this chapter, we briefly introduce some concepts of population genetics as they relate to humans.

The Hardy-Weinberg Law

When the principles of Mendelian inheritance were rediscovered at the turn of this century, it was not understood how recessive alleles could persist in populations. Why, if brown eyes are dominant to blue eyes, do not all persons have brown eyes?

The answers to questions of this nature were provided in 1908. An English mathematician, G. H. Hardy, and a German physician, W. Weinberg, independently developed a mathematical model that showed how the frequency of recessive alleles in a population could be estimated knowing only the frequency of persons homozygous for the recessive allele.

The *Hardy-Weinberg equilibrium equations* apply to diploid, sexually reproducing populations and have five key underlying assumptions:

1. Genetic mutation must not occur.
2. Individuals must not migrate from one population to another.
3. Individuals must mate randomly with respect to genotype.
4. Reproduction and mortality must be the same for all genotypes—that is, *natural selection must not occur.*
5. The population must be very large.

In a mathematical sense, the Hardy-Weinberg equilibrium equations describe the fate of genetic variation in an "ideal" population that is not subjected to disturbing influences. Such ideal conditions virtually never occur in nature, yet many populations seem to conform to the expectations described by the Hardy-Weinberg equations. On the other hand, where populations differ from Hardy-Weinberg predictions, we can use the five underlying assumptions as a guide to understanding the process of genetic change and population divergence.

The Hardy-Weinberg equations describe frequencies of both alleles and genotypes in a population. The following notation is commonly used for a single locus with two alleles (A,a):

$$\text{The frequency of allele A} = p$$
$$\text{The frequency of allele a} = q$$

The frequencies of each allele vary between 0 and 1, and because there are only two alleles, the *first Hardy-Weinberg equation* states:

$$p + q = 1$$

Stated simply, the sum of the frequencies of alleles A and a comprise 100% of the alleles at that locus in the population.

However, the population consists of diploid individuals, meaning that alleles always occur in pairs, which constitute the genotype of the individuals carrying them. Thus, the basic data will always be the frequencies of the various genotypes in a population. Because the individuals are assumed to be mating randomly, the frequencies of the genotypes can be calculated as the product of the frequencies of alleles involved in each combination. The frequency of genotype AA is $p \times p$, or p^2. Thus,

The frequency of AA = p^2
The frequency of aa = q^2
The frequency of Aa = $2pq$

The frequency of heterozygotes, Aa, results from the fact that a heterozygote may be formed in either of two ways, receiving the A from the mother and a from the father, or vice versa (table 12.1).

Because a locus with two alleles can have only three genotypes, we know that their frequencies must also total one. Hence, the *second Hardy-Weinberg equation* states:

$$p^2 + 2pq + q^2 = 1$$

In practice, allele frequencies are calculated from data on genotype frequencies. For example, suppose that MN blood group data were determined for a large population of humans, with the following results:

Genotype	Number
MM	360
MN	480
NN	160

Allele frequencies can be quickly calculated, remembering that homozygotes have two copies of an allele and heterozygotes only one:

$p = [(2 \times$ number of MM$) +$ number of MN$]/(2 \times$ number of individuals)
  $= [(2 \times 360) + 480]/(2 \times 1000)$
  $= (720 + 480)/2000$
  $= 1200/2000 = 0.60$

The frequency of the other allele, q, can be calculated the same way or more simply as:

$q = 1 - p$
$q = 1 - 0.60$
$q = 0.40$

Allele frequencies provide precise data for comparing populations, something we have already seen when comparing blood group data between human populations, for example. The interesting part of the Hardy-Weinberg calculations results from using calculated allele frequencies in one generation, as we have just done, to predict the allele frequencies in the next generation. Using the above data, we can calculate for the next generation:

Table 12.1
The Hardy-Weinberg equations and a Punnett square showing the derivation of the frequencies of each genotypic class

Let the frequency of the dominant allele **A** be represented by **p**
Let the frequency of the recessive allele **a** be represented by **q**
Let the frequency of occurrence of the homozygous dominant genotype $= p^2$
Let the frequency of occurrence of the recessive genotype $\qquad = q^2$
Let the frequency of occurrence of the heterozygote $\qquad\qquad = 2pq$

$$\text{Equation 1:} \qquad p + q = 1.00$$
$$\text{Equation 2:}\ \ p^2 + 2pq + q^2 = 1.00$$

All possible ♂ gametes
(their frequency)

♀ \ ♂	A(p)	a(q)
A(p)	AA p^2	Aa pq
a(q)	Aa pq	aa q^2

All possible ♀ gametes (frequency)

For example, consider a population where the frequency of the homozygous recessive genotype was 1/10,000. How many persons are carriers of the a allele?

$$q^2 = 1/10,000 = .0001$$
$$q = \sqrt{q^2} = .01$$
$$p = 1-q = .99$$

The frequency of carriers in the population is $2pq = 0.0198 \simeq 2\%$

$$p^2 = (0.60)^2 = 0.36 = \text{frequency MM}$$
$$2pq = 2 \times (0.60) \times (0.40) = 0.48 = \text{frequency MN}$$
$$q^2 = (0.40)^2 = 0.16 = \text{frequency NN}$$

These frequencies are precisely those found in the original data. Continuing such calculations would show that, as long as these ideal circumstances apply, the allele frequencies *never* change. Hence, the Hardy-Weinberg Law states: *In a sexually reproducing diploid population not subjected to disturbing influences such as mutation, migration, non-random mating, natural selection, and small population sizes, neither allele nor genotype frequencies change in succeeding generations.* In other words, in a mathematically ideal world, evolution does not occur!

Geneticists know that in reality disturbing influences such as migration, mutation, and natural selection do occur, and that population sizes in nature are often very small. Moreover, biologists accept evolution as the central process in biology. Quite clearly, evolution does occur.

The importance of the Hardy-Weinberg Law is that it provides a basis for interpreting population differences and evolutionary change in the real world. Analysis of gene frequencies in populations over many generations often shows little or no change (a fact which is unsurprising in light of the Hardy-Weinberg Law). But in those cases where allele frequencies are changing, or where genotype frequencies are not those predicted, we can reasonably infer that one or more of the five disturbing influences is acting upon that population. Further study can often reveal the precise forces governing the evolutionary process.

The science of population genetics is an elegant mathematical development of the Hardy-Weinberg equations. The field of **ecological genetics** relates patterns of phenotypic and genetic variation within and among natural populations to principles derived from population genetics, particularly the five assumptions of the Hardy-Weinberg equations.

For the remainder of this chapter, we introduce aspects of genetic diversity and evolution as they relate to immediate human concerns: agriculture, conservation, and human population differences.

Agricultural genetics

Domestication of wild species marked a crucial turning point in the origins of human civilization. Domestication increased the quantity and predictability of food and other important supplies. In turn, some members of developing societies were freed to focus their attention on art, commerce, engineering, law, medicine, and science. Today, domestic species are managed so productively that only a small fraction of the population in developed nations is directly involved in agriculture. Much of the success in removing most members of society from the burden of food production is the result of genetic changes in domestic species during the past few thousand years.

Agricultural productivity, that is the *quantity* of domestic species, largely depends upon genetic *quality*. The essence of nature is variety. Some individuals are docile, easy to manipulate. Others produce more, healthier, or larger offspring. In the past, individuals with particularly desirable traits were chosen as parents of succeeding generations, and the desirable traits thus increased in domesticated populations. In other words, early humans exercised **artificial selection.**

Artificial selection has been used successfully for at least ten thousand years. As the science of genetics has developed in this century, artificial selection has become a precise and sophisticated technology.

Artificial selection

Artificial selection remains the primary basis for crop and livestock management. It has been so successful that almost all domesticated species are quite different, phenotypically and genetically, from their wild progenitors. Traits that permit survival in the wild only complicate management in captivity. Wild cattle and sheep, for example, have long legs for speed and horns for defense, but domestic cattle (figure 1.4) and sheep are bred for short legs, and their horns are polled.

The variety of traits selected artificially is astonishing. Food species have always been bred for features such as size, flavor, and disease resistance. But harvesting and distribution factors favor selection for some unexpected traits. Tomatoes, for example, have been developed with tough skins to withstand mechanical harvesting machines, extremely slow ripening to give long shelf life, and even cubical shape for efficient packaging (figure 12.2).

The specialized requirements of medical research have often been met with unusual strains of experimental animals. Nude mice, lacking a thymus gland, are used widely in cancer and immunological studies. A strain of pigs weighing an average of 140 pounds (65 kilograms) has been developed to provide a same-size model for surgical and other experimental studies relating to humans.

In spite of extraordinary success, artificial selection has serious limitations. Plant and animal breeders cannot produce new varieties on demand. Development of particular varieties is often a protracted, difficult process, with no guarantee of success.

Figure 12.2 Artificial selection in dogs, tomatoes, and roses.

Bloodhound

Collie

Scottish terrier

Pear tomato

Cherry tomato

Beefsteak tomato

Hybrid Tea

Floribunda

Grandiflora

Several factors limit the success of artificial selection. First, artificial selection is slow. A new mutation appears in only one individual. It must then be bred for many generations before a large number of individuals acquire the same trait. The time required to spread a new mutation depends upon the fecundity and generation time of the species, as well as the sex of the mutant individual.

Second, new genetic variation occurs at random. The effectiveness of artificial selection depends totally upon the particular alleles present in a population. Selection cannot change a population that is genetically uniform. Furthermore, most mutations are deleterious, so most new variation is useless to plant or animal breeders, who are forced to wait opportunistically for the rare, useful mutation to appear. If a particular trait never appears, or if it cannot be perpetuated, or if it is not recognized, then it can never be developed in a domestic species.

Third, the most rapid way to develop a strain of organisms bearing a desirable trait is through inbreeding. Consanguineous mating, however, leads to **inbreeding depression.** Inbreeding depression is the genetic price paid for the benefits of rapid selection.

Plant and animal breeders will continue to rely upon time-proven techniques of artificial selection. Modern genetic and reproductive technologies, however, are now providing solutions to the limitations of artificial selection. Coming decades should see the manipulation of captive gene pools become an increasingly precise and effective science.

Creating genetic variation

Artificial selection operates only on existing genetic variation, and selection is increasingly effective at higher levels of variability. *Mutation* is the ultimate source of genetic variation, so geneticists now accelerate the occurrence of new genetic variants using mutagenic agents such as *chemicals* and *radiation.*

A major problem with the application of mutagens is the unpredictability of the results. Valuable mutations usually occur only after many harmful ones. Individuals having desirable mutations may possess other mutations that are harmful.

The use of mutagenic substances is most profitable on species such as plants that produce large numbers of offspring. Each kernel on a stalk of corn, for example, represents a different individual. Therefore, a large number of individuals can be quickly assayed for mutations. Similarly, mutagens can be effective in generating commercially desirable results in bacteria, which have only a single copy genome. Thus, any mutations will be immediately expressed and reproduce extremely rapidly.

Polyploids, resulting from induced mutation, are important contributors to the world's food supply. As described in chapter 10, many commercially important strains of wheat, berries, and other food plants are polyploid. Polyploidy increases the overall size of organisms, especially the size of edible fruit and seed.

Inducing polyploidy in animals has had limited success. Polyploidy is achieved by subjecting recently fertilized eggs to temperature extremes. If these eggs successfully develop, the adults are polyploid and often sterile. Such techniques have been successfully applied to cultivation of salmon, oysters, and abalone.

Combining genes in new ways

Most species possess large amounts of genetic variation. New techniques of cell and gene manipulation are now permitting the transfer of specific genes between individuals as well as across species boundaries.

Grafting Cuttings from different individual plants will often grow together, fusing to form a single new individual. Such **grafting** has long been a successful method of combining the best characteristics of different strains. Many cuttings can be taken from a single individual, so a newly arisen desirable mutant can be effectively cloned by appropriate grafting. All seedless oranges, for example, were propagated by grafting from the original mutant branch, as the seedless fruit is sterile. Much of the French and California wine industries now rely upon the grafting of the best French wine grape stock to insect-resistant American rootstock.

Grafting is also a method of "bulking up," that is, rapidly producing large numbers of individuals of a desirable genetic type. In this regard, grafting parallels the techniques of superovulation and embryo transfer now widely practiced on domestic livestock (box 7.1).

Hybridization Interbreeding either different genetic strains within a species or individuals of different species is called **hybridization.** By definition, a *species* consists of all individuals that can freely interbreed in nature and produce fertile offspring. Commonly, members of different species cannot and do not interbreed, or if they do, the offspring are not viable. Occasionally, however, members of different species not only interbreed, but also produce viable offspring. Often, interspecific *hybrid* offspring have undesirable traits or reduced fertility, but in a few cases the results are spectacularly successful. Interspecific hybridization is more common in plants than in animals.

The best known interspecific hybrid is the mule, a cross between a male donkey and a female horse. Like many hybrids, the mule exhibits **hybrid vigor,** an increase in size, physiological vitality, or other desirable qualities often found in hybrid individuals. The mule is valuable as a work animal precisely because of these traits. Like many interspecific hybrids, however, mules are usually sterile.

Hybrid vigor can also result from the cross of different strains *within* a species. Most grain crops are the F1 progeny from hybridization of inbred strains. Their hybrid vigor is expressed as increased productivity. The F2 generation of these lines does not show hybrid vigor; their productivity is reduced. Hybridization has thus changed farming practices permanently. Farmers no longer save the best seed to produce next year's crop. Modern grain agriculture depends upon the breeding programs of seed companies that supply hybrid seeds annually.

Hybrid corn was developed in the United States in 1917. Since then hundreds of different hybrid combinations have been produced, each uniquely suited to a specific set of growing conditions. Often a single highly productive hybrid is grown over a broad geographical area, **monoculture.** While the productivity achieved with hybrid strains has been remarkable, monoculture increases the potential vulnerability of a crop to genotype-specific diseases.

Protoplast fusion Recent developments in *tissue culture* techniques also involve hybridization. In plant tissue culture, a small segment of plant tissue is grown in a petri dish in the laboratory. With proper chemical stimulation, the cells can sometimes be induced to grow into new plants. A technique known as **protoplast fusion** produces hybrid cells from the union of cells from different species grown in tissue culture. The hybrid cells exhibit properties of both parent species, qualities that in combination may produce an agriculturally superior plant.

Potential applications of tissue culture and protoplast fusion are almost limitless. For example, potatoes resistant to potato blight (the disease that caused the nineteenth-century Irish potato famine) have now been produced by protoplast fusion of potatoes with tomatoes in laboratory culture. Tomatoes belong to the same family as potatoes, but are resistant to the blight. Genes for resistance are incorporated into the hybrid offspring, which then produce normal, but blight-resistant, potatoes.

Protoplast fusion is still in the developmental stage. Mutagenic agents are used to induce fusion of cells, but the results of this process are unpredictable. It is not possible to control which cells fuse and which do not, nor to control which characteristics are expressed in the hybrid. The technique is so rapid, however, that many combinations can be tried. If a successful hybrid is produced, individuals can be cultured from only a few cells. One successful fusion can result, potentially, in a thousand or more offspring.

Gene transfer Recombinant DNA technology, **genetic engineering,** offers great promise for transcending the limits of artificial selection. Using recombinant techniques, a single desirable gene may be introduced into an organism, **gene transfer** (see chapter 13 also). Selective breeding can then spread the gene throughout a population. The advantages of gene transfer are clear. Only a single gene or set of genes is introduced, while the remainder of the genome is unaffected. Furthermore, the desired gene can be removed intact from another individual, even from another species. Or the gene can be constructed precisely, one base at a time. Breeders will no longer need to wait for the random appearance of a valuable trait by mutation.

Domestic crop species are already being subjected to transformation experiments with the goal of increasing their productivity. A major factor affecting productivity is the availability of nitrogen. Although nitrogen is an essential nutrient for plant growth, most plants cannot directly use abundant atmospheric nitrogen. Instead, plants use fixed nitrogen, i.e., nitrogen that has been combined with oxygen to form nitrates and nitrites by nitrogen-fixing bacteria. Nitrates are a primary component of fertilizers. A few plants, particularly legumes such as beans, clover, and alfalfa, are essentially self-fertilizing because they have nodules on their roots that house colonies of nitrogen-fixing bacteria. Such plants actually enrich, rather than deplete, the soils on which they grow. Plant geneticists are attempting to isolate and purify the nitrogen-fixing genes of these bacteria. If successful, gene-transfer techniques could be used to introduce these genes into domestic crop species.

Other recent successful experimentation with gene transfer has involved producing frost resistant strawberry plants by infecting them with "ice-minus" bacteria (normal bacteria from which the gene promoting ice formation has been removed); producing a genetically engineered vaccine (made by removing part of a natural viral gene) to protect swine against the deadly disease pseudorabies; adding herbicide resistance genes to normal food crops; transferring into tobacco plants a bacterial gene that triggers the production of a protein that kills caterpillars that feed on tobacco leaves; and adding extra copies of the gene for the hormone somatotropin to cattle to boost milk production (by 10%–40%).

The price of agricultural success: new genetic problems

Modern agriculture bears little resemblance to farming practices used prior to this century. Single crops are grown over vast areas, and massive doses of chemicals are applied to all crops. In addition to increased production, the new techniques also create a set of problems never before encountered by farmers.

Pesticide resistance

The first widely used pesticide was the chemical DDT. Following the discovery of its insecticidal properties in 1939, DDT was sprayed indiscriminately worldwide for control of mosquitoes, houseflies, and other insect pests. At that time malaria, for which mosquitoes are a vector, was the single greatest cause of human death in the world.

DDT was extremely effective against mosquitoes and houseflies, and for a few years the populations of these insects remained at low levels. Within a decade, however, DDT-resistant populations began appearing. While DDT killed most insects in a population, a few rare individuals possessed genetic traits that made them resistant to DDT. The large population sizes of insects make it likely that a few resistant individuals will exist in any population. The rapid reproductive rate of most insects means that in only a few generations large populations of pesticide-resistant insects will replace the nonresistant populations. Control of these populations consequently requires the application of increasingly larger doses of pesticide to keep pest numbers at a low level, the **pesticide treadmill.**

The result of several decades of use of DDT and other pesticides is that many insect populations in the world are largely resistant to the common pesticides employed against them. For example, between 1970 and 1980 the number of documented pesticide-resistant insects increased from 224 to 428. One response to DDT resistance has been the rapid development of other pesticides. In a sense, an evolutionary arms race is required if pesticides continue to be a primary agent of pest control.

All species have the potential to develop resistance to toxic chemicals. A spectacular example is the Norway rat, a serious pest that consumes human food supplies and transmits disease. Following World War II, the anticoagulant warfarin was used to kill rats in cities. Warfarin was mixed with a food bait that included ground glass. Rats eating warfarin suffered internal hemorrhaging, causing death.

About ten years after the introduction of warfarin, however, warfarin-resistant populations of rats began to appear independently around the world. As with insect pests, genetic mutations had conferred resistance to warfarin giving rise to numerous resistant populations. In fact, some warfarin-resistant rats actively metabolize warfarin, using it to their physiological benefit.

Drug-resistant diseases of livestock

Moderate doses of antibiotics increase the productivity of pigs and cattle. As a result, antibiotics are routinely fed to healthy livestock, especially when they are moved into feed lots, confinement areas where they are fed massive amounts of grain to increase their weight prior to slaughter. The drugs may have a prophylactic effect in preventing the spread of disease when animals are kept in such large concentrations. However, drug-resistant strains of disease organisms are now known to have resulted from these practices, just as antibiotic-resistant strains of human diseases have developed.

Fears that humans might become ill as a result of eating livestock that have been fed antibiotics were realized in 1984 when eighteen people became severely ill with salmonellosis. It was discovered that all eighteen were infected by an antibiotic-resistant strain of *Salmonella* that they had ingested in hamburger meat, which was traced to a single farm where cattle had been fed low doses of antibiotics. It is estimated that 30% of all chickens, 80% of all swine, and 60% of beef cattle marketed in the United States have been fed antibiotics for some part of their lives. The residual effect of the antibiotics on human consumers is uncertain. The FDA estimates that nearly 50% of all antibiotics produced in the United States are fed to livestock.

Monoculture and pest susceptibility

Today fewer than fifty species of plants are used as food sources for most of the human population. The traditional diversity of cultivated strains of these few species has now been replaced by the planting of single, genetically uniform strains over vast geographic areas. Monoculture has many practical advantages over older methods of farming; however, as noted earlier, crops grown in monoculture are dangerously susceptible to epidemic disease.

One serious epidemic has already affected United States agriculture. In 1970, southern corn leaf blight destroyed about 15% of the corn crop, causing about one billion dollars in damages. The blight-susceptible crop was saved only by quick action and good luck. Blight-resistant strains were introduced the following year. Damage from future epidemics, however, may not be controlled so successfully. Maintenance of substantial genetic variation in crop species provides the best long-term insurance against such potential calamities.

Genetic techniques of pest control

In nature, potential pest organisms seldom reach destructively high numbers. Most species are limited by interactions with predators, diseases, parasites, and other factors.

Biological methods offer alternatives to the use of chemical pesticides. Some biological methods use genetic technology to circumvent many problems associated with the use of pesticides. Pesticides, for example, kill both pest and beneficial insects indiscriminately. The loss of beneficial insects may be more serious than the damage done by pests. Furthermore, pesticides such as DDT persist in the environment for years, accumulating in the tissues of living animals, including humans. The damage caused by chemical residues is serious, and the long-term effects are unknown. In contrast, biological control methods usually affect a single target species, for a relatively short period of time.

The sterile male technique

Release of large numbers of sterile males into pest populations can significantly reduce their numbers (figure 12.3). Larval pest insects are grown to adulthood in artificial growth chambers. Adult males are then given large doses of radiation, rendering them sterile. The artificially reared, sterile males are released into nature. When fertile wild females mate with sterile males, no offspring are produced. Some females also mate with fertile wild males, but sterile males greatly outnumber fertile males, so most females mate with sterile males. The **sterile male technique** is most successful in limited pest populations where sterile males are released for several consecutive years. Note, again, that basic research has practical application. In this case, the work of H. J. Muller, who identified the mutagenic and sterilizing effects of radiation, provided a basis for the control of noxious insects.

The sterile male technique has been used successfully against a parasite of domestic cattle, the screwworm fly. The screwworm fly lays its eggs in the open wounds of mammals, particularly cattle. The eggs develop into larvae that feed on the wounds. Severe infection may cause death. In the 1950s, a sterile male program was begun against the screwworm, then causing an estimated twenty million dollars damage to the American cattle industry. Since then, losses to screwworm have been dramatically reduced; in some portions of the United States this pest has been totally eliminated.

Chemicals

All organisms produce chemicals that regulate their own lives. In particular, *hormones* control development and reproduction. Some insect hormones can be used to help suppress pest populations. Hormones, however, usually occur in minute quantities, far too small for commercial use. Genetic engineering techniques provide a method for mass production of insect hormones, much like production of artificial human insulin.

Pheromones, hormones released into the environment as mate attractants, are now used for pest control. Male insects searching for a mate, for example, often follow pheromone trails laid down by females. Scientists bait insect traps with pheromones, attracting males who are then killed. This technique requires many traps over large areas. Many males are still likely to find mates before entering the traps.

A second, more devious, approach involves aerial spraying of pheromones throughout an infested region. Mass application of pheromones confuses the males—females appear to be everywhere. Pheromone trails laid down by females are obscured, reducing the probability of a male finding a mate.

Growth hormones can also control pest species. *Juvenile hormones* suppress sexual maturation. *Anti-juvenile hormones,* on the other hand, produce premature maturation and consequent sterility. Aerial application of either of these hormones greatly reduces the reproductive potential of pest populations.

The advantage of biological control of insects by hormones is obvious. Unlike DDT or other pesticides, hormones are naturally occurring chemicals that are not toxic to humans or other animals. Their primary effect is to disrupt normal mating behavior and reproduction of a target species without the harmful secondary effects of traditional pesticides. The market for insect pheromones and other agricultural chemicals produced by recombinant DNA technology will be substantial.

The Green Revolution: mixed verdict

The **Green Revolution** refers to a series of programs begun during the 1960s to increase agricultural productivity in underdeveloped countries. The development of genetic strains of food crops with high productivity has been a major goal of the Green Revolution. Several international institutes were established for this purpose.

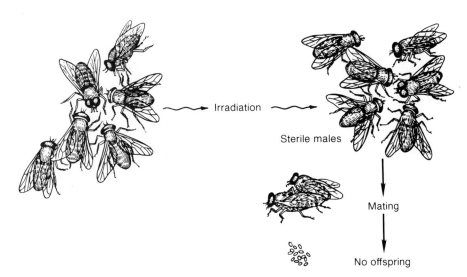

Figure 12.3 The sterile male technique. Large numbers of the insect are raised and exposed to sterilizing amounts of radiation. At the appropriate time, sterile males are introduced into the natural population. These males mate with the normal fertile females, and thus the females contribute no offspring to the next generation.

Irradiation

Sterile males

Mating

No offspring

For example, the Centro International de Mejoramiento de Maiz y Trigo (CIMMYT) developed strains of wheat with relatively large seeds and short stems, thus increasing productivity (figure 1.5). Nearly all wheat grown in Mexico and about 75% of that in India are the short-stemmed or dwarf varieties. Crop yields in India and Mexico have increased three and five fold, respectively. High-productivity strains of maize developed by CIMMYT are now an important food crop in Mexico and Africa.

Geneticists at the International Rice Research Institute (IRRI) in India have developed high-yield strains of dwarf rice. Today, about 70% of all rice planted in the Philippines and 90% of the rice grown in Colombia are these dwarf varieties.

Public attention has focused on the genetic basis of the Green Revolution, but the new strains are successful only when modern techniques of cultivation are used. The crops must be grown in monoculture with massive applications of fertilizers, pesticides, and herbicides—a major change from traditional agriculture. In the Peruvian Andes, for example, native farmers grew fifteen or more different strains of potato in a single garden. The variety of genotypes provided protection against soil depletion and loss of the entire crop to diseases or pests. Native farmers maintained their own seeds from year to year, whereas the hybrid strains must be purchased from the seed companies for each planting.

In short, traditional agriculture is labor intensive and self-sustaining. The Green Revolution requires capital-intensive techniques feasible only on a large scale because of large investments required for seeds, machinery, petroleum, and agricultural chemicals.

Although the goal of the Green Revolution—more food for more people—is laudable, results have been mixed. Productivity has increased greatly in many countries, but the imposition of modern technologies upon underdeveloped societies has sometimes caused significant social and economic stress.

Because small farmers cannot afford initial capital investments for the new techniques, they have often been forced to sell their land to large agribusiness corporations. Thus, the distribution of land has shifted from small farmers to large corporations. Profits made from increased productivity have been channeled into foreign investments rather than reinvested in the local economy. The small farmers, forced off the land, have often moved to urban ghettos, with little prospect of employment. The benefits of technology may increase productivity, but the social consequences are considerable.

Some biological consequences of the Green Revolution have also been unpredictable. The massive applications of pesticides to protect crops grown in monoculture have selected for pesticide-resistant strains of insects, including mosquitoes. In some underdeveloped countries the incidence of mosquito-borne malaria has increased dramatically with the introduction of modern farming practices.

Finally, increased agricultural production has not reduced the problems of world hunger. Increases in food production are being more than matched by increases in world population, most of which are occurring in the poorest countries. Only a few countries are net exporters of food. In addition, displaced small farmers who no longer grow their own crops cannot afford to buy food. Concentration of agricultural wealth in a few large corporations only serves to increase the disparity in wealth among sectors of the population, a factor that has added to political instability in many parts of the world.

Conservation

The results of biological evolution are extraordinary. Since life originated on earth over 3.5 billion years ago, species diversity has increased regularly to produce a fantastic array of living species. No one knows for sure how many species presently exist, but the number of species has recently been estimated to be between ten and thirty million. Most of these live in the tropics and have never been studied—or even described and named—by scientists. Tens of thousands, possibly millions, of species face an uncertain future because of the activities of humans.

Extinction and genetic resources

For the past several centuries, human numbers have increased exponentially. Since World War II, the rate of growth has actually increased and shows no signs of abating. Human populations now double about every thirty-five years. In 1987 the human population passed the five billion mark and by the turn of the next century is likely to number between seven and eight billion.

The unprecedented human population explosion has resulted in massive loss of habitat for all other species. Very few places on earth are undisturbed by human activity. As the amount of space remaining for wild plants and animals has decreased, many species have become extinct. Many other species have seen their natural ranges fragmented. Large populations—and large gene pools—have been reduced to tiny remnants.

The potential disaster posed by massive species extinction is difficult to imagine, but predictions are dire. Of the possible thirty million species now alive, about one million are predicted to become extinct by the turn of this century. If extreme conservation measures are not implemented soon, half of all living species are likely to become extinct by the end of the next century.

The importance of this issue has been recognized by many organizations. The International Union for the Conservation of Nature publishes a series of Red Data Books that list the threatened, endangered, and possibly extinct species in countries throughout the world. The genetic aspects of extinction are explicitly recognized in the World Conservation Strategy of the United Nations, which makes the conservation of genetic resources one of its three primary goals.

Concern for diminishing genetic resources is well founded. Many people feel a strong ethical responsibility for the human role in species extinction, but compelling practical reasons underlie the need for biological conservation. Genetic resources comprise fundamental life support systems for humans. The many genetic varieties of plants and animals are the basis for all human food and agriculture; they are also the source of about one-fourth of all medicines and drugs. Plants play a major role

Figure 12.4 Potatoes for sale at an outdoor market in Ecuador. Several different varieties are grown by South American Indians in the Andes, and all are marketed throughout the growing season.
Source: D. Donne Bryant.

in regulating weather and climate. Fungi and bacteria quickly recycle the remains of dead organisms. The loss of genetic diversity diminishes the quality of human life and ultimately threatens human existence.

Plant genetic resources and agriculture

Human food supplies are narrowly based. Fewer than fifty plant species and about ten animal species are the major food sources for most of the world's population. Obviously, many other species could provide food, but many potential food sources are threatened with extinction before they can be properly evaluated.

Modern agricultural practices are also a serious source of erosion of genetic resources. Most crop species are comprised of a large number of **land races,** locally adapted varieties well known to farmers in a particular region. Traditionally, farmers grew a number of these land races (figure 12.4). Now, farmers plant only one hybrid variety supplied by a seed company. The result is that many land races have become extinct, reducing the genetic resources on which future plant breeders can draw to produce new varieties. Some people also worry that the seed companies—often large multinational firms—have acquired all supplies of some land races. From the companies' viewpoint it represents good business practice, but farmers and the public have lost control of a resource formerly held in common. **Plant patenting laws** now allow plant breeders to own exclusive rights to specific varieties of plants and animals.

Loss of land races reduces the genetic base—the survival resources—of agricultural species, but some domestic crops have always had a narrow genetic base. No better example exists than coffee. The coffee plant originated in Africa, from where a few plants were taken by the Dutch to the East Indies (Indonesia) in the seventeenth century. These few plants formed the basis for large coffee plantations, which provided much of the wealth flowing into Holland from their colonial empire.

In the nineteenth century, the British surreptitiously took a single coffee plant to the Kew Gardens in London. A cutting of this plant was subsequently taken to South America, where it formed the basis for the South American coffee industry, now the largest in the world.

At each step in its history, the domestic coffee plant was subjected to a **genetic bottleneck,** reducing its total genetic variation drastically. The dangers of low variation were realized in the 1970s, when a disease epidemic struck the South American coffee plantations. Only the import of disease-resistant coffee plants from Indonesia—an input of genetic variation—was able to restore the South American plantations to production.

An equally dramatic example of the survival value of genetic diversity is shown by the history of the European wine grape, *Vitis vinifera.* In the late nineteenth century, an infestation of a root-destroying insect, *Phylloxera,* reached plague proportions and totally devastated most European vineyards. Their grape varieties were the finest in the world, the result of centuries and even millenia of captive propagation and selection for desirable characteristics.

Many of these unique and irreplaceable varieties, as well as the European wine industry itself, were saved by grafting cuttings onto rootstock from native American species of grapes, which are resistant to *Phylloxera.* Thus, the best wine grapes in the world survive as a genetic mosaic. Natural genetic diversity in wild species provided crucial information that saved an entire agricultural industry.

Kew Gardens (figure 12.5), perhaps the most famous botanical gardens in the world, are a major tourist attraction for visitors to London. Most visitors are unaware that they were founded entirely as a repository for plants from around the world that might support the British Empire. Kew Gardens formed the center of a major program to obtain by any possible means plants that might be of value to the Empire. The British fully realized the commercial value of genetic resources, their exploitation of which allowed them to dominate colonial economies and accumulate immense wealth for their country.

In a sense, Kew Gardens may be considered a forerunner of plant gene banks, which are now being developed in many countries to preserve vanishing plant species and races. The *International Board for Plant Genetic Resources* (the *IBPGR*) based with the United Nations Food and Agriculture Organization in Rome, coordinates the activities of national plant germ plasm banks in many countries. These gene banks maintain collections of seeds to preserve differing varieties of agriculturally important species and their wild relatives. In the United States, the National Seed Storage Laboratory at Fort Collins, Colorado, administered by the U.S. Department of Agriculture, is the primary storage facility for plant germ plasm.

Although sound in principle, the reality of the task confronting plant germ plasm collections is immense. The collections for a single species may include millions of seeds. Seeds can survive for many years if stored in proper conditions, but they do not last forever. The collections must be renewed continually by planting and growing the seeds and then collecting the next generation of seeds. In addition, millions of wild species are largely ignored because most collections deal primarily with agriculturally or commercially valuable species.

Furthermore, most plant genetic diversity occurs in tropical continents, especially South America, Africa, and Asia. Tropical areas of especially high plant diversity—called *Vavilov centers* (figure 12.6), after the famous Russian geneticist who identified them—are the site of origin of most domestic crop species. They also occur mainly in underdeveloped countries with rapidly increasing populations. Natural diversity is being greatly reduced in these countries due to human population pressures. These countries have urgent conservation problems with only limited resources, if any, to devote to the problem.

Figure 12.5 Kew Gardens in London, England. Source: De Casseres/Photo Researchers.

The IBPGR organizes assistance programs for the conservation of plant genetic diversity in underdeveloped countries, but such efforts sometimes have been highly controversial. The IBPGR is supported financially by seed companies, who are also represented on its board of directors. Other support has come from independent organizations such as the Kellogg Foundation, again with links to agro-industry and based in the developed world. Thus, many underdeveloped countries view germ plasm collections as a form of modern imperialism, which tries to steal from them their biological heritage. National germ plasm collections in developed countries are viewed as a contemporary equivalent to the Kew Gardens, where genetic resources of economic value are stored for national purposes.

Seeds stored in national germ plasm collections that participate in IBPGR programs are supposed to be available to and shared among all nations in the world, yet material is often not shared. The United States, for example, does not freely exchange seeds with communist nations. Following the Nicaraguan revolution in 1979, many seeds stored in the national collection in Nicaragua were removed for their own safety and stored in the United States. Subsequently, and despite many requests from the Nicaraguan government, the U.S. government has not allowed these seeds to be returned to Nicaragua. Some of the seeds represent varieties no longer found in nature that could greatly benefit Nicaraguan agriculture.

Figure 12.6 Regions of genetic diversity and their associated crops. These sites are called *Vavilov centers,* in honor of the Russian geneticist who identified them. They are the sites of origin of most domestic crop species. Source: Jack Kloppenburg, Jr. and Daniel Kleinman, "The Plant Germplasm Controversy," *BioScience,* 37:190–98, Fig. 3, p. 193. Reprinted by permission of American Institute of Biological Sciences and the authors.

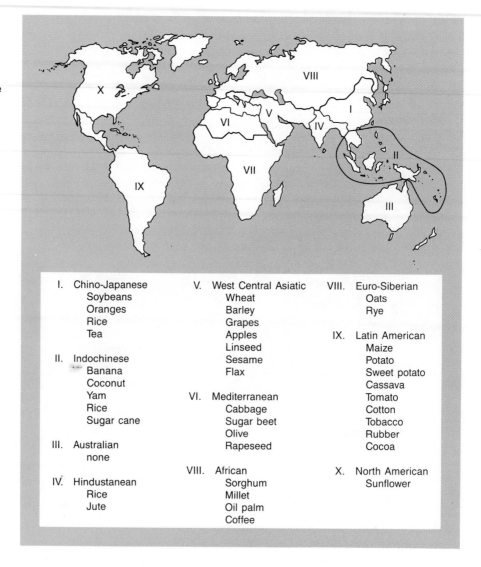

I. Chino-Japanese Soybeans Oranges Rice Tea	V. West Central Asiatic Wheat Barley Grapes Apples Linseed Sesame Flax	VIII. Euro-Siberian Oats Rye
II. Indochinese Banana Coconut Yam Rice Sugar cane	VI. Mediterranean Cabbage Sugar beet Olive Rapeseed	IX. Latin American Maize Potato Sweet potato Cassava Tomato Cotton Tobacco Rubber Cocoa
III. Australian none	VIII. African Sorghum Millet Oil palm Coffee	X. North American Sunflower
IV. Hindustanean Rice Jute		

Thus, politics intrude on the conservation of genetic resources. Developed countries claim they are acting with the best motives, observing correctly that only they have the money and expertise to deal with such a major task. Underdeveloped nations point to the fact that multinational seed companies are likely to be the primary beneficiaries of germ plasm conservation programs, and also that their biological resources are now being plundered just as their mineral resources were in previous centuries. Meanwhile, rapid continuing development of tropical continents, particularly deforestation of the Amazon basin and remnant tropical forests in Africa and Asia, poses an ever-increasing threat to remaining plant genetic resources.

Wildlife conservation

Wild animal species face the same problems confronting plant species: reduced and fragmented habitat, and decreased population sizes—conflict with humans at every turn. These problems are compounded by the large home ranges required by many species; caribou, for example, have migratory routes of several thousand miles. The problems are especially acute for large mammals, the species we often value most.

Figure 12.7 Przewalski's horses, the forerunners of the domestic horse, are bred in a special program at the San Diego Zoo.
Source: Zoological Society of San Diego.

Predicted extinction rates for large mammals in the next century are extremely high because so little habitat is available for nature reserves to meet the needs of these species.

Although some species will survive in reserves, many others can be saved only by extreme measures. Zoos already are an essential part of world conservation programs. Some species, such as Przewalski's horse (the forerunner of the domestic horse) and the European bison, survive only in captivity (figure 12.7). The very small numbers that zoos can support mean that these species have already undergone severe genetic bottlenecks. They are also subject to intense inbreeding. Many zoos practice rigorous breeding programs to reduce the effects of inbreeding depression. ISIS, the International Species Inventory System, is a computer database that records the lineage of animals held in zoos worldwide and coordinates international captive breeding programs.

The limited space in zoos cannot accommodate all threatened species. More extreme conservation measures are under development, such as the use of ultra-cold gene banks to store sperm and embryos from endangered species; ova cannot yet be stored successfully. Techniques such as sperm banking, **superovulation,** *in vitro* fertilization, and **embryo transfer,** which have been used on humans and domestic livestock, are now being used successfully on wild animals by institutions such as the San Diego Zoo.

Unfortunately, such techniques are primarily applicable to mammals, but not to most of the millions of other animal species. Again, limitations of space mean that many species will not receive the conservation attention they deserve. Thus, mammals and birds will be a special focus for wildlife conservation, but many other species of animals will be ignored. As with plants, animal genetic diversity will continue to shrink dramatically in coming decades.

Human genetic diversity

Human genetic diversity has been shaped by the same evolutionary forces affecting all other species. Humans have been subject to reductions in genetic variation similar to those affecting other species. European colonialist expansion in the past five centuries has resulted in the extermination of many isolated populations. In some cases, indigenous peoples were subject to programs of genocide. In others, small local populations were literally consumed by hybridization as a result of intermarriages. In many other cases, local populations were not resistant to European diseases such as smallpox and syphilis. Contact with Europeans exposed the narrow genetic base of these populations to new diseases, and extinction ensued.

We will examine briefly key aspects of the processes guiding human population and evolutionary genetics.

Human origins

Fossils show that life has been evolving on earth for over 3.5 billion years. Vertebrate animals have existed for over 500 million years, and mammals—warm-blooded animals with hair and mammary glands—for perhaps 150 million years. Primates (monkeys and their relatives) appear in the fossil record about 50–60 million years ago. About 5 million years ago, the human lineage diverged from the common ancestor of the Great Ape lineage (see chapter 1), and only one species now survives: *Homo sapiens,* wise man.

Although our activities now dominate the earth, *Homo sapiens* has been an abundant and ecologically significant species for only several hundred thousand years, about one ten-thousandth of 1% of the history of the earth. Perhaps the most notable feature of human evolution has been its rapidity, particularly the rapid rate of increase in brain size. In the brief period of human existence, many racial stocks have also evolved. Presumably, natural selection and the other evolutionary factors that led to the development of land races in plants and animals also produced the diversity of human types.

Races

Genetic variability is the essence of life. One of the great achievements of modern genetics has been to establish the large amount of genetic variation found within natural populations, including humans. Virtually every human is genetically unique. Even identical twins or triplets differ genetically due to somatic mutation. Genetic differences are accentuated by environmental effects, producing a phenotypic array of seemingly limitless diversity. We have no difficulty, for example, identifying by visual appearance alone the thousands of persons we meet in a lifetime.

Most individuals are usually also easily identified as belonging to a particular *race.* White Europeans are easily distinguished from black Africans. Despite the simplicity of this example, race is not a simple or easily defined concept. There is no objective definition of race. Most geneticists would agree that a race is a distinct or identifiable subgroup of a species, but there is no agreement as to *how distinct* a race must be.

The difficulties in defining race explain the extent of disagreement among reputable anthropologists regarding the number of surviving human races. In this century, some anthropologists have suggested only three or four races exist; others have described more than one hundred. Four major racial groups of humans can clearly be identified by skin color and numerous anatomical differences (figure 12.1): Mongoloid, generally yellow-skinned populations including such groups as Chinese,

Figure 12.8 Differences in Negroid characteristics: a tall Dinka woman (left) in contrast to a short San man (right). Note that the anthropologist standing next to the San is 5′6″ tall. Source: (a) J. F. E. Bloss/ Anthro Photo; (b) Irvin DeVore/Anthro Photo.

Japanese, Malayans, Eskimos, and American Indians; Caucasoid, white- to brown-skinned populations originally found from Europe to Asia and northern Africa; Negroid, blacks of African descent; and Australoid, Australian aborigines.

The common practice of characterizing races by skin color—as we have just done—almost certainly is a gross oversimplification, concealing both the large amounts of variability within each race and the extensive interbreeding that has occurred whenever racial groups meet. Among the Negroid populations, for example, are the very tall Dinka and very small pygmies and San tribesmen (figure 12.8). Differences perhaps less striking but equally significant distinguish an Apache Indian from a Cantonese, and both from a Vietnamese. The terms Swedish, French, and Sicilian arouse stereotypic images that reflect in part genetic differences among peoples native to these regions. To deal with such diversity, many anthropologists have felt that the genetic variation within the human gene pool was best recognized by the designation of dozens of races, rather than three or four.

The concept of races probably reflects the obvious phenotypic effects of only a few loci. Alleles of some genes have marked phenotypic effects that correlate with commonly designated racial groups. The most obvious example is the set of genes involved in production of skin pigments. As few as four loci may encode varying amounts of pigment production, leading to apparently obvious racial differences in pigmentation. To realize the dramatic effect that only a small number of genes may have, recall from chapter 6 how few loci are involved directly in controlling human sexual dimorphism. When considering the role of skin color in racial identification, it is worth noting how many environmental factors also affect skin color: age, health, emotional state, exposure to the sun, exercise, and even diet.

Just how genetically different are the major human races? And how do genetic differences between members of different races compare to genetic differences between members of the same race? In the past twenty years biochemical techniques have been developed to identify genetic variation at many loci. Fifty or more loci can be examined from the proteins in a single blood sample. Studies using these techniques show the existence of large amounts of variation within humans. Overall, about 34% of all loci examined within a population are **polymorphic,** that is, have two or more alleles. The remaining 66% are **monomorphic,** having only one identifiable allele at the locus. A particular individual, of course, has only a fraction of this variation. On average, most individuals are heterozygous at about 3%–5% of their loci. Polymorphic genes are invaluable tools in medicine, paternity cases, and in evolutionary studies of human populations.

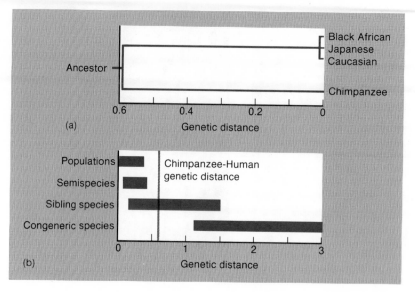

Figure 12.9 (a) Genetic distances among human races compared to genetic distances between humans and our nearest evolutionary relative, the chimpanzee. (b) The ranges of genetic distance commonly found between closely related taxa. Genetic distances are estimated on the basis of proteins assayed by gel **electrophoresis.**

Electrophoretic comparisons among individuals show that genetic differences between races comprise a relatively small fraction of total variation. Most of the total human genetic variation occurs within each of the major races. For example, on average, two randomly chosen American blacks differ at about 12% of their loci; two Japanese at about 12% of their loci; and two American whites at about 14% of their loci. Thus, variation *within* each race is about equally large.

Comparisons *between* races, however, reveal only *limited additional variation.* Beyond the level of differences normally found within races, for example, American blacks and American whites differ at only an additional 1.6% of their loci; Japanese and American blacks at 1.5%; and Japanese and American whites at only 0.5%. In other words, the average genetic difference between members of different races is only slightly greater than that between members of the same race.

The genetic differences between human races are so small that a prominent population geneticist observed: "If the holocaust comes and a small tribe deep in the New Guinea forests are the only survivors, almost all of the genetic variation now expressed among the innumerable groups of four billion people will be preserved." This statement does not deny racial differences, but places them in proper perspective. The genetic differences that underlie characteristic racial phenotypes occur at a small minority of all loci. Conversely, even a small population contains most of the genetic variation found in the total human gene pool.

Genetic comparisons between species provide another framework for assessing racial differentiation. Average genetic differences between populations or species can be expressed in terms of a statistic called **genetic distance (D).** Genetic distance is a measure of the number of different alleles per gene locus between two populations, taking into consideration the frequency of these alleles in both populations. Genetic distances range from 0 to infinity. When about 40% of the loci compared have different alleles, the distance is 0.5; if 63% of the loci differ the distance is 1.0, and 92% differ at a distance of 3.

Estimates of genetic distance between humans and their closest evolutionary relative, the chimpanzee, vary from 0.39–0.62, depending upon the laboratory conducting the analysis (figure 12.9). This value is small compared to the genetic distance found between other closely related species; it is smaller than the genetic distance between domestic dogs and the coyote, for example. In fact, the genetic differentiation of humans and chimpanzees is so small that the average human polypeptide is more than 99% identical to its chimpanzee counterpart. Chromosome analyses also confirm the close evolutionary relationship, and thus the recent common ancestry, of humans and chimpanzees (box 12.1).

On the other hand, the average genetic distance between races of humans is only about 0.01–0.02, trivial in comparison to the interspecific comparison. Interracial genetic distances are so small because, as noted earlier, genetic differences between races are normally *quantitative,* not qualitative. That is, populations differ in the frequencies of alleles, but most populations contain all the alleles found in other populations. This pattern is shown clearly by population frequencies for various blood groups (table 4.4 and table 12.2).

Although genetic differentiation between human races is limited, characteristic allelic differences between populations often allow us to identify the genetic history of peoples as well as their historical geographic history. Recall our discussion in chapter 4 of the distribution of ABO alleles and the origin of American Indians.

Racism

Human races and populations can be characterized by their allelic frequencies at some loci. Persons with racist viewpoints frequently claim that such genetic differences support their views and warrant differential treatment of persons of particular racial, ethnic, geographic, religious, or economic groups. Modern genetics, however, does not provide any support for **racism.**

More than any other science, genetics provides evidence that refutes racist claims of inherent superiority of certain races. It is hardly surprising that both ignorance and malice have led racists to pervert, misconstrue, or falsely represent the findings of modern genetics. Because racist misrepresentations are common, it is well worth clarifying the implications of modern genetics for racist assertions.

Consider the problem of assigning a particular *individual* to one of the major racial groups. Suppose you are given, one at a time, the individual's genotype at a number of loci. You first learn that the ABO genotype is OO. From table 12.2, you might predict that the most likely possibility is that the person is an American Indian, but you could not specifically assign him or her to any race. You could also not *exclude* the person from membership in any race on the basis of ABO genotype. If you were told the person was genotype AB, you would probably feel confident in saying the person was not an American Indian from Lima, but he or she might also be an Indian from Bolivia. Are they different races?

Suppose next you learn that the person's earwax genotype is WW, giving a wet earwax phenotype (chapter 4). That genotype is most common in both black and white populations, but also occurs at low frequencies in Mongoloid populations. In some American Indian populations, the wet allele is reasonably common. So the addition of a second locus still does not allow assignment of the individual to a specific race.

Now consider that you are given information on the same loci for a second person, whose genotypes are AA and ww. As with the first person, you cannot with any certainty whatever assign that individual to a race. Beyond that, however, you cannot even make a reasonable guess as to whether the two persons belong to the same or different races, even though they share no alleles whatever at those loci.

As you acquire more information on each person, you might be able to improve your guesses. You might eventually be able to conclude with reasonable certainty that each person belonged to a particular race, but as with paternity testing, even a thousand loci might not *exclude* the person from belonging to other races. After all, virtually all alleles occur in all populations—only the frequencies vary.

BOX 12.1

Chromosomal relationships of humans and Great Apes

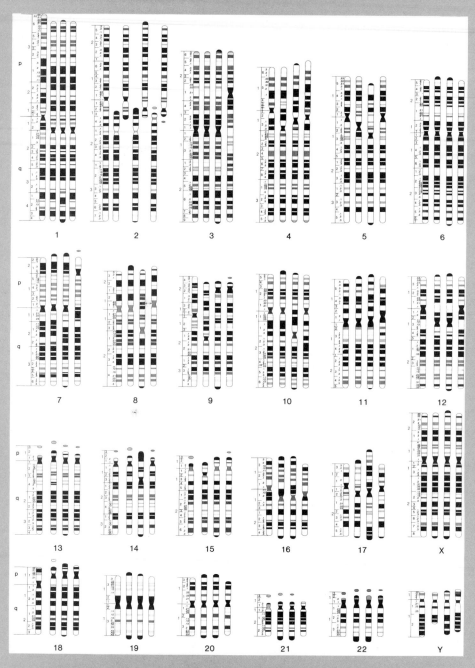

Figure 12.10 Comparison of late-prophase chromosomes of (left to right) human, chimpanzee, gorilla, and orangutan. Drawings represent the 1,000-band stage. Note: human chromosome 2 is a fusion of two chromosomes found in the Great Apes.
Source: J. J. Yunis and O. Prakash, "The Origin of Man: A Chromosomal Pictorial Legacy," *Science,* Vol. 215, Fig. 2, pp. 1525–30, March 19, 1982. Copyright 1982 by the AAAS.

BOX 12.1 (*Continued*)

Chimpanzees, gorillas, and orangutans all have a diploid number of 48 chromosomes. A fusion of one chromosome pair in a human ancestor, producing the large, submetacentric chromosome 2, reduced the diploid number in humans to 46. High resolution G-banding techniques now permit the identification of about one thousand bands in humans and in their Great Ape relatives. Comparative studies reveal virtually total homology of these one thousand bands.

Two major types of differences are found among the chromosomes of these species. First, variations in the amount of nongenic heterochromatin are common. Many chromosomes of the chimpanzee and gorilla, for example, have terminal bands, heterochromatic regions not present in the other two species. The Y chromosomes also show differences in heterochromatin that do not reflect differences in genic content. If these nongenic sequences are disregarded, many chromosomes are essentially identical among the various species: thirteen pairs for human and chimpanzee (chromosomes 3, 6, 7, 8, 10, 11, 13, 14, 19, 20, 21, 22, and XY); nine pairs for human and gorilla (3, 6, 11, 13, 19, 20, 21, 22, and XY); and eight pairs for human and orangutan (5, 6, 12, 13, 14, 19, 21, and 22).

The other type of chromosomal differences are structural rearrangements. The most common, pericentric inversion, has occurred between chromosomes 4, 5, 9, 12, 15, and 16 of human and chimpanzee. A paracentric inversion can be seen in chromosome 7 of chimpanzee and gorilla, and in several instances both paracentric and pericentric inversions have occurred (chromosome 16 of human and gorilla; chromosomes 3 and 17 of human and orangutan). A reciprocal translocation can be seen between chromosomes 5 and 17 of the gorilla.

These comparisons show that essentially the same one thousand nonheterochromatic bands are present in these species. In other words, they all have the same genes. The primary differences are in the order of those genes along the chromosomes. These karyological differences agree with other genetic studies that have found extraordinarily high levels of genetic similarity among humans and Great Apes—over 99% identity in the amino-acid sequence of proteins among humans and chimpanzees, for example. Such findings suggest that structural rearrangements of chromosomes may be associated with regulatory alterations of the genes that produce the phenotypic differences between species.

Such high levels of genetic similarity also indicate an exceptionally close evolutionary relationship of humans to Great Apes. Only a few decades ago, the lines leading to humans and the Great Apes were believed to have diverged perhaps forty million years ago. The accumulating evidence of high genetic similarity, however, combined with the discoveries of key fossils such as Lucy, suggests that divergence of the human–chimpanzee lineages occurred far more recently, perhaps five million years ago, and divergence from the other Great Apes not too many million years earlier. In evolutionary time, humans and Great Apes are separated by only a few ticks of the clock.

Table 12.2
Frequencies of phenotypes and alleles

Population	Total	Phenotype frequencies				Allele frequencies		
		O	A	B	AB	A	B	O
Iowa City, U.S.A.	49,979	.45	.42	.09	.03	.26	.07	.67
Nord, France	103,242	.44	.43	.09	.03	.27	.06	.66
Dublin, Ireland	36,879	.54	.32	.11	.03	.19	.07	.73
Armenians, U.S.S.R.	44,632	.29	.50	.13	.08	.35	.11	.54
Moscow, U.S.S.R.	63,549	.34	.38	.21	.07	.26	.15	.58
Bihar, India	9,257	.31	.22	.38	.09	.16	.27	.56
Pureblood Cherokee Indians, North Carolina, U.S.A.	166	.95	.04	.01	0	.02	.01	.97
Eastern Islanders, Oceania	754	.33	.65	.01	.01	.42	.01	.57
		M	MN	N		M	N	
New York City, U.S.A.	3,268	.32	.50	.18		.57	.43	
Danes, Copenhagen	4,319	.29	.50	.21		.54	.46	
Eskimos, Greenland	569	.83	.16	.01		.91	.09	
Papuans, Oceania	240	.02	.18	.80		.11	.89	

As you acquired similar information on more individuals, the pattern that would emerge would be a continuum of virtually limitless variation. Some groups of individuals would share a few more alleles with each other than with members of other groups, but the key finding would be diversity. Essentially every individual is genetically unique. An understanding of genetics tells us that we all differ, not that we are better or worse than each other. Racism must be seen as a perversion of morality, not as a dictate of nature or our genes.

Population differences in allele frequencies have come about slowly over evolutionary time. Ultimately all new alleles arise through mutation. An allele is incorporated into the population either because it confers a selective advantage, such as in the case of the sickle-cell hemoglobin allele, or because it is effectively neutral and is randomly incorporated into the population. Migration of persons between populations and assortative mating also must have contributed to the present-day distribution patterns of alleles throughout the world. Although current patterns are the result of biological evolution, we are becoming able to actively interfere in our biological heritage and influence our future genetic potential.

Manipulating the human gene pool

Humans have been altering their own evolution in very direct ways throughout recorded history. Assortative mating, by increasing the phenotypic expression of certain genes, may produce combinations of alleles favored by natural selection. Humans mate preferentially for many traits, some genetic (such as height, skin color, earlobe length, and intelligence) and some environmental or cultural (religious or economic status, for example). The imposition of culture upon the selective process can have exceptionally strong effects. Social structures allowing chieftains and kings dozens or even hundreds of wives have greatly increased frequencies of genes carried by these families, especially within small populations.

A more widespread effect may be associated with improving medical and hygienic practices. Many persons with genotypes that would have been fatal in previous generations now live essentially normal lives and have children, thus transmitting some once-lethal genes. Are improving health practices leading to genetic deterioration of our species? And what positive steps can we take to improve the genetic quality of our species?

Dysgenics

Factors increasing the frequency of deleterious genes in a population are called **dysgenic.** Among the genes that may be increasing in frequency in the human gene pool due to the potentially dysgenic effects of modern medicine are those causing diabetes, bad eyesight, PKU, and many other genetic diseases. Should we fear such changes?

Probably not. The effect of any gene depends upon the environment in which it is expressed. Contemporary medical practice will be able to provide an environment in which individuals bearing genes deleterious to survival in primitive conditions will have perfectly normal phenotypes. If—and it is not a certainty—an increasing proportion of the population develops diabetes, the provision of insulin and other treatments will not be an unsupportable burden, particularly in comparison to any systematic programs to eliminate such genes, i.e., persons, from the population. The expansion of medical knowledge is so rapid that treatments for genetic diseases will be developed far more rapidly than deleterious genes, however defined, accumulate in the gene pool.

Virtually every individual harbors four or five, perhaps many more, deleterious recessive genes. We could not remove all deleterious genes from a population without exterminating that population. An equally intractable problem lies in defining deleterious. An allele harmful in some sense in one environment may be beneficial in another.

Euphenics and eugenics

Genetic engineering is a term heard frequently in the popular media. Any human interference that alters allelic frequencies in populations, or alters the genome, or alters the expression of genes can be considered a form of genetic engineering. There are at least two major different approaches to human genetic engineering: euphenics and eugenics.

Euphenics (good appearance) involves direct manipulations of the phenotype in an attempt to nullify an undesirable genetic trait and ameliorate or cure persons of their genetic affliction. Examples of euphenic measures include wearing glasses, providing daily injections of insulin to a diabetic, keeping phenylalanine *out* of the diet of a person with phenylketonuria, and removing the cancerous eye in persons with retinoblastoma. Over twenty-two hundred inherited disorders have been identified by The National Foundation–March of Dimes. Theoretically a euphenic treatment could be devised for each of these.

Eugenics (good birth) is a term coined by Francis Galton, a cousin of Charles Darwin. Eugenics refers to any effort to improve the genetic quality of the human population through selective breeding. Galton started an active eugenics movement in the 1880s with a book pointing out (quite incorrectly) the hereditary component of genius. Galton was very interested in improving the overall mental abilities of the population by breeding programs. Galton's ultimate goal was to improve the genetic composition of the human gene pool.

Eugenics programs have been common in plants and animals and can take one of two directions. *Positive eugenics* involves breeding individuals with good genes to increase these genes in the population. Using a prize bull to father an entire herd of dairy cattle is an example of a positive eugenics program. The sperm bank for Nobel prizewinners (chapter 7) is a small-scale voluntary scheme of positive eugenics whose population effects will be negligible, however, and represents no threat to human rights, whatever the motivation of participants.

Negative eugenic programs prevent individuals with undesirable genes from leaving any progeny. Considerable enthusiasm was generated for negative eugenic programs in the first several decades of this century. Numerous persons incarcerated in institutions for the mentally defective and criminally insane were sterilized to prevent their reproducing. In 1913 Wisconsin passed a law requiring that male applicants for a marriage license pass a stringent physical examination. Various other states passed laws forbidding marriage of insane, epileptic, or otherwise physically or morally unfit persons. A very sad chapter in American social history involves racism and the controversy concerning inheritance of intelligence.

Some aspects of negative eugenics are presently practiced and require an informed public to debate the issues involved. Therapeutic abortion of genetically defective fetuses is a negative eugenic measure. Another form of negative eugenics is birth control, whether sterilization or otherwise, for victims of Down's syndrome and other diseases to prevent them from reproducing.

An insoluble problem in any eugenic program is how to decide what constitutes good and desirable genes and what does not. Who is to make these very important decisions? Dare we even entertain the notion that a eugenics program could work in a democratic society? A major problem with attempts at a negative eugenics program ridding the population of deleterious genes is that most of these genes are recessive. Preventing reproduction of affected persons is a slow way of reducing the frequency of recessive alleles in the population; the allele is carried not only by the affected person, but by all heterozygotes as well. In a population where affected persons (aa) account for one in ten thousand of the population, $q^2 = 0.0001$, the frequency of the allele a would be 0.01, and an estimated 2% of the population, 2pq, would be carriers (Aa). In a population with an initial allele frequency of a = 1% (.01), if all aa persons were not allowed to reproduce, it would take five hundred generations to reduce the allele frequency to 0.0017, and over five hundred generations to reduce the allele frequency by a factor of ten. A negative eugenics approach is *not* a quick, effective means of decreasing the frequency of a recessive deleterious allele in a population.

Many eugenic proposals and genetic and reproductive technologies raise serious ethical, legal, and scientific questions. A eugenic measure such as therapeutic abortion may be accepted by a broad spectrum of our society, but it also raises legal and moral questions that will be debated emotionally in the coming decades. Societal resolution of these issues will require participation by lawyers, lawmakers, theologians, philosophers, scientists and—importantly—informed citizens.

Summary

1. The Hardy-Weinberg equilibrium equations permit estimation of allele and genotype frequencies in populations.
2. Ecological genetics relates patterns of phenotypic and genotypic variation within and between populations.
3. Human life depends upon all other organisms with which we share the Earth.
4. Humans have long practiced artificial selection in breeding high-quality plant and animal species.
5. New genetic combinations of domestic species are created by mutagens, grafting, hybridization, protoplast fusion, and gene transfer.
6. Inadvertent selection by humans has resulted in the emergence of pesticide-resistant and drug-resistant strains of plant and animal disease organisms.
7. Genetic knowledge has led to varied methods of pest control, including sterile male releases and pheromone "trickery."
8. The interplay of politics and agriculture is exemplified by the Green Revolution of the 1960s.

9. It is imperative for humans to conserve the remaining genetic diversity of plants and animals around the world.
10. Genetic bottlenecks have greatly diminished the evolutionary potential of many species and driven others to extinction.
11. Various governmental agencies around the world are now conserving valuable plant and animal germ plasm.
12. The gene pool consists of all alleles contained in a group of interbreeding individuals. About one-third of all human genes are polymorphic, having more than one allele. Races can be characterized by allele frequency differences. Over 85% of the allelic variation in human races is seen within populations, not between races. Electrophoresis is a rapid way of visualizing allelic differences in structural genes. These allelic differences between populations can be expressed as a genetic distance.
13. Dysgenic factors lead to an increase in the proportion of harmful genes in a population. Modern health practices are unlikely to have significant dysgenic effects.
14. Euphenics involves manipulation of the phenotype to ameliorate genetic conditions. Insulin injections for diabetics are a good example.
15. Eugenics involves encouraging persons with favored genes to reproduce (positive eugenics) or discouraging persons with deleterious genes from reproducing (negative eugenics).

Key Words and Phrases

artificial selection	genetic distance (D)	monoculture
biological evolution	genetic diversity	monomorphic
dysgenics	genetic engineering	pesticide treadmill
ecological genetics	genome	pheromones
electrophoresis	grafting	plant patenting laws
embryo transfer	Green Revolution	polymorphic
eugenics	Hardy-Weinberg equilibrium equations	population genetics
euphenics		protoplast fusion
gene pool	hybrid vigor	race
gene transfer	hybridization	racism
genetic bottleneck	inbreeding depression	sterile male technique
	land races	superovulation

Questions

1. What are some of the problems associated with using mutagens to produce new genetic variants? Are plants or animals more likely to yield favorable results? Why?
2. To what domestic mammals other than cattle could superovulation and embryo transfer be most profitably applied? What are your criteria for selecting these species?
3. Until the twentieth century all farmers saved seed from one harvest to plant the succeeding year's crop. In advanced countries, farmers tend to purchase their seed annually from seed companies. What changes have caused farmers to purchase seeds annually? Can you think of any disadvantages arising from seed companies controlling seed stock?

4. Suppose a seed company developed a perennial variety of corn or wheat (all present varieties are annuals). What dilemma might such a development pose to a seed company? Why would a perennial corn be advantageous to the farmer?

5. Compare and contrast the advantages and the disadvantages of crop monoculture.

6. What characteristics shared by organisms such as bacteria, insects, and rats result in their ability to evolve rapidly in response to attempts to limit their populations with chemical pesticides and antibiotics?

7. Could the sterile male technique be applied to control rats or other noxious mammals? What problems might arise from such attempts?

8. Many triploid (or other polyploid) plants are of commercial value because they produce exceptionally large fruit or seeds. Induced triploid animals have, in some cases, increased commercial value due to their lack of reproductive tissues. Comment on the different phenotypic results of polyploidy in plants and animals.

9. Recombinant DNA techniques permit the transfer of DNA between species. Does this new technology change the potential commercial value of the millions of nondomesticated species of plants and animals? How?

10. If 84% of a random-mating population express the dominant phenotype of being able to taste the bitter compound PTC (taster phenotype), what is the frequency of the recessive allele, which when homozygous makes its owner unable to taste this compound (a nontaster)?

11. If the frequency of a recessive allele is 30%, in a population of one thousand persons how many would you predict would be carriers of this allele, but would not express the recessive phenotype?

12. Of what significance is the fact that the proteins of humans and chimpanzees are so alike that any two homologous proteins chosen at random have a 99% chance of being identical in their sequences?

13. Distinguish between euphenics and eugenics. Which program has the potential for effecting the greatest genetic change in the human population?

14. What immediate measures for the genetic health of the human population can you think of that could be implemented now without raising serious moral issues? Are there associated economic costs?

15. How many specific human traits can you think of that are the likely result of natural selection?

16. Compare genetic variability in the population of melting-pot countries such as Canada and the United States with genetic variation in countries that are less racially diverse.

For More Information

Cann, R. L.; M. Stoneking; and A. C. Wilson. 1987. "Mitochondrial DNA and Human Evolution." *Nature,* 325:31–36.

Cohen, M. L., and R. V. Tauxe. 1986. "Drug-Resistant *Salmonella* in the United States: An Epidemiologic Perspective." *Science,* 234:964–69.

Ehrlich, P. R. 1977. "Ecologists, Ethics, and the Environment." *Bioscience,* 27:239.

Elkington, J. 1985. *The Gene Factory: Inside the Genetic and Biotechnology Business Revolution.* Carroll & Graf Publishers, New York.

Gould, S. J. 1983. *The Mismeasure of Man.* W. W. Norton and Co., New York.

Hardin, G. 1974. "Living on a Lifeboat." *Bioscience,* 24:561–68.

Harris, M. 1973. "The Withering Green Revolution." *Natural History,* 82(3):20–22.

Jukes, T. H. 1986. "Frost Resistance and *Pseudomonas.*" *Nature,* 319:617.

King, M. C., and A. C. Wilson. 1975. "Evolution at Two Levels: Molecular Similarities and Biochemical Differences between Humans and Chimpanzees." *Science,* 188:107–17.

Knorr, D., and A. J. Sinskey. 1985. "Biotechnology in Food Production and Processing." *Science,* 229:1224–29.

Lewin, R. 1987. "The Unmasking of Mitochondrial Eve." *Science,* 238:24–26.

Lewin, R. 1987. "My Close Cousin the Chimpanzee." *Science,* 238:273–75.

Miller, J. A. 1984. "The Mating Game." *Science News,* 126:232–35.

Miller, J. A. 1986. "Barnyard Biotech: Dissent on the Farm." *Science News,* 129:213.

Miller, J. A. 1986. "First Live Gene-Splice Release: It's Already History." *Science News,* 129:228.

Motulsky, A. G. 1974. "Brave New World?" *Science,* 185:653–63.

Motulsky, A. G. 1983. "Impact of Genetic Manipulation on Society and Medicine." *Science,* 219:135–40.

Ruse, M. 1981. *Is Science Sexist? And Other Problems in the Biomedical Sciences.* D. Reidel Publishing Company, Boston, London.

Sun, M. 1986. "Engineering Crops to Resist Weed Killers." *Science,* 231:1360–61.

Tatum, L. A. 1971. "The Southern Corn Leaf Blight Epidemic." *Science,* 171:1113–16.

Wade, N. 1974. "Green Revolution, I and II." *Science,* 186:1093–96, 1186–92.

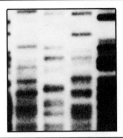

Genetic engineering: the promise of the future

Figure 13.1 (a) A portion of a human DNA sequencing gel showing sequenced DNA from three different individuals. The sequence of individual number 18, reading from the bottom up would be GGTCTATCCTCA-TTTTCGGCGAGT; (b) polymorphic human DNA fragments from six unrelated British Caucasians (1–6) and twelve members from a large British Asian-Indian pedigree (7–18).
Source: (a) Dr. Jack Griffith; (b) Dr. A. Jeffreys, *Nature* 314:67–73.

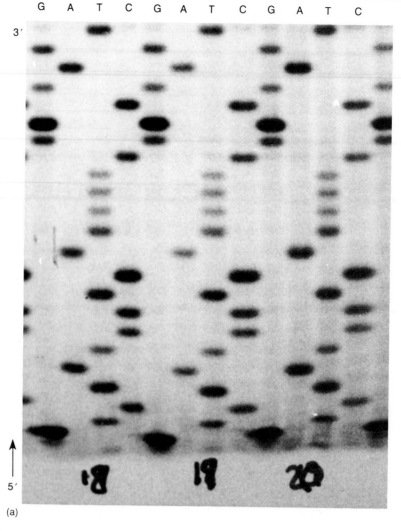

(a)

Modern genetics has advanced in three major steps. First, the findings of Mendel and his followers established the foundations of genetics as a true science. Mendelian genetics has been of greatest practical value as the basis for breeding programs and artificial selection to improve domesticated agricultural plant and animal species.

The second stage began in 1953 with the identification of the physical structure of DNA, confirming it as the genetic material. Subsequently, enormous amounts of money have been spent on research programs in molecular genetics, attempting to understand the biochemical basis of the genetic mechanism and human disease. These programs have greatly advanced the knowledge of the molecular basis of many medical conditions, but in most cases did not directly lead to improved health care. As we have previously seen, diagnosis is not the same as treatment.

The third level of development of modern genetics began in 1969 with the isolation, for the first time, of a single gene. Since then, techniques have developed rapidly, allowing manipulation of DNA. Virtually any segment of DNA in any organism can now be isolated, sequenced, and transferred to other organisms for many possible uses. (See figure 13.1.)

A new set of terms has quickly entered the language to describe these new scientific and technological methods. **Genetic engineering** and **gene splicing** are phrases commonly used to describe any type of manipulation of DNA for any purpose. Both

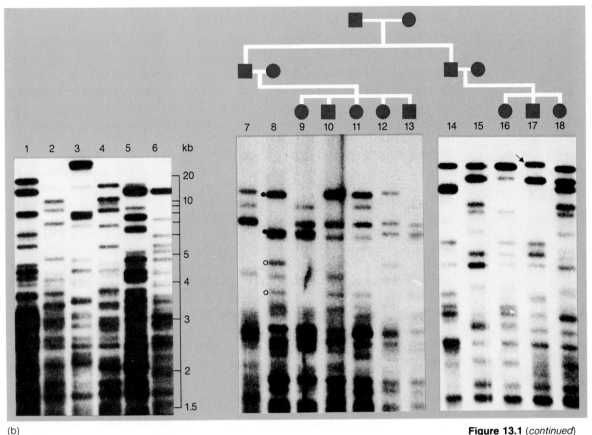

(b)

Figure 13.1 (*continued*)

imply the ability to cut segments of DNA from one or more organisms and recombine these segments to create novel genetic forms. A more precise and less emotive term for these methods is **recombinant DNA technologies.**

Another long-used word has acquired new life and added meaning: **biotechnology.** Biotechnology is the industrial use of biological systems and processes. Biotechnology is as old as human civilization; fermentation, for example, is a biological process used to manufacture such products as wine, beer, alcohol, cheese, vinegar, and bread. Other types of biotechnology allow the production of soap and medicines and the treatment of waste products.

To many people, the term biotechnology has become synonomous with genetic engineering because recombinant DNA technologies offer so much promise for industrial use. Recombinant DNA methods have been used commercially for over a decade and already affect the lives of most of us. Fortunes have been made many times from various aspects of recombinant DNA work, and in the future recombinant DNA methods will be a routine and pervasive part of human life. The rapid expansion of these techniques is difficult to comprehend. Japan, for example, had no commercial application of recombinant DNA techniques in 1975, but by the year 2000 Japan plans to generate 10% of its enormous gross domestic product directly from recombinant DNA methods.

In this chapter we introduce briefly some aspects of these methodologies, then consider some of the applications of these techniques to everyday human life. Additionally, we note some of the hazards and ethical concerns raised by this rapid technological advancement.

Recombinant DNA technology

Two obvious problems confront anyone wanting to manipulate and recombine DNA. The first problem is how to identify and isolate a single potentially useful or interesting segment of DNA. An average bacterium such as *E. coli* has a genome about four million nucleotide bases long, and a complex organism such as a human has about three billion base pairs in the genome. To scan those billions of nucleotides, recognizing their functions and plucking out only the relevant segments, seems an impossible task and led to the early pessimism about the possible practical use of molecular techniques.

The second problem is concentration. A single cell is visible only under a microscope, and most cells contain only two sets of chromosomes, which constitute a small fraction of its mass. How could one ever hope to study specific segments occurring in such tiny concentrations in a single cell?

These problems were solved surprisingly easily by using reasonably well-known biological processes in a method now called **molecular cloning.** These processes had been discovered in prior research into molecular biology conducted purely for its own rather than applied interest, but proved essential to the development of recombinant DNA technologies. Molecular cloning is the key process in recombinant DNA technology; it allows precise identification and examination of specific segments of DNA in a particular species and the subsequent transfer of those segments to other species for useful purposes. Molecular cloning is now conducted routinely in research and industrial laboratories around the world. Its applications seem limitless.

Molecular cloning

Molecular cloning begins with DNA from a test organism (figure 13.2, step 1), a cloning vehicle, a host bacterium, and a set of enzymes. The test organism is chosen because of special interest in its DNA. For disease diagnosis in humans, for example, DNA might be obtained from a liver or bone marrow biopsy, or from cells in amniotic fluid. Alternatively, DNA might be taken from a plant that is believed to have a trait that might be of agricultural importance such as, for example, resistance to an inspect pest.

Second, a **vector** or **cloning vehicle** is required. Two types of vectors are commonly used, **plasmids** and **phages.** A plasmid is a small circular segment of extra-chromosomal DNA, about five thousand base pairs long, found in the cytoplasm of bacteria (figure 13.2, step 2). Plasmids reproduce independently of the DNA of the bacterium and have long been known to carry genes for antibiotic resistance. Furthermore, plasmids are transmissible between bacteria, meaning that antibiotic resistance has now spread to virtually all disease-causing bacteria of humans.

Phages, more properly called bacteriophages, are viruses that attack bacteria. Like plasmids, phages move between bacteria, and they reproduce independently of the bacterium itself. The key trait of both plasmids and phages is their ability to incorporate short segments of foreign DNA into their own genomes. Thus, both can accept DNA from another organism, reproduce it along with their own DNA, and transmit or move that DNA to a host bacterium. Neither plasmid nor phage is considered to be alive, however, because neither is capable of independent self-reproduction. Each uses the enzymes and cellular apparatus of a bacterium to reproduce itself. Whether a plasmid or a phage is chosen for molecular cloning experiments usually depends upon the test species being examined. A phage will incorporate much larger segments of foreign DNA than will a plasmid and is thus often chosen for working with DNA from species with a large genome, such as humans.

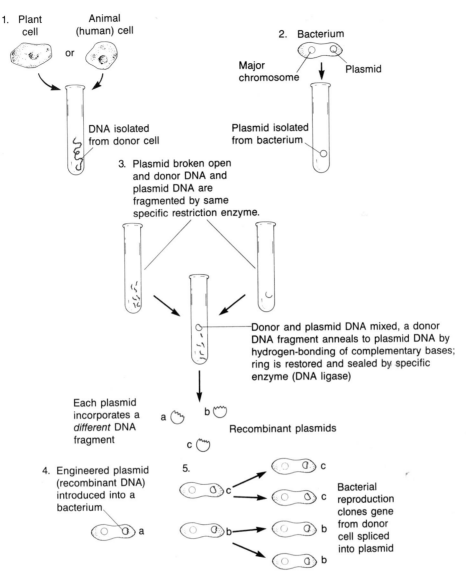

1. Plant cell   Animal (human) cell   or

DNA isolated from donor cell

2. Bacterium

Major chromosome   Plasmid

Plasmid isolated from bacterium

3. Plasmid broken open and donor DNA and plasmid DNA are fragmented by same specific restriction enzyme.

Donor and plasmid DNA mixed, a donor DNA fragment anneals to plasmid DNA by hydrogen-bonding of complementary bases; ring is restored and sealed by specific enzyme (DNA ligase)

Each plasmid incorporates a *different* DNA fragment

a   b
c

Recombinant plasmids

4. Engineered plasmid (recombinant DNA) introduced into a bacterium.

a

5.

c → c
c → c
b → b
b → b

Bacterial reproduction clones gene from donor cell spliced into plasmid

Figure 13.2 An outline of molecular cloning. Plasmids are isolated from bacteria and cleaved by special enzymes into small fragments. DNA is simultaneously isolated from plant or animal donor cells. Plasmid DNA and donor DNA are incubated together, and a special DNA ligating enzyme restores the ring plasmid. The recombinant engineered plasmid is introduced into a bacterium that then reproduces clonally making more recombinant DNA.

The third ingredient in molecular cloning is a host bacterium such as *E. coli.* The bacterium is the key to amplification of DNA from other species. When the bacterium reproduces, it reproduces its own genome plus the DNA incorporated into the bacterial genome by a lysogenic phage (see chapter 11 to review aspects of phage biology). If the bacterium contains plasmids, they will have been reproduced independently in the cytoplasm; daughter bacteria will receive plasmids in their cytoplasm also. Thus, tiny amounts of DNA can be rapidly turned into very large quantities by bacterial reproduction.

The final key ingredient is one or more of a set of enzymes called **restriction endonucleases** or **restriction enzymes.** These enzymes were first identified as a key component of the defense mechanism of a bacterium against viral attack. Their name derives from their ability to restrict the entry of foreign DNA (such as that of a phage) into a cell by enzymatically cutting it into harmless fragments. Each species of bacterium produces its own set of restriction enzymes, against which its own genome is protected. Furthermore, some of these enzymes cut only at specific four- or six-base sequences of DNA.

Thus, a large group of these enzymes is available to use in the first step in molecular cloning—cutting DNA into manageable fragments at specific sites. Because the DNA of each species has a unique sequence, a particular restriction enzyme will cut DNA from one species into a set of fragments that differs from that of all other species.

In step 1 of molecular cloning (figure 13.2), purified test DNA from a source such as human bone marrow is incubated with a specific restriction enzyme, which cuts it into thousands of segments. Usually, each segment is either about 250 base pairs long (for an enzyme that recognizes a four-base sequence; any random four-base sequence of the four nucleotide bases will occur roughly every $4^4 = 256$ bases) or about 4,000 base pairs long ($4^6 = 4,096$, for an enzyme recognizing a six-base sequence).

At the same time (step 2), a cloning vehicle such as a plasmid is also cut in exactly one place with a restriction enzyme so that the circular DNA is "opened up" but not fragmented. Next, step 3, a small quantity of the test DNA is mixed with a large number of broken plasmids, and an enzyme called **DNA ligase** is added. DNA ligase, a naturally occurring enzyme, functions to ligate, or join, broken segments of DNA. The excess of plasmids over fragments ensures that the reformed plasmids will incorporate no more than one fragment of test DNA. This step results in many **recombinant plasmids,** containing their own DNA combined with a fragment derived from another organism.

These plasmids are then incubated with their host bacterium (step 4), which incorporates them. The result is often referred to as a **DNA library,** meaning that the bacterial culture contains a set of bacteria that has incorporated fragments of DNA from the test species. All the DNA from the test species may have been incorporated, but a single bacterium will have only a short fragment of the entire test genome.

Often, the next step is simply to allow the bacterial culture to incubate and reproduce itself, thus *amplifying* the quantity of bacteria *and* the test DNA. Once bacteria are in a library, essentially infinite quantities of DNA can be produced—cloned—for subsequent testing. This means that particularly interesting DNA from only one source can be used in laboratories all over the world, and that the problem of tiny concentrations of DNA is overcome by standard microbiological techniques. If a phage is used as a cloning vehicle, the phage colony itself is directly amplified by culturing in a medium containing its bacterial host.

Probing

The next part of the procedure is like searching for a needle in a haystack. The haystack, or DNA library, consists of a set of clones, and the object is to select efficiently only the one set of interest from the tens of thousands of clones in the library. This is achieved by the use of a specific genetic **probe.**

A probe is a sequence of nucleotide bases complementary to any segment of a particular gene and long enough to hybridize uniquely with that gene. Construction of a probe is the key aspect of gene isolation and one of the most difficult parts of recombinant DNA technology.

Often, a probe is created by collecting large amounts of the messenger RNA specific to the gene to be isolated. For example, mRNA for human hemoglobin can be obtained in large quantities from red blood cells; few other mRNAs are found in red cells. Other tissues and cell types would be specific for other types of mRNA.

Next, reverse transcriptase from retroviruses is used to construct a segment of DNA complementary to the sense strand of the gene of interest from the isolated mRNA. Radioactive nucleotide bases are added to the culture where the complementary DNA (or **cDNA**) is produced. Thus, genetic probes normally are radioactive cDNA segments.

1. Make radioactive probe:

   (a) Isolate hemoglobin mRNA from red blood cells

Red blood cells ⬭ ⬭ ——— m-RNA

   (b) Use reverse transcriptase and radioactive DNA bases to make cDNA* (= probe for hemoglobin gene)

          cDNA

           m-RNA

Figure 13.3 How a probe works.
Source: (agar plate) Dr. Jack Griffith.

2.

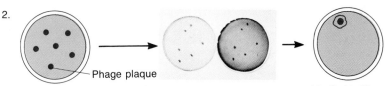

     Phage plaque

(a) Grow recombinant phage on bacteria on agar plate   (b) Make nitrocellulose replica of agar plate   (c) Probe filter with cDNA* under conditions allowing cDNA* to hybridize with *any* complementary DNA sequences on filter; expose filter to X-ray film

3. Pick colony off original agar plate; grow phage on new bacterial culture

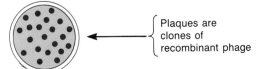

    Plaques are clones of recombinant phage

4. Isolate DNA from culture; will contain the cloned hemoglobin gene

        hemoglobin-DNA

The first step in the use of a probe in a phage library is to "plate" the phage containing the DNA library on an agar medium covered with a bacterial lawn, a continuous growth of bacteria (figure 13.3). The result is a set of phage colonies—clones—each originating from a single recombinant phage. Each phage clone is called a *plaque,* which looks like a clear hole in the bacterial lawn, where the bacteria have all been lysed by the replicating phage.

The problem is identifying which of these colonies, if any, contains the gene of interest. A nitrocellulose filter is then applied to the surface of the agar medium; the filter absorbs some of the phage from each colony and thus becomes a mirror image of the original plate. The original plate is saved for future use.

Next, the nitrocellulose filter is heated to separate, or *melt,* the two strands of DNA in the phage in each plaque. DNA melts at about 85°–100°C. The filter is now incubated in a solution containing the radioactive probe under conditions favoring *hybridization,* or rejoining, of the complementary but separated strands of DNA. In the phage clones containing the gene that matches the cDNA of the probe, hybridization will occur to the probe. Those colonies can be identified by exposing the agar plate to X-ray film, then matching the colonies that exhibit radioactivity to the original plate. Those clones can then be cut out of the original plate for further study and use (figure 13.3).

The most difficult part of these procedures is developing the probes. Many genes may not produce a protein or mRNA in sufficient quantities to allow production of a probe. Additionally, a single agar plate can grow only a few dozen plaques; screening enough plates to ensure coverage of the fifty thousand or more functional genes in the entire genome of a complex organism is an arduous task, which is not always feasible.

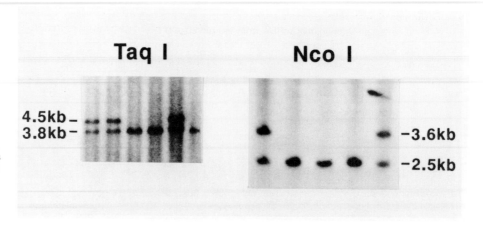

Figure 13.4 Gel electrophoresis allows visualization of DNA fragments. RFLP patterns are shown for the restriction enzymes Taq I and Nco I from six and five unrelated individuals respectively. The probe used is a human muscle creatine kinase cDNA which maps to the long arm of chromosome 19. Molecular sizes in thousands of bases are indicated in the right and left margins. Source: Dr. Fielding Hejtmancik, Institute for Molecular Genetics, Baylor College of Medicine, Houston.

Nonetheless, the general techniques described above have proven extremely successful in offering access to the genome of virtually any species. Molecular genetic techniques are continuing their rapid development and improvement, but molecular cloning and probing are the basis for the developing recombinant DNA industry. They are usually used in conjunction with the technique of gel electrophoresis, which allows fragments of DNA created by restriction enzymes to be visualized directly and compared between individuals and species (figures 13.1 and 13.4).

Safety

When recombinant DNA experimentation began in earnest in the early 1970s, many scientists and laypersons expressed serious concern about the potential dangers. The possibilities for a technological disaster seemed very real. Moving genes between species seemed to offer the opportunity for creation of new and possibly dangerous organisms. The use of *E. coli* as the amplification host for recombinant vectors seemed especially threatening. Humans are the natural host of *E. coli;* recombinant forms might produce toxins or initiate infections that could not be controlled and could lead to disease epidemics. Release of new genetic combinations, even new species, into the environment might have disastrous effects. With some justification, the public imagination ran wild.

Fears seemed to culminate in a situation in Cambridge, Massachusetts, where the City Council passed laws to prohibit recombinant DNA research within the city limits, which include Harvard University and Massachusetts Institute of Technology. This placed some scientists in the equivocal position of seeing their research and careers threatened by the actions of a public who had only a limited knowledge of their work. However, the scientific community itself recognized the novel nature of recombinant DNA research and the potential for danger.

A series of scientific meetings in the early 1970s eventually led to a major international conference at Asilomar, California in 1975, which established a set of voluntary guidelines for scientists engaged in recombinant DNA research. These guidelines prohibited certain types of research entirely. Some potentially dangerous experiments could be conducted only in facilities designed for maximum containment of pathogens; only a few such facilities existed and were used for studies with the most dangerous disease organisms such as smallpox and anthrax. Many other types of experiments were considered acceptable if standard laboratory protocols were followed.

Businesses and private individuals could not be forced to comply with these guidelines. However, the guidelines were adopted by the National Institutes of Health, who fund most public research of this type in the United States. Failure to adhere to proper safeguards could result in an investigator losing future funding, as well as other penalties.

Experience in the past decade has led to relaxation of these guidelines. For several reasons, recombinant DNA research now appears to be considerably less dangerous than imagined in the early 1970s. Recombinant DNA itself is not dangerous. It may become so only after transfer to another living cell, which is not a simple procedure. Even in a species such as *E. coli,* which might be of potential danger to humans, laboratory strains are usually selected for special traits that restrict their infectiveness outside the laboratory. Such strains have been employed for decades. No disease epidemic in humans has ever resulted from escape of an infectious agent from a laboratory.

Even more compelling is the information discovered since the early 1970s on the interindividual and interspecies transfer of DNA. It is now quite clear that genes are naturally transferred between individuals, even of different species, and have always been throughout evolutionary history. Oncogenes appear to represent widespread evidence of such transfer. New genetic combinations are always arising as a result of such natural processes. Such discoveries have been made primarily as a result of recombinant DNA research, and they argue convincingly that our initial fears of the dangers of recombinant DNA research were, largely but not totally, unfounded.

Thus far, recombinant DNA research has proven to be safe, but there is no way to guarantee that all such experiments will be safe. Opponents of this research argue that, although the odds of a disaster may be small, even a tiny risk is not worth taking. If recombinant DNA research results in a new epidemic-creating organism, the costs will be borne by the public, who have not been given the chance to consent to being exposed to such dangers. Many ecologists still express concern over the potential damage made possible by the release of recombinant organisms into nature.

Furthermore, the safety protocols for recombinant DNA research, still administered by the National Institutes of Health, are totally voluntary. Several serious violations of these protocols have occurred, and the perpetrators were reputable and well-intentioned scientists. If even the scientific community does not conform to safety guidelines, many people despair of compliance by businesses motivated by profits. In addition, governments have shown few limits on their behavior during times of war, and actions of unscrupulous individuals can seldom be restricted. Opponents of recombinant DNA research argue that these technologies should never be developed because they are certain to be misused.

On the other hand, the safety record thus far is good, and as noted above, increasing knowledge seems to be diminishing our perceptions of danger. The techniques are simple enough and already so widely used that it would be virtually impossible to forbid recombinant DNA research entirely. The actual and potential benefits to be derived from this technology are so high that for most people the risks seem worth taking.

Cloning of whole organisms

Although molecular cloning is new, cloning of entire organisms is not. Clones (genetically identical copies) of various agricultural species of plants have long been produced by asexual (vegetative) reproduction. Navel oranges, lombardy poplars, dates (figure 13.5a), and many other cultivated species all comprise clones.

Some animals produce clones as well. Identical twins and triplets of any species are clones. Armadillos commonly give birth to identical quadruplets, making them extremely useful experimental animals (figure 13.5b).

In the 1950s the first clone of a complex animal was produced experimentally by a technique known as **nuclear transplantation** (figure 13.6). Nuclei from adult frog or tadpole cells were introduced into enucleated eggs. About 20% of eggs treated this way later develop into a frog that was an exact replica of the tadpole that donated the transplanted nuclei.

In 1978 a writer named David Rorvik published a book entitled *In His Image: The Cloning of a Man.* It purported to document the secret cloning of a wealthy man. Published as nonfiction, *In His Image* received a flurry of media attention. Rorvik, however, was unable to verify his claims, and in 1982 the publisher publicly admitted, following a court ruling, that no attempt had been made to confirm Rorvik's claims and that his book was a fraud and a hoax.

The successful cloning of mammals from somatic cells of embryos or adults has not been accomplished. It appears that normal mammalian development requires specific contributions from both the male and female parent, unlike the situation observed in amphibians. Although it may be possible in the future, cloning of humans is generally regarded as ethically unacceptable.

Biotechnology

The uses of recombinant DNA technology seem as limitless as the imagination. In this section, we briefly discuss some of the applications that are presently affecting human life or that will soon do so.

Genetic pharmacology

The amplification capabilities of molecular cloning are of great pharmaceutical and industrial value. Once a biological substance, particularly a protein, has been identified, it is often possible to produce large quantities relatively cheaply by recombinant DNA methods. This technique is especially useful for producing biochemicals that occur in small concentration in the body.

For example, **interferon** is a naturally occurring antiviral agent believed to be of value in combatting cancer. Only tiny quantities can be purified from experimental animals, and then only at enormous cost. An adequate quantity of interferon to conduct experiments on only one or two cancer patients costs tens of thousands of dollars. Recombinant DNA methods, however, allow mass production at a price suitable for mass marketing.

The first substance produced by recombinant DNA methods for clinical use was somatostatin, a pancreatic hormone. The market for this chemical is relatively limited, but the market for synthetic **human insulin** for the treatment of diabetes is huge and commercially lucrative. Diabetes results from inadequate production in the pancreas of insulin, which is required for normal carbohydrate metabolism. Diabetics

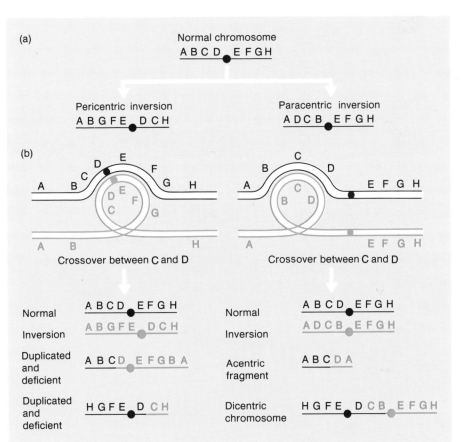

Plate 16 Chromosome inversions: (a) production of inversions by two breaks in a single chromosome followed by healing; (b) homologous pairing in meiosis of individuals who are heterozygous for an inversion. Notice that in both inversions only two of the possible four gametes will contain an entire chromosome, one normal and one inverted. The remaining two meiotic products will be missing some genes, have additional genes, be missing a centromere, or even have two centromeres. Both of the latter conditions interfere with normal meiosis.

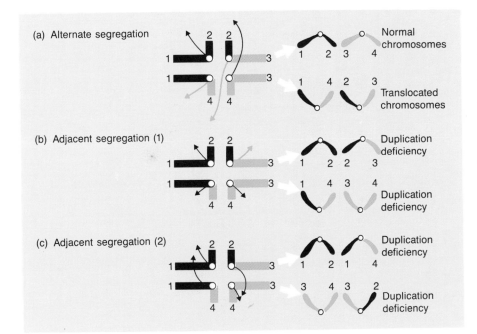

Plate 17 Meiosis in a cell that is heterozygous for the reciprocal translocation depicted in figure 10.8b results in different products depending upon whether segregation at anaphase I is alternate (a) or adjacent (b and c). (a) In alternate segregation both gametes have a balanced genome, although half have normal chromosomes and the other half have translocation chromosomes. (b and c) In both types of adjacent segregation, homologous centromeres go to the same (c) or opposite (b) poles, resulting in all gametes being duplicated and deficient for portions of the genome.

Plate 18 Human insulin is a protein molecule composed of two polypeptide chains: the A chain consists of twenty-one amino acids; the B chain consists of thirty amino acids.
Source: Leland G. Johnson, *Biology*, 2d ed. Copyright © 1987 Wm. C. Brown Publishers, Dubuque, Iowa. All Rights Reserved. Reprinted by permission.

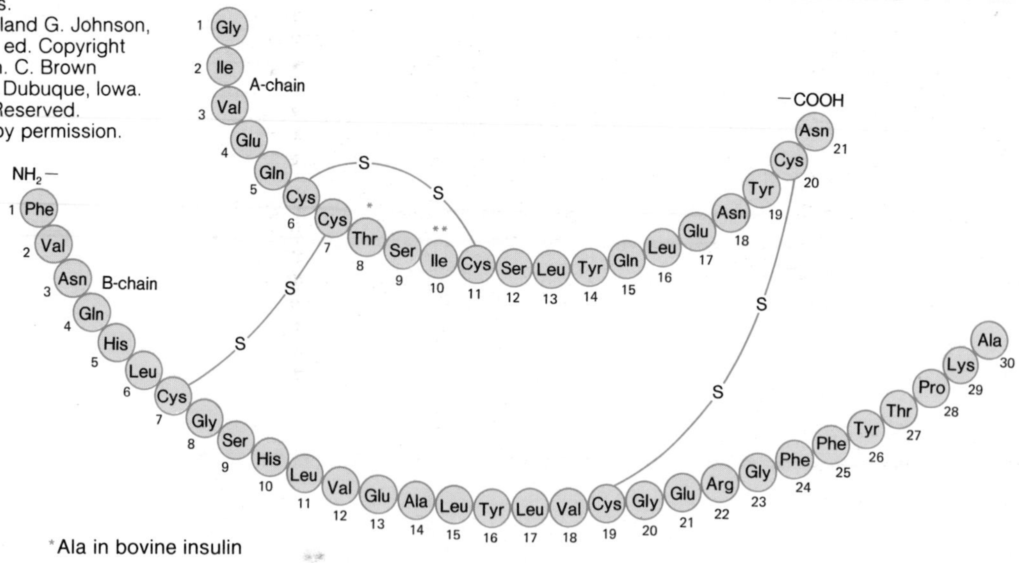

A-chain: 1 Gly, 2 Ile, 3 Val, 4 Glu, 5 Gln, 6 Cys, 7 Cys, 8 Thr, 9 Ser, 10 Ile, 11 Cys, 12 Ser, 13 Leu, 14 Tyr, 15 Gln, 16 Leu, 17 Glu, 18 Asn, 19 Tyr, 20 Cys, 21 Asn —COOH

NH_2—

B-chain: 1 Phe, 2 Val, 3 Asn, 4 Gln, 5 His, 6 Leu, 7 Cys, 8 Gly, 9 Ser, 10 His, 11 Leu, 12 Val, 13 Glu, 14 Ala, 15 Leu, 16 Tyr, 17 Leu, 18 Val, 19 Cys, 20 Gly, 21 Glu, 22 Arg, 23 Gly, 24 Phe, 25 Phe, 26 Tyr, 27 Thr, 28 Pro, 29 Lys, 30 Ala

*Ala in bovine insulin

**Val in bovine insulin

Plate 19 Genetic engineering of human insulin. The synthetic insulin A gene is introduced into the lac operon of one group of *E. coli*. The synthetic insulin B gene is introduced into another batch of *E. coli*. Both cultures are induced by lactose. The resulting A and B polypeptides are purified and allowed to combine to form active human insulin.

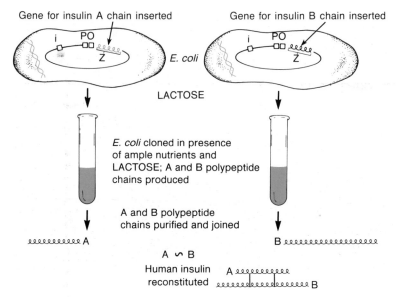

Gene for insulin A chain inserted

Gene for insulin B chain inserted

E. coli

LACTOSE

E. coli cloned in presence of ample nutrients and LACTOSE; A and B polypeptide chains produced

A and B polypeptide chains purified and joined

A

B

A ∾ B

Human insulin reconstituted

(a)

Figure 13.5 Examples of clones from nature: (a) date trees grown in the Middle East; (b) mother armadillo and identical quadruplet offspring.
Source: (a) C. J. Hylander/ Photo Researchers; (b) Karl H. Maslowski/Photo Researchers.

(b)

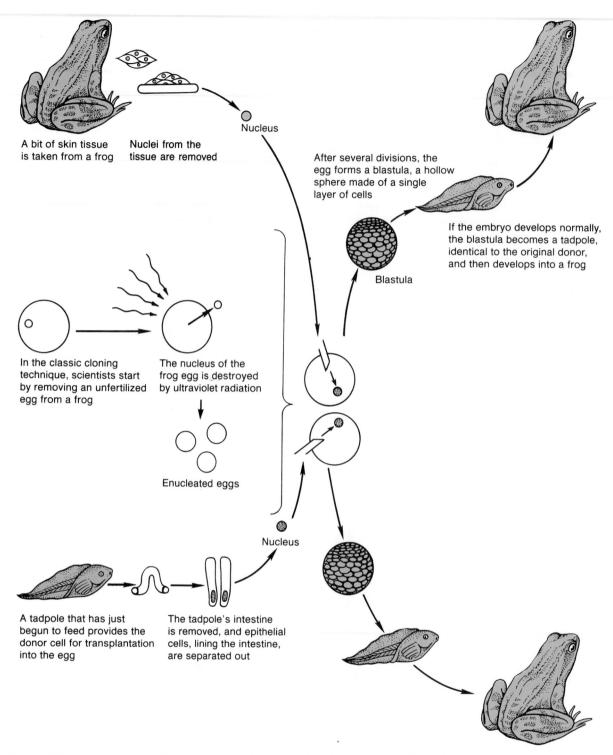

Nucleus

A bit of skin tissue
is taken from a frog

Nuclei from the
tissue are removed

After several divisions, the
egg forms a blastula, a hollow
sphere made of a single
layer of cells

If the embryo develops normally,
the blastula becomes a tadpole,
identical to the original donor,
and then develops into a frog

Blastula

In the classic cloning
technique, scientists start
by removing an unfertilized
egg from a frog

The nucleus of the
frog egg is destroyed
by ultraviolet radiation

Enucleated eggs

Nucleus

A tadpole that has just
begun to feed provides the
donor cell for transplantation
into the egg

The tadpole's intestine
is removed, and epithelial
cells, lining the intestine,
are separated out

Figure 13.6 Cloning of frogs. In the upper example the nucleus of a skin cell from an adult frog is used to produce a frog identical to the donor. With the use of multiple eggs and multiple nuclei, a clone of frogs is produced. In the lower example, the nuclei used come from the intestinal cells of a tadpole. The success rate of cloning from adult tissues is much lower than when using nuclei from tadpoles.

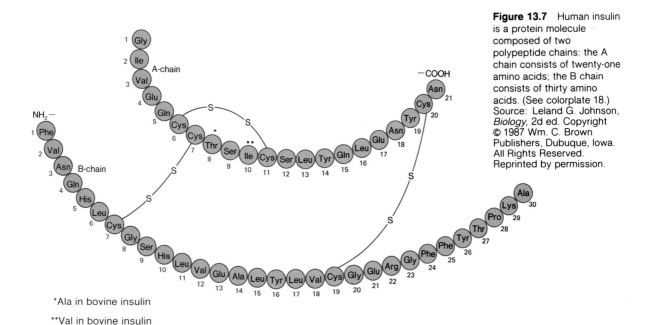

Figure 13.7 Human insulin is a protein molecule composed of two polypeptide chains: the A chain consists of twenty-one amino acids; the B chain consists of thirty amino acids. (See colorplate 18.) Source: Leland G. Johnson, *Biology,* 2d ed. Copyright © 1987 Wm. C. Brown Publishers, Dubuque, Iowa. All Rights Reserved. Reprinted by permission.

*Ala in bovine insulin

**Val in bovine insulin

do not properly metabolize sugar, which then accumulates in the blood and results in many harmful effects. Diabetes is the fifth leading cause of death and second leading cause of blindness in the United States.

Until recently, the primary treatment for diabetics was insulin purified from the pancreases taken from cattle in slaughterhouses. Bovine insulin, however, is expensive to produce and has one serious side effect. Chemically, bovine insulin differs slightly from human insulin (figure 13.7), a difference sufficient to induce an immune response in about 5% of all diabetics.

Recombinant DNA methods now allow relatively cheap mass production of synthetic human insulin (figure 13.8). Insulin is composed of two polypeptide chains, designated A and B, each encoded by a separate gene (figure 13.7). The A chain is twenty-one amino acids long, and the B chain is thirty amino acids long. To produce human insulin artificially, geneticists first determined the amino-acid sequence of both chains, and then constructed synthetic DNA sequences encoding those amino-acid sequences. In other words, they created artificial A and B genes of human insulin.

Next, the artificial genes were inserted into plasmids of *E. coli.* These particular plasmids insert themselves and the artificial insulin genes into the *E. coli* genome immediately adjacent to the Z gene of the lac operon (see figure 9.2), a key feature. Recall that the lac operon is normally turned off. However, it can be turned on simply by adding to the bacterial growth medium the inducer substance, lactose. This in turn induces transcription of bacterial mRNA, *as well as* the mRNA for the inserted A or B gene of human insulin. The system is extremely efficient; a single bacterium produces about one hundred thousand copies of one of the chains. As a final step, either the A or the B chain is purified from each separate culture, and the purified products then combined. Between 10% and 40% of the chains of each type bond to chains of the opposite type, forming entire molecules of synthetic human insulin.

Figure 13.8 Genetic engineering of human insulin. The synthetic insulin A gene is introduced into the lac operon on a plasmid in one group of E. coli. The synthetic insulin B gene is introduced into another batch of E. coli. Both cultures are induced by lactose. The resulting A and B polypeptides are purified and allowed to combine to form active human insulin. (See colorplate 19.)

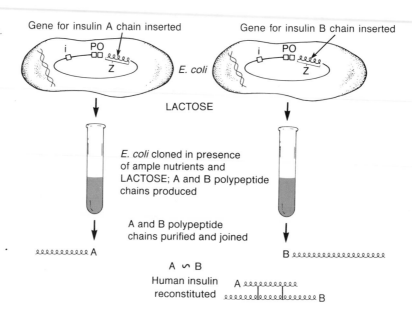

Synthetic human insulin is identical to natural human insulin and therefore does not induce allergic responses in diabetics. It is also produced far more cheaply than bovine insulin. Additionally, the commercial success of synthetic insulin shows clearly the impossibility of separating so-called "pure" from "applied" research. When the lac operon was first proposed in 1961, it seemed little more than an interesting academic hypothesis that explained something about how nature functions, but had no obvious commercial value. In less than twenty years, knowledge of the regulatory mechanism for lactose metabolism in bacteria proved to be a key ingredient in an industrial process.

Pest control

Another use for genetic engineering of valuable biochemicals lies in the area of insect pest control. Present practice relies primarily on pesticides, which have many disadvantages. Most are synthetic chemicals that are not species-specific, killing valuable as well as target pest species. Most also remain in the environment—like DDT—and accumulate in the tissues of rare species like hawks and sea birds.

Recombinant DNA manufacture of naturally occurring chemicals has none of these or other disadvantages. For example, species-specific control can be achieved by the use of **sex pheromones,** chemicals that insects use to find partners during the brief mating season. In many species of insect pest, the female releases a pheromone that attracts males and identifies her not only as a female, but also as a member of the proper species. These chemicals have been used as a form of pest control by saturating fields with them, overwhelming the pheromones released by females and so confusing the males that they cannot find females with which to mate.

Like interferon, most sex pheromones occur in such tiny quantities that purifying them from natural sources is prohibitively expensive. However, once even a tiny amount of pheromone has been obtained, recombinant DNA methods can allow mass production for agricultural use at reasonable prices. Residues of such chemicals pose no long-term health hazard to other species because they are species-specific and also because as naturally occurring biochemicals they rapidly degrade in nature.

Prenatal diagnosis

Nearly four thousand Mendelian loci have been identified in humans, many of which are associated with disease or congenital abnormality. By conventional means, prenatal diagnosis is possible for only about one hundred conditions. Virtually all of these conditions are identified either by chromosomal abnormality, as in Down's syndrome, or because of the presence of a biochemical marker such as α-fetoprotein associated with neural tube defects. Many conditions are not susceptible to early diagnosis, however, because they either involve a point mutation not visible karyotypically or do not produce a marker protein.

Analysis of fetal tissues by recombinant DNA techniques offers a means of direct examination of fetal DNA for genetic anomalies very early in pregnancy. Cells obtained by either chorion villus biopsy or amniocentesis can be amplified by tissue culture and then subjected to one of several recombinant DNA procedures for diagnostic purposes. In some cases, cDNA probes have been developed to assay deleterious mutations directly, but because probes are so difficult to construct, a powerful "genetic-marker" test known as **restriction fragment length polymorphism (RFLP)** mapping (figure 13.9) has come into prominence. RFLP mapping uses restriction enzymes to cut DNA, which is then separated into size fragments on an electrophoretic gel and visualized by staining and/or use of radioactive probes. Usually the disease-causing gene is *not* visualized, but the mutation changes the RFLP pattern.

Thus, RFLPs serve as linked markers to mutant genes. One of the most exciting results of this analysis has been diagnosis (with over 99% accuracy) of the gene for Huntington's disease, both prenatally and in children of victims of Huntington's disease, prior to average age of first onset. Cystic fibrosis, phenylketonuria, muscular dystrophy, and sickle-cell anemia are among the genetic diseases that can now be diagnosed by RFLP pattern analysis.

As more of the human genome is explored, direct DNA examination should become available for virtually all genetic diseases. However, these procedures are expensive and time-consuming. Thus far, with the possible exception of prenatal sex identification in some societies, prenatal diagnosis has been used retrospectively, that is, to test fetuses in families where affected children have occurred previously. Although recombinant DNA techniques will quickly expand the number of conditions subject to prenatal detection, it is unlikely they will be used in shotgun fashion to test all fetuses for a wide range of possible genetic conditions.

DNA fingerprinting

Within a species, most sections of the genome show little or no variation between individuals. There is, however, one region of the genome of humans and some other vertebrate species that is genetically unique in every individual (except identical siblings, clones, etc.). Called **satellite DNA,** this segment of the genome can be isolated by a cDNA probe and then examined in further detail by RFLP analysis.

The odds of two individuals having an identical set of electrophoretic bands by chance are greater than one in one billion. Attachment sites for restriction enzymes are transmitted as Mendelian alleles, so parents should share half their bands with each offspring, and siblings should share half their bands with each other.

This high level of Mendelian variability is of clear **forensic** (legal) importance. Paternity testing can now be conducted with a very high level of certainty. Unlike paternity tests with ABO and other blood groups, where possible fathers can only be excluded but never confirmed by genetic analysis, the odds of mistakenly identifying a man as the father of a child by DNA fingerprinting are negligible. DNA

Figure 13.9 (a) Restriction fragment length polymorphism (RFLP) patterns from three generations in one family. The genotypes are shown directly below pedigree symbols; M is a marker lane. (b) Disease analysis by RFLP analysis. DNA is cut by restriction enzymes and the resultant fragments are separated on gels and visualized by appropriate methods.
Source: Nakamura, Y., *Science,* March 27, 1987, fig. 3, 235:1616–22. © 1987 by the AAAS.

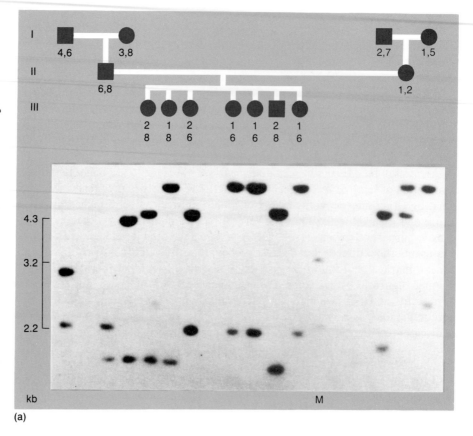

(a)

fingerprinting could also be used in maternity testing, for example, to identify the true mothers in rare cases where babies are switched or mistakenly labelled in maternity hospitals. The first official use of genetic fingerprinting has been to verify maternity in the cases of children claiming citizenship in Britain on the basis of maternal relationship.

In domestic livestock like cattle and sheep, calves and lambs sometimes get separated from their mothers. The purchaser of the offspring of especially valuable mothers might want genetic confirmation of maternity. In the case of thoroughbred racing horses, DNA fingerprinting could verify both parents, information that can be added to the pedigree of a colt as a form of title insurance to confirm its high value.

DNA fingerprinting is likely to become an important tool for identifying criminals, as well. Criminals often leave some small quantity of tissue at the scene of the crime, especially in the case of murders and sexual offenses. Sufficient DNA for fingerprinting can be obtained from blood, semen, or hair and later matched with samples taken from suspects. In its first successful use in a criminal case in Britain, DNA fingerprinting was used as a last resort in a case of sexual assault and murder of a young girl. After many weeks the police had no clear leads, but they had been able to obtain DNA samples from blood and semen left on the victim. Using traditional ABO and other blood group tests, blood from about four thousand men in the community was compared with blood left on the victim. All but about four hundred men could be excluded. Samples from these four hundred were then compared by DNA fingerprinting with the samples on the victim. A perfect match was found and a conviction obtained on a man who lived near the scene of the crime and who had not been on the list of suspects. The DNA evidence also confirmed the innocence of a second man who had confessed to the crime. Thus, one person was shown guilty and, equally important, hundreds of others were proven innocent.

DNA cleaved by restriction enzyme

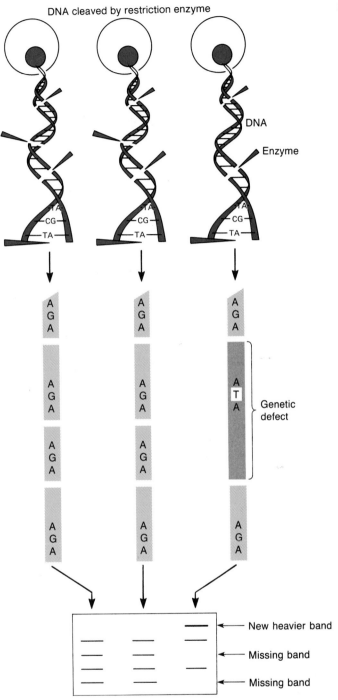

(b)

New heavier band

Missing band

Missing band

Figure 13.9 (continued)

BOX 13.1

Genetic surprises: mapping the human genome

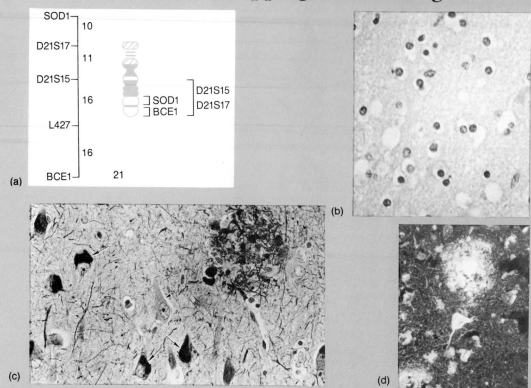

Figure 13.10 (a) Current genetic map of chromosome 21 where genes involved in Down's syndrome and Alzheimer's disease are located. (b) Section of tissue from the brain of a normal person; dark staining areas are the cell nuclei. (c) Section of tissue from the brain of a forty-two-year-old Down's syndrome patient showing a neuritic plaque (upper right), neurofibrillary tangles (arrows), and a normal neuron (arrowhead). (d) Section of tissue from the brain of a patient with Alzheimer's disease, with several neuritic plaques (lower left corner and center with prominent central core) and neurofibrillary tangles.
Source: (b) Martin M. Rotker/Taurus Photos; (c) Wisniewski H. M. and Wen G. Y.; (d) Robert Terry/Alzheimer's Disease and Related Disorders Association, Inc.

Sequencing the human genome

Recombinant DNA methods can be used to obtain sequences of segments of DNA. Scientific journals now regularly report the base sequences of genes of particular interest from many species. The U.S. National Institutes of Health now directs an international computer database to store and disseminate for scientific research all known DNA sequences. Information is being produced so rapidly that the database is already several years behind in entering newly identified sequences into its system.

Some scientists believe that an all-out effort should be made to determine the entire sequence of the human genome as rapidly as possible. This would be a massive task because the genome is so large and because the sequencing procedure is still laborious and costly. Normally, only a few thousand bases—a few kilobases—can be sequenced at one time by "walking along" the DNA. Sequencing the entire genome would require participation by many laboratories around the world, each of which would undertake examination of specific short segments. Additionally, the level of coordination and standardization of procedures would be very high, and a massive computer data processing operation would be essential.

BOX 13.1 (*Continued*)

Application of recombinant DNA methods has allowed new access to the human genome and new ways of searching for genetic disease. Many surprises have emerged; one of the most extraordinary is a possible relationship between two well-known conditions: Down's syndrome (DS; see chapter 10) and Alzheimer's disease (AD), a form of senile dementia affecting several million older Americans. Down's syndrome has long been associated with trisomy of a small region on chromosome 21, and brains of patients with DS who live into their thirties and forties show physical changes (plaque formation) similar to those seen in persons with AD (figure 13.10). Both patients with DS and AD show an abnormally high level of deposition of a protein (β amyloid) in the brain. Using a cDNA probe, the gene encoding β amyloid was localized on chromosome 21, very close to the region associated with DS.

Findings such as these are medically valuable and are being pursued by many research laboratories. Scientists opposing funding for programs to sequence the human genome point out that a more productive approach might be to focus on specific instances such as this where a base sequence of a particular gene can be interpreted directly in terms of its phenotypic or pathogenic effects. In other words, examine only those genes that appear from medical and biochemical data to warrant further detailed analysis.

In October 1987 the world's first genetic linkage map of the entire human genome was announced. Researchers used polymorphic loci and RFLPs as gene markers; 403 such markers, known to be inherited along with specific disease traits, were placed on the 23 human chromosomes. The resolution of the map will increase with the identification of additional markers. The current map allows geneticists to localize genes within five to ten million bases, rather than among the three billion that comprise the human genome. All chromosomes are not equally well mapped. There are 60 markers on chromosome 7, where the gene for cystic fibrosis is found. Workers estimated that with this map there is a 95% chance of being able to map any disease gene to an approximate chromosome location.

This map of the human genome will not only allow the placement of additional markers, but will facilitate efforts to clone other disease genes, assist in prenatal or presymptomatic diagnosis of individuals at risk for genetic disease, and aid efforts to completely sequence the human genome.

Most of all, sequencing the entire genome would be very expensive. Although many geneticists agree that it would be very interesting to know the entire human genomic sequence, there is considerable disagreement as to the immediate value of such an endeavor. Many argue that the funds required to sequence the human genome would be better spent on other types of projects. It might be more productive, for example, to engage in major programs to identify specific gene products that are of medical importance and then to construct cDNA probes to them. The location of these specific genes could then be mapped to particular chromosomes. Genes of special importance, where precise information about their structure and function is required, could then be sequenced selectively and efficiently. This more opportunistic approach would allow efficient exploitation of the many surprising findings likely to appear in the next several decades as the human genome is explored (box 13.1).

Opponents of a major sequencing program also point to another problem: interpretation. Lists of DNA sequences several billion base pairs long will not be easily assessed. It is now clear that much of the human genome does not code for any protein. Does this DNA have any function at all? No one knows, but merely having a list of base sequences does not guarantee that its importance can be deciphered and understood.

Despite these concerns, efforts to sequence major parts of the human genome are continuing. It is likely that within the coming generation, geneticists will be able to describe humanity in terms of its DNA content.

Gene therapy

Techniques are in development to allow transfer of single genes between higher organisms. Such gene engineering techniques offer hope for curing diseases caused by defective genes by inserting a properly functioning gene into the affected individual. The defective genes themselves are not removed—it might some day be possible to inactivate them—but addition of a normal gene may facilitate normal metabolism.

Such procedures, if effective, are simply a step beyond traditional therapies. For example, diabetes has long been treated by administering insulin to supplement the small amounts produced by the victim. Insertion of a gene to produce insulin would simply allow the additional insulin to be produced by the diabetic's own metabolic processes.

Gene therapy is most likely to be successful in treating diseases caused by a single gene mutation. Among the candidates for such therapy are sickle-cell anemia, thalassemia, Lesch-Nyhan disease, hemophilia, and phenylketonuria, diseases whose abnormal gene products are well known. Other conditions, such as Huntington's disease, cystic fibrosis, and muscular dystrophy, whose abnormal products, if any, remain unknown, may be subject to gene therapy when better characterized.

The feasibility of gene therapy has already been shown by tests on experimental animals. Rabbit hemoglobin genes, for example, have been transferred to and expressed in mice. More dramatically, genes encoding rat growth hormone have been transferred to and expressed in mice (figure 13.11). By the end of the 1980s, gene therapy will likely have been used in human experimental tests for several conditions in which the disease is life-threatening and no other treatment exists.

The most likely conditions for first treatment by gene therapy in humans are the various hemoglobinopathies. At least one trial was conducted surreptitiously in the early 1980s. A doctor removed bone marrow from two victims of β-thalassemia and incubated the marrow cells (where red blood cells are formed) in culture with purified quantities of normal β-globin gene. β-thalassemia is caused by defective β-globin genes, and the hope was that the normal genes would be incorporated into the cells of the affected individuals. Subsequently, the marrow was reimplanted into the bones of the victims. In neither case did the patient show any evidence of accepting or expressing the normal genes.

Unfortunately, the physician conducted the experiment in secrecy, seriously violating the ethical guidelines of the National Institutes of Health, who had sponsored his research. He was punished severely for his actions by removal of his funding and other limitations on future research, as well as by the censure of his professional peers.

Despite the unfortunate aspects of this case, it illustrates how close gene therapy is to becoming a reality. Mendel's discoveries were recognized at the beginning of the twentieth century. It is quite likely that by the end of the twentieth century, geneticists will have the capacity to manipulate some portions of the human genome for therapeutic purposes.

Environmental release of recombinant organisms

Gene transfer techniques have applications far beyond human health. Transfer of useful genetic traits to domestic plant and animal species can supplement and far

Figure 13.11 The mouse on the right is normal. The mouse on the left carries a transferred rat gene coding for growth hormone. The genetically engineered mouse had hormone levels up to eight hundred times higher than its companion and was nearly twice the weight of the normal mouse. Source: Ralph Brinster, *Nature* 16 Dec. 1982.

surpass traditional reliance on mutation and artificial selection. As with gene therapy, single genes will be the first to be used, but complex genetic traits involving many genes are also likely.

Some uses of gene transfer are obvious. Genes for increased production, shortened time to maturity, and disease resistance would be valuable in any plant or animal used for food. Some gene transfer programs now undergoing research and trials may fundamentally change agriculture. Among the most valuable plants are the legumes (beans and peas) and some other species that harbor *nitrogen-fixing bacteria* in nodules on their roots. These bacteria combine atmospheric nitrogen with elements from the soil to produce their own nutrients, a form of natural fertilizer. Thus, these plants enrich the soil, unlike most other species, which require large quantities of fertilizers.

Genetic engineers are developing strains of recombinant bacteria that can live symbiotically with many nonleguminous plants and that are capable of nitrogen fixation. If successful, such research could greatly reduce reliance on the massive amount of fertilizer needed by modern agriculture.

Another program, now in the stage of outdoor trials, has produced a strain of bacteria that confers frost resistance on host plants. If successful, such research could expand agriculture for many species into areas too cold or with too short a growing season to be commercially feasible at present.

Release of recombinant organisms into nature has not met with universal approval. Stringent government controls in the United States have delayed field trials. Protest groups pulled up the first plants on which the frost-resistant bacteria were to be tried. Court action by environmental lobbyists further delayed government approval.

Such delays can exacerbate an already complex and trying situation. One scientist who had worked for many years to produce a recombinant bacterium conferring resistance to Dutch elm disease, which has largely exterminated American elm trees in North America, began releases of the bacterium without government approval after lengthy, frustrating delays. His motivations were good, as with the physician who tried gene therapy for β-thalassemia, but he saw the delays as an unnecessary impediment to achievement of a goal he perceived as good. Unfortunately, flaunting of regulatory systems only increases public suspicion of recombinant DNA research. Valuable research programs, about which there is reasonable public concern and caution, may then be evaluated more on emotion than on their intrinsic merits.

Ethical concerns

We noted earlier that recombinant DNA research and technologies cannot be guaranteed to be risk-free. Certainly, these methods can be used for making war and other unhappy human activities, just as can many other new technologies.

The public is especially concerned over gene manipulation. Genetic engineers "tinker" with the essence of life, which some people and religions view as arrogant and intrinsically offensive. Other people compare the risks posed by genetic engineering to the threats of nuclear accident. In both cases, the activities of scientists are changing nature forever. Nuclear experiments (and certainly nuclear bombs and power plants) create isotopes that persist in the environment for tens of thousands of years. The release of recombinant organisms confronts nature with new species, the effects of which cannot be guaranteed. Once released, some organisms may not be subject to control, and the consequences cannot be totally predicted. Whatever the effects, nature may be irreversibly altered.

Clearly, caution is required. Most people accept that some types of genetic engineering research are little different than other long-accepted types of research, particularly in medicine and biotechnological production of valuable chemicals. Artificial selection is a form of genetic manipulation, as is the use of yeast for fermentation of alcohol and food. Ethicists generally accept that gene transfer to treat human illness is no different in kind than other traditional therapies, so long as it affects only somatic cells.

Manipulation of germ cells, however, is viewed as unacceptable, as are efforts to clone individual humans. But many gray areas exist, and bioethics has become a true growth industry as society slowly tries to deal with the many moral aspects of genetic engineering research. As with other developing technologies, moral consensus will be reached only slowly, and different societies may arrive at different views.

We noted earlier that governments and individuals may often be tempted to use genetic engineering for evil purposes, but we have also shown how the thicket of regulatory procedures, combined with legal challenges and complex ethical questions, can compel informed and well-meaning scientists and physicians to break the rules. Modern genetic technologies will likely continue to develop a checkered history—many exciting and extraordinary advances will be combined with well-known (and doubtless some secret) infringements of accepted standards.

For certain, recombinant DNA technologies are with us to stay. Their potential for good and evil is as limitless as the human mind. Quite likely, our lives will be improved in countless ways by new genetic technologies. We can only hope that reason will prevail in the applications of genetic methods, and also that the risks continue to be minimal.

Modern genetic technologies will continue the long tradition of genetics: teaching us about ourselves. Genetics has much to tell us about the meaning of *humanness*— What are we? What does it mean to be a human? As we learn more about ourselves and our world, that knowledge may help us to act with greater wisdom and prudence.

Summary

1. Genetic engineering, gene splicing, and recombinant DNA technologies all refer to modern ways of manipulating DNA.
2. Biotechnology is the industrial use of biological systems and processes.
3. Molecular cloning uses recombinant DNA techniques to amplify specific genes or gene sequences.

4. Either plasmids (for small pieces of DNA) or phages (for larger DNA) can be used as cloning vectors.

5. Restriction endonucleases are bacterial enzymes that cut DNA at specific sequences. They are used in gene cloning and also to produce characteristic restriction fragment length polymorphisms (RFLPs) in DNA. These RFLPs are used as markers for a variety of medical and forensic purposes.

6. DNA ligase is an enzyme used to produce recombinant plasmids; it rejoins broken segments of DNA.

7. DNA libraries are cultures of bacteria, with plasmids or phages, that carry genes or gene fragments from recombinant DNA studies.

8. cDNA probes are DNA sequences complementary to the sense strand of the gene to be studied. They are often produced using the enzyme reverse transcriptase and mRNA for the gene in question.

9. Cloning of many cells and genes can be done routinely; cloning of entire organisms is more problematic. Although some plants and frogs can be cloned, mammals cannot.

10. Genetic engineering methods have been used successfully to produce proteins and hormones of great medical importance, such as insulin and growth hormone.

11. Genetic engineering techniques have also been used in prenatal and presymptomatic diagnosis of individuals at risk for genetic disease.

12. DNA fingerprinting, using RFLP analysis is becoming extremely useful in medicine and forensics because each human appears to have a distinct DNA fingerprint.

13. Although a sequence of the entire human genome is a dream of some geneticists and a costly project, genetic maps are now available that are extremely useful and less costly.

14. Gene therapy for humans is still not a reality, but is being experimented with successfully in plants and some mammals.

Key Words and Phrases

biotechnology	human insulin	recombinant plasmids
cDNA	interferon	restriction endonucleases or restriction enzymes
DNA library	molecular cloning	
DNA ligase	nuclear transplantation	restriction fragment length polymorphism (RFLP)
forensic	phage	
gene splicing	plasmid	satellite DNAs
gene therapy	probe	sex pheromones
genetic engineering	recombinant DNA technologies	vector or cloning vehicle
genetic pharmacology		

Questions

1. Discuss the genetic potential of the cloning of single genes, cells, organs, and entire organisms. Distinguish between plants and animals.
2. Do you believe it is important to sequence the entire human genome? Why?
3. Should there be government regulation of research using recombinant DNA methodologies?

4. Should prenatal diagnosis be available to all pregnant women? Or, should it only be available to those at risk for known genetic disease?
5. Why would geneticists use plasmids rather than phages for molecular cloning?
6. How can RFLPs be used in prenatal diagnosis?
7. How can RFLPs be used in paternity testing? How does this genetic information compare to data from blood groups?
8. Why has gene therapy not been successfully applied to human genetic disease?

For More Information

Anderson, W. F., and E. G. Diacumakos. 1981. "Genetic Engineering in Mammalian Cells." *Scientific American,* 245:106–21.

Caskey, C. T. 1987. "Disease Diagnosis by Recombinant DNA Methods." *Science,* 236:1223–29.

Clark, M.; S. Begley; and M. Hager. 1980. "The Miracles of Spliced Genes." *Newsweek,* March 17:62–71.

Cooper, D. N., and J. Schmidke. 1986. "Diagnosis of Genetic Disease Using Recombinant DNA." *Human Genetics,* 73:1–11.

Dodd, B. E. 1985. "DNA Fingerprinting in Matters of Family and Crime." *Nature,* 318:506–7.

Donis-Keller, H. et al. 1987. "A Genetic Linkage Map of the Human Genome." *Cell,* 51:319–37.

Drlica, K. 1984. *Understanding DNA and Gene Cloning: A Guide for the Curious.* John Wiley & Sons, New York.

Higuchi, R. et al. 1988. "DNA Typing from Single Hairs." *Nature,* 332:543–46.

Kolata, G. 1986. "Genetic Screening Raises Questions for Employers and Insurers." *Science,* 232:317–19.

Lewin, R. 1988. "Genome Projects Ready to Go." *Science,* 240:602–4.

Lewis, R. 1987. "Genetic-Marker Testing: Are We Ready for It?" *Issues in Science and Technology,* 4:76–82.

Marx, J. L. 1982. "Building Bigger Mice through Gene Transfer." *Science,* 218:1298.

Marx, J. L. 1987. "Role of Alzheimer's Protein Is Tangled." *Science,* 238:1352–53.

Miller, J. A. 1987. "Mammals Need Moms and Dads." *Bioscience,* 37:379–82.

Murray, M. 1987. "Battling Illness with Body Proteins." *Science News,* 131: 42–45. (recombinant DNA technology).

Patterson, D. 1987. "The Causes of Down's Syndrome." *Scientific American,* 257:52–60.

Roberts, L. 1987. "Flap Arises over Genetic Map." *Science,* 238:750–52.

Wurtman, R. J. 1985. "Alzheimer's Disease." *Scientific American,* 252:62–75.

Appendix A

Classification of Life

Species are groups of organisms that can interbreed to produce new organisms of the same kind. Humans comprise one species, dogs another, and dandelions, onions, spruce budworms, brook trout, great white sharks, and starlings yet others. At least one million and perhaps ten million species now exist on earth.

Aristotle (384–322 B.C.) is often considered to be the first biologist, and among his many works is a classification of organisms known to the Greeks. Aristotle's system was used as the basis of biological classification by Western societies for almost two thousand years, but most other societies had some form of classification. Perhaps the most common classification was to separate the living world into two categories of species: plants and animals. This was a commonsense approach. After all, plants are similar in being green, not ingesting other organisms, and not being able to move. Animals, on the other hand, are capable of movement and eat other organisms. Within the animal kingdom, smaller groups of organisms could be identified on the basis of increasingly precise characteristics.

In the mid-eighteenth century, a Swedish biologist, Carolus Linnaeus (1707–1778), invented the *binomial system,* which is the basis of modern biological nomenclature. In the Linnaean system, each species is assigned its own unique set of two latinized names. The first is the *generic* name, and the second is the *specific* name. (Genus is Latin for birth or race; species is Latin for kind.) For example, the scientific name of the domestic dog is *Canis familiaris.* The generic name (*Canis* meaning dog) is the more inclusive category and refers to a group of very similar (which Linnaeus knew) and also evolutionarily closely related (which Linnaeus did not know) species. For example, the genus *Canis* also includes *Canis latrans,* the coyote, and *Canis lupus,* the wolf.

A genus may include only one species, or it may include dozens or even hundreds of species. Species that are very closely related are assigned the same generic designation. On the other hand, the same specific epithet may be given to many different species, but only if they are in different genera. Figure A.1 gives the classification of two distantly related frogs with the same specific epithet, the spring peeper (*Hyla crucifer*) and the South American toad (*Bufo crucifer*). Both the peeper and the toad are assigned the descriptive name *crucifer* because each has a mark on its back resembling an X. *Crucifer* means crossbearing in Latin. No two species are ever assigned the same generic and specific names together.

The binomial system permits all scientists to speak one language. A species may be called by many common names in different places. A tree frog in England is a grenouille in France and a rana in Spain, but scientists in all countries know it as *Hyla arborea.* A squirrel in the United States is an ardilla in Spain and an ecureuil in France, but it is *Sciurus niger* to all scientists. Groups of species are organized into hierarchical series of increasingly less-similar groups. Genera are combined into Families, Families into Orders, then Classes, Phyla, and finally Kingdoms. These major categories (or *taxa*) may be further divided if necessary. The biological system of classification is designed both to organize the immense diversity of life into groupings of organisms according to similar or shared characteristics and to reflect the degree of evolutionary relationship among them.

Three kingdoms of life

A five-kingdom classification of life, introduced in 1969 by R. H. Whittaker, enjoyed widespread use until quite recently. In the late 1970s C. R. Woese demonstrated that major genetic differences exist between two groups comprising Whittaker's Prokaryotes and that these differences are equal to or exceed the genetic differentiation between the plants and

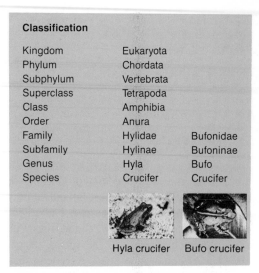

Classification		
Kingdom	Eukaryota	
Phylum	Chordata	
Subphylum	Vertebrata	
Superclass	Tetrapoda	
Class	Amphibia	
Order	Anura	
Family	Hylidae	Bufonidae
Subfamily	Hylinae	Bufoninae
Genus	Hyla	Bufo
Species	Crucifer	Crucifer

Hyla crucifer   Bufo crucifer

Figure A.1 Example of classifying animals, using frog and toad having same specific name because both species have markings resembling a cross on their back.
Source: (left) Don Valenti; (right) W. R. Heyer.

animals. Woese and his students therefore recognize three kingdoms, dividing the Prokaryotes into the Archaebacteria and the Eubacteria, and combining all remaining taxa in the Eukaryotes.

Kingdom Archaebacteria

Archaebacteria are prokaryotes that live in conditions hostile to most life, but possibly similar to the harsh conditions under which life began. Although in external appearance they differ little from Eubacteria, genetically they are extremely different.

Many of the Archaebacteria are methanogenic (meaning methane producing). They live only where oxygen is absent, such as the decaying vegetation at the bottom of a swamp. A by-product of their metabolism is methane, a combustible gas that is known as swamp gas when seen burning in a swamp at night. Methanogens are also found in hot springs, stagnant water, sewage treatment plants, the intestinal tracts of animals, and the bottom of the ocean. Other types of Archaebacteria are halophiles (meaning salt loving), which inhabit highly saline waters such as the borders of evaporation ponds, the Dead Sea, and the Great Salt Lake, and the thermoacidophiles (meaning heat-acid loving), which occupy acidic hot springs, such as at Yellowstone National Park, and burning coal tailings.

Kingdom Eubacteria

The other group of prokaryotes, the Eubacteria, are ubiquitous and essential members of the living world. They occupy most habitats and have many different modes of life. Many are free living and function in the decay and breakdown of dead organic matter. If the Eubacteria did not serve as recyclers to the world, we would soon be buried under the carcasses of plants and animals.

Other Eubacteria occupy the bodies of living organisms. *Symbiotic* (meaning living together) bacteria benefit their host organisms: in return for a place to live, the gut bacterium of vertebrates, *Escherischia coli,* helps digest our food. *E. coli* is an important organism in genetic research because it is easy to culture in the lab and because it reproduces so rapidly—one generation every twenty minutes (figure A.2). Some bacteria are parasitic or pathogenic (meaning disease causing). The agent of common food poisoning is *Salmonella,* and *Staphylococcus* causes a variety of serious infections in humans.

Blue-green algae are another type of Eubacteria. Most algae are eukaryotes, but blue-greens are clearly prokaryotic. They commonly inhabit stagnant water and many harsh environments, such as hot springs. Blue-green algae, as well as many bacteria, are capable of *photosynthesis.* They use the energy from sunlight to combine carbon dioxide and water into carbohydrates such as glucose, a simple sugar from which they then derive energy.

Kingdom Eukaryota

Protists

Protistans are single-celled eukaryotes. They occur in a great variety of forms, but they are larger and more complex than any prokaryotes. The two basic types of protistans are the protozoa (meaning before animals), which exhibit many features of the multicelled animals, and the single-celled algae, which are much like the multicelled higher plants. The ancestors of the multicelled eukaryotic kingdoms probably originated from organisms much like living protistans.

Most protozoa are heterotrophic (meaning other food), i.e., they derive their energy from eating other organisms. Like the animals,

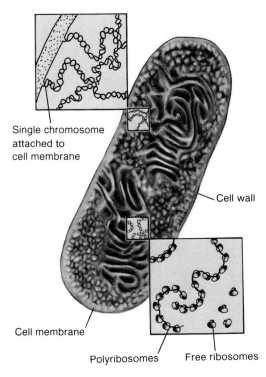

Single chromosome
attached to
cell membrane

Cell wall

Cell membrane

Polyribosomes   Free ribosomes

Figure A.2 The common gut bacterium, *E. coli,* is a favorite organism of geneticists. The average cell is rod shaped and about 2u long by 1u in diameter. Under optimal conditions of food and temperature it reproduces by fission every twenty minutes to produce two identical daughter cells per division. In twenty-four hours the descendants of a single bacterium can number up to one billion.

many have a primitive ability to detect and respond to external stimuli. They occupy many habitats, and both free-living and parasitic, disease-causing species are known. Familiar protozoa include *Paramecium* and *Amoeba* (both found in freshwater ponds), *Trypanosoma* (the cause of African sleeping sickness), and *Plasmodium* (the cause of malaria).

One-celled algae are protistans that are primarily aquatic autotrophs (meaning self-feeding). They use photosynthesis to produce their own food, but like multicelled plants and unlike photosynthetic bacteria, they possess specific organelles, *chloroplasts,* in which photosynthesis occurs. Photosynthesis requires the presence of one of a variety of pigments called *chlorophylls.* Common chlorophylls are green and cause the green color of most plants. In photosynthetic eukaryotes, chlorophyll occurs in chloroplasts.

Single-celled algae are extremely important because they form the primary source of food for many aquatic organisms. Common

types of these algae are the diatoms (whose beautiful silicon shells have been used in toothpaste), dinoflagellates, and *Euglena.* One species of dinoflagellate causes the well-known *red tides.* Red tides occur when outbreaks of *Gonyaulax catanella* poison huge numbers of fish and other marine life.

With minor exceptions, the remaining members of the Eukaryota are all multicellular.

Fungi

Toadstools and mushrooms are familiar to us all. Some exist as single-celled forms, but most are multicellular. Fungi used to be considered degenerate plants because they lacked chlorophyll, but the differences between fungi and all other organisms are so great that the fungi are now considered a major independent line of evolutionary descent.

Fungi are composed of a filamentous cell called a *hypha.* Often, hyphae form a multicelled structure called a *mycelium,* which is the body of the fungus. Fungal cells differ from plant cells in having neither chloroplasts nor chlorophyll (which is why they are often white in color) and in having a supportive cell wall of *chitin,* a polysaccharide that is also found in the skeletons of insects but never in plants. Unlike either plants or animals, fungi use extracellular digestion; digestive enzymes are secreted onto food, and the partially digested mixture is then absorbed by the fungal cells.

Fungi are important decomposer organisms. Like bacteria, they are one of the agents of decay of dead organic matter, particularly in forest communities. Many fungi such as molds and mildews are nuisance organisms, while others are serious disease organisms of crops (rusts and smuts of wheat and rye, for example) or humans (ringworm—not a worm at all—and thrush). Humans eat many species of fungi, but others such as the *Amanita* are highly toxic. Some are hallucinogenic. Fungi are the source of penicillin, an important antibiotic drug, and are used to produce Roquefort and other cheeses. Yeasts are a type of fungi whose metabolic waste products are used by humans: carbon dioxide from their respiration causes bread to rise, and fermentation involves the production of alcohol from the breakdown of sugars.

Figure A.3 The oldest bristlecone pine, from eastern Nevada, has been dated at forty-nine hundred years of age by analysis of its annual growth rings. These pines are among the world's oldest organisms and remain standing long after death.
Source: USDA Forest Service.

Plants

Plants are multicellular autotrophs. They range in size from unicellular algae to the largest organisms now on earth, the giant sequoias in California. The oldest living organisms are also trees. Some bristlecone pine trees in California and Nevada are almost five thousand years old (figure A.3). They have been alive for almost the entire recorded history of human civilization.

Plants are the dominant organisms in most terrestrial ecosystems. They provide structure to the living community, a place for many organisms to live, and food for all. Organisms that do not eat plants nonetheless eat other organisms that do eat plants. Among the most important types of plants are mosses and liverworts, ferns, horsetails, gymnosperms (such as pine trees), and the familiar angiosperms or flowering plants.

Plants are able to attain great size because of their extraordinarily tough and strong cell walls. The cell walls of plants consist of cellulose, a polysaccharide. Cellulose is not only strong, but also resistant to decay. The cellulose trunks of some bristlecone pines that have been dead for four thousand years have yet to decay. Most heterotrophs are incapable of digesting cellulose. In a living tree, most of the mass consists of cellulose that once surrounded living cells. As the tree grew, the cells died, but the cellulose remained and continued to provide support for the surviving structure. Most of the living tissue of a tree consists of new leaves and twigs and branches and roots, plus a layer of growing tissue surrounding the trunk just under the bark.

Although plants now dominate terrestrial habitats, they originated in the oceans. Multicelled algae evolved from single-celled algal protistans, which then invaded dry land four hundred million to five hundred million years ago. Algae need not be small; many seaweeds like kelp are more than one hundred feet long.

Plants provide our primary source of food; civilization probably began as a result of the development of agriculture. But plants have a variety of less obvious but equally important uses to humans. Plants hold the soil in place, inhibiting erosion. Plants help cool the surface of the earth (compare a forested park to paved cities in summer). Plants are the source of hundreds of drugs. Plant products are used to build our homes and to furnish them, to write on, to write with, and to clothe us.

Plants also play a vital role in regulating the composition of our atmosphere. Plants consume carbon dioxide and release oxygen. As industrial pollution continues to pump carbon dioxide into the atmosphere, the capacity of plants to control levels of carbon dioxide is being strained. A particularly serious problem is the destruction by humans of the huge tropical forests, especially in the Amazon, which further reduces the mechanism for abating the levels of carbon dioxide. The potential buildup of carbon dioxide may induce a *greenhouse effect,* in which the surface temperatures of the earth will begin to increase. The effects of such an increase would be extremely serious, and there is good evidence that such an increase has begun.

Animals

Animals are multicellular heterotrophic eukaryotes. They differ from fungi (which are also multicellular and heterotrophic) in many ways: animal cells have no cell walls; digestion is usually within the body; animals usually are capable of active movement; and animals usually possess a nervous and sensory system.

The diversity of animal life is astonishing. Most of us are familiar with a few types of animals: mammals, birds, frogs, lizards, snakes, fish. All those are animals with backbones (or *vertebrates*), like humans. Over one million species of animals are now alive, but only about forty thousand are vertebrates. There are probably at least one million species of insects alone, and three hundred thousand of those are beetles. Other types of *invertebrates* (animals without backbones) include sponges, corals and jellyfish, flatworms, roundworms, segmented worms, ticks and spiders, crustaceans, sea squirts, and mollusks—by the tens of thousands.

Animals occur everywhere on the surface of the earth, beneath the surface, on the bottom of the ocean, and even in the air above the earth. Many species of insects and spiders have been caught in nets in wind currents thirty thousand feet in the air, as well as at the top of Mt. Everest. Others have been brought up from wells in excess of one thousand feet below the surface. Many species survive by parasitizing other animals and plants, and parasites often have their own parasites.

Plants cannot live deeper in the ocean than light penetrates, but animals survive on the rain of debris of all kinds that reaches many thousands of feet below the surface. In the 1970s, undersea expeditions discovered thriving invertebrate communities (including many species never before seen by humans) living around vents of hot water at depths of fifteen hundred to two thousand meters in the eastern Pacific Ocean. These communities survive by eating the dense growths of Archaebacteria surrounding the vents.

While plants usually determine the basic character of a terrestrial community, animals can also change the face of the earth massively. Humans have touched and changed nearly every spot on earth, but other animals can be equally formidable. Insect plagues can turn forests and croplands into deserts. A small colony of beavers can turn a valley into a series

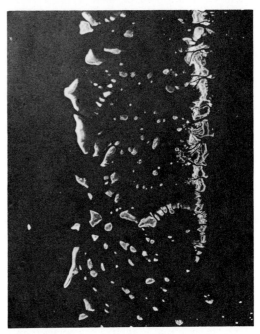

Figure A.4 Two of the largest structures produced by living species—the Great Wall of China and the Great Barrier Reef—as seen from outer space. Source: NASA.

of lakes and meadows. (Some beaver dams are more than one-half mile long.) The largest individual organisms may be trees, but the largest structures are built by animals. Two are large enough to be visible to astronauts (figure A.4): the Great Wall of China, thirty-two hundred kilometers long and built by thousands of humans; and the Great Barrier Reef, two thousand kilometers long off the coast of Australia and built from the bodies of billions of corals.

Human uses for animals are both mundane and imaginative. We use animals for food, for clothing, for shelter, and for transportation. They provide us with companionship. Carrier pigeons communicate for us. Mules pull our plows. And in every facet of biological, medical, and psychological research, animals provide a source of data through which we can learn about our world and ourselves.

Viruses: what are they?

Viruses are extremely small particles that possess some, but not all, of the characteristics of normal living organisms. They seem to do little more than infect a living cell and use the organelles of that cell to produce more viruses. Thus, they are capable of reproduction, but only with the assistance (not freely given) of true cells. They are not capable of other life processes such as metabolism. Because they must invade another organism to reproduce, they are parasitic, but their effects are not always harmful to the host.

The structure of viruses is remarkably simple (figure A.5). They consist of an outer protein coat surrounding a segment of nucleic acid, either DNA or a related molecule, RNA. Often, a short tail is used to attach to the surface of a host cell. Once attached, the virus injects its DNA or RNA into the host. The virus nucleic acid then commandeers the host system and causes the host cell to begin manufacturing more viruses, at the expense of the host cell's metabolic processes.

Thus, a virus may be viewed as a length of parasitic DNA or RNA—a sort of freelance genetic material. Its only occupation is infection and self-reproduction. Biologists have not been able to decide if a virus is a living organism or what its evolutionary origins might be. Is a virus an extremely simple structure that might be like the very first organisms or precursors of life on earth? Or is a virus a degenerate bacterium, a piece of feral nucleic acid?

Whatever the answers to such questions, viruses have been of great importance to genetics. Because a virus is little more than an

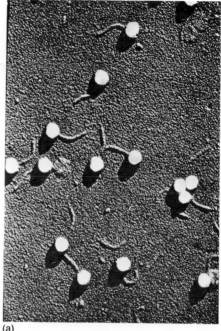

(a)

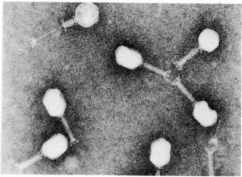

(b)

Figure A.5 The simplicity of viruses can be seen in these electron micrographs of T-2 and T-4 viruses. Viruses consist of an outer shell, with a head containing its genome and a tail for attaching to the bacterium it infects. Source: (a) E. Kellenberger; (b) Carl Zeiss, Inc.

independent piece of genetic material, its function has taught us much about the ways in which DNA and RNA function.

Evolutionary relationships of the three kingdoms

All species represent a branch tip of one large evolutionary tree. How do the branches fit together? Relationships among organisms are increasingly being assessed on the basis of molecular differences. Because genes carry encoded information, and because each gene is a

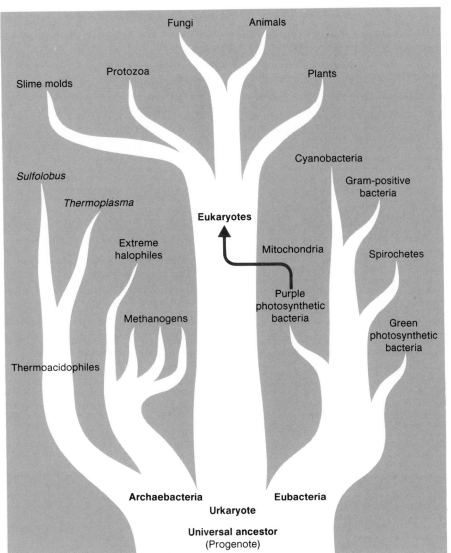

Figure A.6 Three primary kingdoms proposed by Carl Woese at the University of Illinois, Urbana–Champaign. The former prokaryotes are grouped in the *Eubacteria* and in the newly described *Archaebacteria*. The eukaryotes include the plants, animals, fungi, protozoans, and slime molds.

descendant of millions of generations of previous genes, each gene carries information regarding its own history. Thus, we can hope to understand the relationships of organisms by studying their molecular structure and function.

It is hypothesized that three primary kingdoms represent the fundamental types of life: the Archaebacteria, the Eubacteria, and Eukaryotes (figure A.6). These three groups are equally different from each other at the molecular level, and their ancestors probably diverged about 3.5 billion years ago. All three groups originated more or less directly from the first type of life on earth, the Progenotes. Progenotes appeared between 3.5–4.0 billion

years ago, and it was in the Progenotes that a simple genetic mechanism and metabolism developed. Perhaps while these complex processes were still in an experimental stage, the three primary kingdoms went their evolutionarily independent ways.

At first, the Archaebacteria were probably the dominant group, as their metabolism is well suited to the conditions of primitive earth. As oxygen became a major component of the atmosphere and as temperatures decreased, they became restricted to their present habitats. The Eubacteria probably became increasingly abundant as oxygen levels increased. The oldest known fossils are much like today's blue-green algae and are found in the

Fig Tree Series of rocks in Africa (3 billion years old) and from rocks in Western Australia (3.5 billion years old).

Because eukaryotic cells are larger and more complex than prokaryotes, their origin has always been a puzzle. How did they get so large? How did their organelles originate? Recent hypotheses favor a symbiotic origin of the eukaryotes. Perhaps the primitive eukaryote was a slightly larger-than-normal cell capable of ingesting others. If this cell consumed cells that were resistant to digestion, sooner or later the consumed cell might establish a semi-independent existence within its predator. If the two cells survived better together than alone, the ancestor of today's eukaryote could have been formed.

Chloroplasts and mitochondria (chapter 2) are organelles that may have originated in just this way. Each is about the size of a bacterium. Both have their own DNA and apparatus for protein production independent from that of the rest of the cell. Both also reproduce independently from the rest of the cell. Great similarities have been found between chloroplast RNA and the RNA of blue-green algae, a group of the Eubacteria. Some eukaryotic genes also appear to have originated from Archaebacterial genes. Whatever the origins of eukaryotes, they are certainly a very old group with an extremely complex origin. Archaebacteria, Eubacteria, and Progenote all may have participated in the formation of the first eukaryote.

The two prokaryotic types each represent a single kingdom, but the eukaryotes include four major lineages. The evolutionary divergences within the eukaryotes are certainly more recent than those between the primary kingdoms. The most ancient eukaryote fossils are protistans about 1.2–1.4 billion years old. Only a few fossils of multicelled organisms are older than six hundred million years. Terrestrial life began in earnest about four hundred million years ago, the date of the earliest plant fossils. Animals such as amphibians and arthropods soon followed the plants onto the land. Terrestrial habitats have been the site of the greatest diversity of both plant and animal life. The two most abundant terrestrial groups, the flowering plants and the insects, have experienced an immense diversification during the past one hundred million years.

For More Information

Whittaker, R. H. 1969. "New Concepts of Kingdoms of Organisms." *Science,* 193:150–59.

Woese, C. R. 1981. "Archaebacteria." *Scientific American,* 244:98–122.

APPENDIX B

PROBABILITY

A major goal of all science is *prediction*—the ability to foretell events. In genetics, one important type of prediction is the likely genotype of the offspring of specific matings. In some cases, prediction is easy, and the results are certain. For example, if homozygous round-seeded pea plants are bred (RR × RR) we know that, barring mutation, all offspring will be RR and produce round seeds. This result is certain because each parent has only R alleles to contribute to all offspring.

When heterozygotes are mated, however, several genotypes may occur in the offspring: RR, Rr (or its equivalent, rR), and rr. Some types of predictions cannot be made with certainty regarding the results of an Rr × Rr cross. For example, what will be the genotype of a particular offspring? Or, will there be more RR offspring than rr?

The mathematical science of *probability* has developed a language for expressing predictions regarding uncertain outcomes of particular events. A probability is a measure of the likelihood, the chance, that an event will have a particular outcome. Usually, a probability is expressed as a fraction between, and including, 0 and 1. An outcome with a probability of 1 is certain to occur; that with a probability of 0 is certain not to happen. For example, the probability that any one person will live forever is 0. We know that each of us, sooner or later, will die. The probability that a particular person will die is 1. The sum of all possible outcomes of a single event must always equal 1.

In all other cases, the chance of a particular outcome increases as the probability approaches 1, and decreases as 0 is approached. Thus, an outcome with a probability of 0.95 is highly likely, while that with a probability of 0.02 is very unlikely.

Many events in genetics are analogous to a series of coin-tossing experiments. A fair coin has two sides, heads (H) and tails (T), each of which has an equal chance of appearing on a single toss. Thus, the probability of getting heads is $\frac{1}{2}$, as is that of getting tails. The total probability is $\frac{1}{2} + \frac{1}{2} = 1.0$. In other words, the result of one toss must be either heads or tails. No other outcome can occur, and one of the two must occur. Furthermore, the outcome of a single coin toss is said to be *random*, because although each event has a specific probability, the outcome of the toss cannot be precisely predicted beforehand.

Note that the best prediction or educated guess that can be made regarding the outcome of a single coin toss is that heads or tails is equally likely, each with a probability of $\frac{1}{2}$. Knowledge of this single probability, however, makes possible other less-obvious predictions. For example, if the probability of H is $\frac{1}{2}$ for a single toss, what is the probability of getting heads on each of two consecutive tosses (H,H)? Because each toss is *independent* of (does not affect) the other, the latter probability is simply the product of the two individual probabilities: $\frac{1}{2} \times \frac{1}{2} = \frac{1}{4}$. The complete list of all possible outcomes shows that four equally likely outcomes may occur.

$$H,H = \frac{1}{2} \times \frac{1}{2} = \frac{1}{4}$$
$$H,T = \frac{1}{2} \times \frac{1}{2} = \frac{1}{4}$$
$$T,H = \frac{1}{2} \times \frac{1}{2} = \frac{1}{4}$$
$$T,T = \frac{1}{2} \times \frac{1}{2} = \frac{1}{4}$$
$$\text{Total} = 1.0$$

For three tosses, a similar list can be made, showing eight possible outcomes, each of equal probability.

$$H,H,H = \frac{1}{2} \times \frac{1}{2} \times \frac{1}{2} = \frac{1}{8}$$
$$H,H,T = \frac{1}{2} \times \frac{1}{2} \times \frac{1}{2} = \frac{1}{8}$$
$$H,T,H = \frac{1}{2} \times \frac{1}{2} \times \frac{1}{2} = \frac{1}{8}$$
$$H,T,T = \frac{1}{2} \times \frac{1}{2} \times \frac{1}{2} = \frac{1}{8}$$
$$T,T,H = \frac{1}{2} \times \frac{1}{2} \times \frac{1}{2} = \frac{1}{8}$$
$$T,H,T = \frac{1}{2} \times \frac{1}{2} \times \frac{1}{2} = \frac{1}{8}$$
$$T,H,H = \frac{1}{2} \times \frac{1}{2} \times \frac{1}{2} = \frac{1}{8}$$
$$T,T,T = \frac{1}{2} \times \frac{1}{2} \times \frac{1}{2} = \frac{1}{8}$$
$$\text{Total} = 1.0$$

Note that these sets of probabilities can be used to answer other questions. For example, considering two tosses, regardless of the order, what is the probability of obtaining one H and one T? This can occur from either result H,T or T,H. The sum of these two probabilities is $\frac{1}{4} + \frac{1}{4} = \frac{1}{2}$. In other words one-half the time a fair coin can be expected to produce one H and one T. For three coins, the probability of obtaining at least one H and one T is $\frac{3}{4}$.

An offspring's genotype at one locus is analogous to a coin-toss experiment. An offspring's maternally derived allele may be regarded as the result of the toss of a coin, whose two possible outcomes represent the alleles possessed by the mother. The other allele represents a coin toss to choose among the two alleles of the father. A homozygous parent is analogous to tossing a coin with two heads; only one result is possible. On the other hand, the meiotic process insures that heterozygous individuals produce gametes of two types (R and r, for example) in equal proportions. Precisely which gamete of the mother unites with which gamete of the father is random, or totally unpredictable.

Thus, if we wish to predict the genotype of a single offspring of monohybrid cross (Rr × Rr) we may apply the same logic as above. The probability of tossing an R for the first allele is $\frac{1}{2}$; of an r is also $\frac{1}{2}$, and so on. Four possible genotypes may occur: RR, Rr, rR, and rr. However, there is no order to the toss of the genetic coin, so Rr and rR are phenotypically and genotypically identical. Note that 50% of the time the result will be a heterozygote, just as 50% of all two-toss experiments produces one H and one T.

Two rules of probability need to be remembered: the rule of addition and the rule of multiplication. If we wish to calculate the probability of a particular event happening and there is *more than* one way in which it can occur, we *add* the independent probabilities. Thus the probability of obtaining a heterozygous offspring in the above monohybrid cross is the *sum* of the two probabilities, $\frac{1}{4}$ for Rr and $\frac{1}{4}$ for rR $= \frac{1}{2}$.

If, on the other hand, we wish to calculate the probability of two (or more) independent events occurring at the same time (or in succession), we then must *multiply* the probabilities of the independent events. The probability of the above monohybrid cross producing two offspring, one of which is RR and the other of which is rr, becomes $\frac{1}{4} \times \frac{1}{4} = \frac{1}{16}$.

A good rule of thumb in calculating probabilities of genetic events is to first calculate the probability of each independent event. Then state the question to determine whether you are after the probability of event X *or* event Y (add), or whether you are after the probability of event X *and* event Y (multiply).

For example, consider a family with six children. What is the probability that all six children will be girls? The probability of any child being a girl is $\frac{1}{2}$. Our question is concerned with the chance that child number one is a girl, *and* number two is a girl, *and* number three is a girl, and so on. Thus we must multiply $\frac{1}{2} \times \frac{1}{2} \times \frac{1}{2} \times \frac{1}{2} \times \frac{1}{2} \times \frac{1}{2} = \frac{1}{64}$. The probability that the six children will be all boys or all girls becomes the chance that all six are girls ($\frac{1}{64}$) *plus* the chance that all six are boys ($\frac{1}{64}$), or $\frac{1}{32}$. Table B.1 depicts a generalized way to calculate probabilities using the binomial expansion $(p + q)^n$ where p is the probability of one event, q is the probability of an alternative event, and n is the number of events considered.

Table B.1
Expanding the binomial

$(p + q)^n$

p and q represent the probabilities of occurrence of two alternative events
(p = boys, q = girls) and n is the size of the family (group) involved.

$$4! = 4 \text{ factorial} = 4 \times 3 \times 2 \times 1 = 24$$

The probability of a particular combination can be calculated from the following formula:

$$(p + q)^n = \frac{n!}{s!t!} p^s q^t$$

Where n = the number in the group
Where s = the number in one class (i.e., boys)
Where t = the number in the other class (i.e., girls)

Example
If six babies are born in a given hospital on the same day, what is the probability that two will be boys and four will be girls?

$$p = \tfrac{1}{2}, q = \tfrac{1}{2}, n = 6, s = 2, t = 4$$

$$\frac{6!}{2!4!} p^2 q^4 = 15(\tfrac{1}{2})^2(\tfrac{1}{2})^4 = \frac{15}{64}$$

Example
If a couple has three children, what is the probability that two are girls?

$$p = \text{probability of a boy} = \tfrac{1}{2}$$
$$q = \text{probability of a girl} = \tfrac{1}{2}$$

Order of arrival: Different ways in which children may be born in a family of three.
Three boys
$$p\ p\ p = \tfrac{1}{8} = p^3$$

Two boys and one girl
$$\left. \begin{array}{l} p\ p\ q = \tfrac{1}{8} \\ p\ q\ p = \tfrac{1}{8} \\ q\ p\ p = \tfrac{1}{8} \end{array} \right\} = \tfrac{3}{8} = 3p^2q$$

Two girls and one boy
$$\left. \begin{array}{l} q\ q\ p = \tfrac{1}{8} \\ q\ p\ q = \tfrac{1}{8} \\ p\ q\ q = \tfrac{1}{8} \end{array} \right\} = \tfrac{3}{8} = 3pq^2$$

Three girls
$$q\ q\ q = \tfrac{1}{8} = q^3$$

$$(p + q)^3 = p^3 + 3p^2q + 3pq^2 + q^3$$

Glossary

ABO blood group A well-known blood group of humans, characterized by the presence or absence of A and B antigens on the surface of red blood cells.

acentric A fragment of a chromosome lacking a centromere.

achondroplasia A type of dwarfism caused by an autosomal dominant allele that produces abnormal growth in the long bones.

acquired immune deficiency syndrome (AIDS) An infectious viral disease that destroys the immune system; first identified in 1981.

acquired immunity Specific defense or resistance to a particular infective agent, which develops only after exposure to that agent.

acrocentric chromosome A chromosome whose centromere is located very near one end.

active immunity The production of antibodies by an individual's own immune system.

AD. *See* **Alzheimer's disease (AD)**

agammaglobulinemia A sex-linked recessive condition in which individuals do not produce B cells, and therefore cannot produce antibodies.

agglutination The clumping together of red blood cells in serum.

AIDS. *See* **acquired immune deficiency syndrome (AIDS)**

albinism Absence of normal pigmentation in skin, hair, and eyes.

allele An alternate form of a gene; one of the differing sequences of nucleotide bases that occupy a specific genetic locus on a chromosome.

allergy Immunological hypersensitivity to certain foreign antigens.

Alzheimer's disease. (AD) A form of senile dementia.

Ames test A laboratory procedure that directly measures the mutagenicity of a chemical; because of the high correlation of mutagenesis and carcinogenesis, the Ames test is used as an indirect test for carcinogenicity.

amino acid Small biological molecule; twenty common ones used in protein synthesis.

amniocentesis A technique for sampling a small quantity of fetal cells from the amniotic fluid surrounding a fetus for tests.

amplification (gene amplification) The production within the genome of many copies of a single gene.

anaphase The third phase of mitosis, when sister chromatids separate forming two daughter groups of chromosomes.

anaphase I The third phase of meiosis I, in which duplicated homologous chromosomes move towards opposite poles of the cell.

anaphase II The third phase of meiosis II, in which sister chromatids move towards opposite poles of the cell.

aneuploid A cell or individual varying from the normal diploid number; aneuploid individuals have one or more chromosomes too many or too few.

antibody A large protein composed of four polypeptide chains produced by the body in response to exposure to a foreign antigen; antibodies protect the body from invading bacteria and other foreign substances.

anticodon The sequence of three nucleotide bases in a tRNA molecule that is complementary to a codon of an mRNA molecule.

antigen A molecule, often a protein, or a microorganism that elicits the production of antibodies in an individual.

antigenic determinant The portion of an antigen that is recognized by the body's immune mechanisms and, thus, to which the immune mechanisms respond.

antigenic drift A slight alteration in the antigenic determinants of influenza viruses.

antigenic shift A major alteration in the antigenic determinants of influenza viruses.

Archaebacteria One of the two major kingdoms of prokaryotes; a very unusual type of bacteria, including methanogens, halophiles, and thermoacidophiles.

artificial insemination The collection (by agriculturalist, physician, or scientist) of semen from a male that after short-term or long-term storage, is placed in the reproductive tract of a female; used in animal husbandry and medicine.

artificial selection The controlled breeding of domestic plants and animals so that desirable or valuable traits increase within a population or species.

asexual reproduction Any form of reproduction such as fission, budding, or parthenogenesis, in which one parent produces offspring genetically identical to itself.

Australopithecus A genus in the family Hominidae; a possible ancestor of the genus *Homo*.

autoantibody An antibody produced to one's own tissues.

autosomal chromosomes All of the chromosomes in eukaryotes other than sex chromosomes.

auxin A growth hormone in plants.

bacteriophage A virus that parasitizes bacteria.

Barr body A darkly staining structure visible in the nucleus of cells of female mammals; an inactivated X chromosome; in individuals with more than one X chromosome all but one are inactivated, so an individual may have several Barr bodies.

base analogue A chemical whose structure resembles that of a nucleotide base; sometimes base analogues are incorporated in a DNA molecule causing a mutation.

B cell A type of lymphocyte that produces antibodies.

bioengineering The ability to directly alter the genetic material of an organism; also *see* **biotechnology.**

biotechnology The industrial use of biological systems and processes.

bivalent A synapsed pair of homologous chromosomes in meiosis I; each homologue consists of two sister chromatids. Thus, a bivalent consists of four chromatids.

blood group. *See* **blood type (blood group)**

blood type (blood group) A phenotypic trait defined by the type of antigens on the surface of red blood cells.

Burkitt's lymphoma A cancer of the lymphatic tissue, associated with Epstein–Barr virus.

carcinogenic A cancer-causing substance.

carcinoma One of the many types of cancer primarily affecting skin, glands, gastrointestinal and reproductive tracts, and urinary system.

cardiovascular system The larger of the two circulatory systems of the body; including the heart, arteries, veins, and capillaries; providing circulation of the blood.

carrier An individual heterozygous for a deleterious recessive allele.

C band A darkened region on or near the centromere of a chromosome in a cell treated with stain.

cDNA A copy of the DNA made from RNA, using the enzyme reverse transcriptase; used extensively in recombinant DNA techniques.

cell The basic structural and functional unit of life; consisting of cytoplasm, containing organelles and required molecules enclosed by a membrane.

cell cycle (mitotic cell cycle) The complete life history of a single cell, from its origin by mitosis to its own subsequent mitotic division; divided into two major segments—interphase and mitosis.

cell theory The concept that all organisms are composed of cells, and that all living cells arise from other living cells.

cellular immunity A type of specific immunity resulting from the action of T cells, which directly attack foreign antigens.

central dogma The concept describing the flow of hereditary information; within one generation information flows from DNA to RNA to protein, to produce the ultimate phenotype; between generations, DNA replicates itself precisely.

centriole An organelle appearing to serve as an organizing structure for the formation and movement of the spindle during cell division in animal cells.

centromere The primary constriction on a chromosome to which spindle fibers attach during cell division.

cerumen Earwax; the yellow waxy secretion from the glands of the external ear.

Chargaff's rules Within DNA, the amount of adenine equals the amount of thymine, and the amount of cytosine equals the amount of guanine.

chiasma An X-shaped structure formed between chromatids of homologous chromosomes during synapsis; chiasmata are the mechanisms for recombination.

chimera An individual consisting of two or more distinct cell lines; may result from the fusion, early in development, of two or more different zygotes.

chloroplast A chlorophyll-containing organelle of plant cells and some protozoan cells; contains its own DNA; conducts photosynthesis.

chorion villus sampling A rapid and painless way of sampling fetal tissue for detection of genetic disease during the first trimester of pregnancy. This methodology samples fetal tissue directly and results can be obtained with twenty-four to forty-eight hours of sampling.

chromatid One of the two strands that comprise a single chromosome in the early stages of mitosis and meiosis; each chromatid, attached to its sister chromatid by a centromere, forms one-half of the chromosome. *See also* **sister chromatid exchange (SCE)**

chromatin Chromosomes during interphase, the most common configuration of nuclear DNA, combined with histones and non-histone nuclear protein.

chromosome A linear sequence of genes; in prokaryotes, a single circular chain of DNA; in eukaryotes, one of many (usually ten to one hundred) rodlike structures, usually in pairs, formed by complexes of DNA, histones, and non-histone nuclear proteins.

chromosome puffs Expanded regions on a polytene chromosome.

chromosome set Consists of one member of each chromosome pair; in humans there are twenty-three chromosomes in a set.

clone A group of genetically identical genes, cells, or individuals; and, used as a verb, to produce a set of genetically identical genes, cells, or individuals.

cloning vehicle. *See* **vector**

codominance The expression of both alleles at a locus in an heterozygote; a type of allelic interaction at a locus.

codon A sequence of three nucleotide bases in the sense strand of a DNA molecule that specifies which amino acid is to be added to a polypeptide chain; also, the sequence of three nucleotide bases in a messenger RNA molecule that has been transcribed from and is complementary to the DNA codon; a few codons do not specify an amino acid, but rather the initiation or termination of protein synthesis.

coefficient of relationship (r) The proportion of alleles in two individuals that are identical by descent; a measure of relationship.

colchicine A chemical that induces polyploidy in a dividing cell; used for treatment of gout in humans.

colinear The precise, one-to-one correspondence between the sequence of nucleotide triplets in a DNA molecule and the sequence of amino acids in the protein encoded by the DNA.

color blindness The inability to distinguish certain colors.

colostrum A secretion of a mother's mammary glands shortly after birth; colostrum is rich in antibodies and provides a nursing infant with short-term, passive specific immunity.

complementary (complementary base pairing) The precise pattern of bonding of nitrogenous bases in opposing strands of nucleic acids; in DNA, adenine always pairs with thymine, and cytosine with guanine; when RNA is transcribed from DNA, uracil (rather than thymine) in the growing RNA chain pairs with adenine on the opposing DNA.

complete dominance. *See* **dominance (complete dominance)**

consanguineous "Of the same blood;" sharing genes derived from a common ancestor; related by descent.

contact inhibition A phenomenon exhibited by cells grown in culture, limiting their growth to formation of a single layer within a culture dish; growth stops when a barrier, such as the wall of the dish or another cell, is encountered.

cri du chat syndrome "Cry of the cat" syndrome; a serious genetic disease caused by a deletion in the short arm of chromosome 5.

crisscross inheritance The pattern of transmission shown by X-linked alleles in a pedigree; an X-linked allele is transmitted from male-to-female, then female-to-male, in succeeding generations.

crossing-over (crossover) The exchange of segments between homologous chromosomes during meiosis; usually is "reciprocal," involving the same set of homologous loci on each chromosome, but need not be; produces chromosomes with allelic combinations not found on either parental chromosome.

cystic fibrosis A disease caused by an autosomal recessive allele; most common genetic disease of Caucasians.

cytoplasm The viscous fluid within a cell; in eukaryotes, cytoplasm is external to the nucleus and contains many organelles and molecules.

degenerate With reference to the genetic code, redundant; the specification of some amino acids by more than one codon.

delayed onset The absence of expression of a particular gene until later in life.

deletion Loss of a segment of a gene or a chromosome.

deoxyribonucleic acid (DNA) A long molecule comprised of two chains of nucleotides containing the sugar deoxyribose; the hereditary material in all organisms except some viruses.

deoxyribose A five-carbon sugar; a chemical constituent of DNA.

development The process by which a single unspecialized cell becomes a highly complex organism of many specialized cells; the result of growth and differentiation.

diabetes A disease resulting from an inability to metabolize carbohydrates properly, often due to a deficiency in insulin production.

differentiation The structural and functional specialization of a cell or tissue.

dihybrid cross Mating between individuals homozygous for alternate alleles at two loci.

dioxin An extremely toxic mutagen and carcinogen that occurs as a contaminant of some herbicides and other chemicals.

diploid Having two sets of chromosomes; one set derived from the mother, the other set from the father.

diploid number The total number of chromosomes found in species in which chromosomes occur in pairs; often referred to as 2N, where N refers to the number of pairs of chromosomes; usually specific and constant for an eukaryotic species.

dispermy The fertilization of one egg by two sperm.

dizygotic twins Twins derived from the separate fertilization of two different eggs; fraternal twins; dizygotic twins share one half of all alleles, just as do normal siblings.

DNA. *See* **deoxyribonucleic acid (DNA)**

DNA library An extensive series of bacterial cultures that have incorporated different pieces of DNA for the purposes of gene cloning.

dominance (complete dominance) With reference to the interaction of two alleles at a locus, the expression by a heterozygote of a phenotype identical to that of one of its alleles (the dominant allele) when homozygous; the phenotype exhibited by the other allele (the recessive allele) when homozygous is suppressed.

dosage compensation An adjustment (or compensation) in the number (dosage) of active alleles or chromosomes, usually by turning one chromosome off; usually refers to X inactivation.

double helix The shape of the DNA molecule; often used as a synonym for DNA. *See also* **deoxyribonucleic acid (DNA)**

Down's syndrome A set of phenotypic anomalies caused by trisomy of chromosome 21 in humans.

Drosophila melanogaster The common fruit fly, used in many genetic experiments.

Duchenne muscular dystrophy. *See* **muscular dystrophy (Duchenne muscular dystrophy)**

duplication A type of mutation in which a nucleotide sequence that repeats (duplicates) a nucleotide sequence already present is added to a chromosome or genome.

dysgenic Anything that acts to increase the frequency of deleterious genes in a population.

EBV. *See* **Epstein-Barr virus (EBV)**

E. coli. See **Escherichia coli (E. coli)**

electrophoresis A biochemical technique that uses differences in electrical charge, size, and shape to separate a mixture of proteins or nucleic acid fragments.

embryo An early developmental stage of a plant or an animal; descendant of a zygote; in humans, usually refers to period between weeks two and eight of gestation.

embryo transfer The removal of a developing embryo from its genetic mother and its placement for the remainder of gestation in a second female, a surrogate mother.

endoplasmic reticulum A complex network of membranes within the cytoplasm of eukaryotes, on which are found ribosomes.

environment The total of all conditions that impinge upon an individual cell or genome; genes and the environment interact to produce the phenotype of an individual; genes and gene products may comprise a genetic environment for other genes.

enzyme A protein that catalyzes (increases the rate of) biochemical reactions.

epididymis A long, convoluted tubule on the surface of each testis; the site of sperm storage and maturation.

epistasis Interaction of alleles at different loci; the control or masking of the expression of alleles at one locus by alleles at a second locus.

Epstein–Barr virus (EBV) A relatively common DNA virus producing coldlike symptoms, but also associated with Burkitt's lymphoma and infectious mononucleosis; one of the herpes viruses.

erythroblastosis fetalis A type of hemolytic anemia in a fetus or newborn infant, resulting from Rh incompatibility of mother and fetus; potentially life-threatening; anti-Rh antibodies produced by an Rh-negative mother destroy red blood cells of an Rh-positive fetus; most serious for second and subsequent Rh-positive children.

Escherichia coli (E. coli) The common gut bacterium of humans and other mammals, widely used in genetic experimentation.

estrogen One of several types of female sex hormones.

ethics The study of the standards of morality and human conduct.

Eubacteria One of two major kingdoms of prokaryotes; includes most common bacteria and blue-green algae.

euchromatin Lightly staining chromatin; believed to be relatively uncoiled DNA and capable of transcription and other genetic activity.

eugenics The study or practice of improvement in the genetic quality of a population.

Eukaryota The third kingdom of life, including all of the nucleated organisms.

eukaryotic cell (eukaryote) One of two basic types of cells; about 1,000 times larger in volume than prokaryotic cells; well-defined nucleus with rodlike chromosomes; many organelles visible within extra-nuclear cytoplasm.

euphenics The manipulation of the phenotype in an attempt to nullify an undesirable genetic trait.

evolution Change with time; the biological concept that all life on earth changes through time; continuing genetic change may result in one species giving rise to new species.

exons Those portions of the sense strand of DNA that are the template for protein synthesis; exons are interspersed with introns, both of which may be transcribed, but only exon sequences are translated.

expressivity The extent to which an individual demonstrates particular phenotypic effects of a gene; a quantitative measure of expression of a gene.

F_1 The designation for progeny produced by the first mating in a pedigree; subsequent generations are labelled F_2, F_3, etc.; F refers to filial generation.

fallopian tube The duct through which an ovum passes from the ovary to the uterus; oviduct.

fetal alcohol syndrome (FAS) A set of teratologies produced in a fetus by maternal alcohol consumption during gestation.

fetoscopy A procedure for the visualizing of fetal blood vessels that permits removal of blood directly from a fetus.

fetus The developing offspring of live-bearing animals; in humans, refers to developing individual from about age eight weeks until birth.

fixation Where only one allele occurs at a particular locus in a population; that allele is said to be fixed, and all individuals are homozygous for it.

forensic Relating to legal matters.

frameshift A type of genetic mutation in which one or two bases are added or deleted to the nucleotide sequence, thus shifting the reading frame and changing all subsequent codons.

fraternal twins. *See* **dizygotic twins**

gamete A reproductive cell produced in the gonads; either spermatozoa or ova; gametes are usually haploid.

gametogenesis Gamete formation; includes meiosis and the continued development of the haploid daughter products of meiosis to produce a mature gamete.

G band A transverse band on a chromosome in a cell treated with Giemsa stain.

gene A specific location on a chromosome that encodes hereditary information in which sense it is synonymous with the terms *locus* or *genetic locus;* also used to refer to a specific molecular variant occupying a genetic locus, in which sense it is synonymous to *allele*.

gene amplification. *See* **amplification (gene amplification)**

gene mutation (point mutation) A type of mutation affecting only one of a few nucleotide bases; not microscopically visible.

gene pool All the genes that occur within one species; all the genes possessed by all members of a species.

gene splicing A phrase used to describe manipulation of the genome by insertion and/or deletion of DNA into or from a gene.

gene therapy Replacement of disease-causing genes or nucleotide sequences with normal or non-pathological sequences; or inserting additional genetic material that can provide a normal phenotype.

genetically significant dose (GSD) The amount of radiation reaching the gonads.

genetic code The triplet sequences of DNA bases that encode information for protein synthesis.

genetic counseling Guidance provided by a physician or geneticist to a family or individual regarding genetic aspects of pathological conditions; genetic counselors may advise parents concerning the risks of producing genetically defective children or the treatment of individuals with genetic defects.

genetic distance A measure of the average number of allele differences per locus between populations; distances range from zero to infinity.

genetic engineering Deliberate alterations of individual genotypes; any technique that creates new combinations of genetic material or that alters expression of preexisting combinations; recombinant DNA techniques.

genetic load A measure of the proportion of deleterious alleles within a population.

genetics The study of the structure and function of genes and their transmission between generations; the study of inheritance and sources of organismal variation.

genome All the genes possessed by a cell or individual.

genotype The genes possessed by an individual; may refer either to a pair of alleles at a specific locus, or to the entire genome.

germplasm bank Repository for genetic material, cells, or tissues, where they are placed in a dormant but living state for long-term preservation and later use.

gonad Gamete-forming organ; testis in male and ovary in female.

grafting Agricultural process of fusing cuttings from two (or more) different plants to grow as a single new (hybrid) individual.

Great Apes Primates of the family Pongidae; the closest evolutionary relatives of humans; includes gorillas, chimpanzees, and orangutans.

Green Revolution A series of international programs begun during the 1960s to increase agricultural productivity throughout the world, but especially in underdeveloped countries.

GSD. *See* **genetically significant dose (GSD)**

guevodoces Individuals from a village in the Dominican Republic born with a 46, XY karyotype, but as children, they possess ambiguous genitalia; at puberty, they undergo masculinization of the external secondary sex structures, becoming phenotypically normal males.

haplodiploid A system of sex determination in bees, wasps, and termites, in which males are haploid and females are diploid.

haploid Having only one chromosome of each homologous pair (a haploid set).

haploid number The number of pairs of chromosomes in a cell; one-half the diploid number; often designated N; usually constant for a eukaryotic species.

Hardy–Weinberg equilibrium equations A set of mathematical equations describing the relationship between allele and genotype frequencies in successive generations.

Hayflick Limit The number of times normal cells will undergo mitosis in cell culture.

helper T cells A type of T cell that participates with B cells in the immune response.

hemizygous Genes or chromosomes that occur without a homologous partner; most commonly, refers to genes on the X and Y chromosomes of males.

hemoglobin A protein consisting of four iron-bearing groups (heme) and four polypeptide chains (globin); found in vertebrate red blood cells and transports oxygen from the lungs to all body tissues.

hemophilia A genetic disease characterized by the inability to coagulate blood; bleeder's disease; two forms are controlled by X-linked loci, one by an autosomal locus.

heritability The proportion of phenotypic variation within a population attributable to genetic rather than environmental variation.

hermaphrodite An individual possessing both male and female gonads, testes and ovaries (true hermaphroditism); a pseudohermaphrodite possesses some secondary traits of both sexes, but gonadal tissue of only one sex (or none whatever).

herpes virus One of many types of DNA viruses that infect humans and other vertebrates.

heterochromatin Darkly staining chromatin; probably densely coiled and incapable of transcription.

heterogametic sex The sex with only one sex chromosome, or two non-homologous sex chromosomes, which consequently produces two different types of gametes; in humans, the XY male is the heterogametic sex; XO males in some insects are also heterogametic.

heterogeneous nuclear RNA (HnRNA) The initially transcribed product of the sense strand of DNA, found only in the nucleus; mRNA is produced by the excision of many segments of HnRNA.

heterozygous Having two different alleles at a specific genetic locus.

hibakusha A Japanese word meaning explosion-affected persons; refers to survivors of the atomic bomb explosions in Japan at the end of World War II (August 1945), who have not married or reproduced.

highly repetitive sequences DNA sequences that have more than 100,000 copies in a genome.

histamine A chemical released by mast cells that helps produce the allergic response.

histone One of five types of nuclear proteins; they are involved in DNA coiling in chromosomes and regulation of gene activity.

HLA (HLA antigens) Human leukocyte antigens; a set of antigens on the surface of white blood cells.

HnRNA. *See* **heterogeneous nuclear RNA (Hn RNA)**

homogametic sex The sex having two identical sex chromosomes; in humans, the XX female is the homogametic sex.

homologous Chromosomes that are similar in appearance and structure, possessing the same sequence of genetic loci; or, structures in different organisms that share a common evolutionary origin, such as the wing of a bat and the arm of a human.

homozygous Having two identical alleles at a specific genetic locus.

hormone A molecule that transmits information between cells; a chemical produced by one cell to regulate function of another cell.

HTLV. *See* **human T-cell leukemia virus (HTLV)**

human T-cell leukemia virus (HTLV) A class of retroviruses that includes the causative agents of leukemia and AIDS.

humoral immunity A type of specific immunity resulting from the action of B cells, which produce antibodies, which in turn destroy foreign antigens.

Huntington's disease A dominant genetic disease with delayed onset; the average age of onset of symptoms being twenty-five to forty-five years. The disease is characterized by progressive loss of motor control and mental deterioration. There is no known treatment or cure.

H-Y antigen An antigenic protein found on all cells of males, but not females; the product of one or more Y-linked loci; essential for normal spermatogenesis.

hybrid A cell or individual resulting from the mating of genetically different forms; a hybrid may be the result of a cross of genetically different individuals within a species, or the progeny of a cross of individuals of different species.

hybrid vigor An increase in desirable phenotypic traits such as size, strength, and physiological health, which often occurs in hybrid individuals.

identical twins. *See* **monozygotic twins**

immune response The combined activity of T cells, B cells, and macrophages, in destroying or neutralizing the pathological effects of foreign antigens.

immunoglobulin Antibody.

immunoglobulin E (IgE) A type of antibody involved in the allergic response.

immunoglobulin G (IgG) The most common type of antibody, providing protection against bacteria, viruses, and toxins.

immunologically privileged site One of several structures or locations in the body, including the corneas, bone, cartilage, heart valves, and testes, which are isolated from the circulating T cells; these sites are thus not subject to normal immune surveillance and defense.

inborn error of metabolism An inability to produce an enzyme needed for a particular metabolic pathway; many genetic diseases are due to inborn errors.

inbreeding The mating of related individuals; consanguineous mating.

inbreeding depression The reduction in phenotypic vigor, survival, and reproduction resulting from consanguineous mating.

incomplete dominance (partial dominance) With reference to two alleles at a locus, where a heterozygote exhibits a phenotype intermediate to the two homozygous types.

independent assortment. *See* **Mendel's Law of Independent Assortment**

index case An individual whose phenotype is the cause for the study of his or her pedigree.

inducible With reference to an operon, a type of control in which protein (enzyme) synthesis occurs only when the substrate upon which those enzymes are active is present.

influenza Viral disease caused by one of three different classes of RNA viruses; also called flu.

intelligence quotient (IQ) A standardized measure of some aspects of mental capacity, probably related, but in an unknown way, to intelligence.

interferon A naturally occurring protein that has antiviral properties.

interphase Usually the longest portion of the cell cycle, divided into the G_1 (Gap 1), S (synthesis), and G_2 (Gap 2) phases; a time of vegetative growth for most cells; DNA replication occurs during interphase; also refers to the brief pause between meiosis I and meiosis II.

introns Segments of DNA interspersed between exons that do not contain information to be translated; introns are apparently transcribed to HnRNA, but the intron-coded portions are excised from HnRNA to produce mRNA.

inversion A type of chromosomal mutation in which a nucleotide sequence is removed and then reinserted in reverse sequence to which it normally occurs.

in vitro "In glass;" refers to experiments conducted in the laboratory, often in a flask or culture dish.

in vitro **fertilization** Fertilization of an ovum in the laboratory rather than within the mother's body; a child produced by *in vitro* fertilization is misleadingly called "test tube baby" by the public news media. *See also* **test tube baby**

IQ. *See* **intelligence quotient (IQ)**

karyotype The total chromosomal complement of an individual; or, a formal arrangement of photographed chromosomes arrested at metaphase of mitosis from a single cell.

Klinefelter's syndrome A set of phenotypic anomalies associated with males having an extra X chromosome; the genotype is designated 47,XXY.

lac operon The functional and structural unit of DNA of *E. coli* that controls the synthesis of lactose-metabolizing enzymes.

lampbrush chromosome A type of chromosome found in oocytes of some species that has many loops of DNA extending out from the main axis.

latent period The time interval when latent viruses are inactive and remain sequestered in the nervous system; characteristic of herpes viruses, AIDS virus, and others.

latent virus One of many viruses that retreats to the tissues of the nervous system, where it may persist for years without producing pathological symptoms; may leave the latent stage at any time, causing the normal symptoms of disease.

LD50 "Lethal dose fifty percent;" the amount of a toxic substance or treatment which kills fifty percent of a test population.

leukemia Cancer of the blood and blood-forming tissues, characterized by abnormally high production of white blood cells.

leukocyte Another name for a white blood cell.

linkage The occurrence of two or more genes on the same chromosome; such genes are said to be *linked* and are transmitted as a group to gametes; closely linked genes do not obey the law of independent assortment.

linkage group All genes located on the same chromosome; humans have twenty-three linkage groups.

locus A position or location on a chromosome; synonymous with *gene;* a locus is occupied by one of numerous molecular variants called *alleles.*

Lucy An important fossil, *Australopithecus afarensis,* found in Ethiopia.

lymphatic system The smaller of the circulatory systems of the body; consists of a network of lymph vessels that collect lymph fluid from the tissues and transport it to the cardiovascular system, which it enters near the heart; the lymphatic system also includes lymph nodes, tonsils, spleen, and thymus gland, which participate in specific immunity.

lymphocyte A type of white blood cell, which differentiates to become either a T cell or a B cell.

lymphokines Chemicals released by T cells that have antiviral and antibacterial properties.

Lyon hypothesis The concept that all but one of the X chromosomes are inactivated in each cell of placental mammals.

lysogenic infection A type of viral infection in which viral DNA attaches to the DNA of a host cell, where it is replicated with every division of the host; this type of infection does not kill the host cell.

lytic infection A pattern of viral life history in which the host cell ruptures following viral replication.

macrophage A large phagocytic cell that attacks foreign antigens and also interacts with T cells in producing specific immunity.

mast cell A type of cell responsible for histamine release in the allergic response.

meiosis A process of cell division occurring only in the gonads; usually consists of two successive divisions of a diploid parent cell to form four haploid daughter cells.

meiosis I The first meiotic division; a *reduction division,* in which the chromosomal complement is reduced from duplicated pairs of chromosomes to unpaired chromosomes, each still consisting of two sister chromatids.

meiosis II The second meiotic division; an equational division in which each chromosome (consisting of two sister chromatids) splits equally through the centromere; meiosis II begins with a cell containing one haploid set of duplicated chromosomes, which divides to produce two daughter cells, each containing one haploid set of non-duplicated chromosomes.

membrane A thin layer of lipid and proteins that encloses a cell and also penetrates the interior of the cell, providing internal structure.

membrane budding virus A virus that reproduces by packaging viral RNA or viral DNA into the viral coat at the cell wall and occasionally including a piece of host cell membrane in the mature virus.

memory cell A type of T cell or B cell that provides long-term recognition of an antigen to which the body has previously been exposed.

Mendel's Law of Dominance A law of inheritance, articulated by Mendel, which describes one type of relationship between homologous alleles in a heterozygote; traits governed by dominant alleles are expressed in heterozygotes, while those governed by recessive alleles are expressed only in homozygotes.

Mendel's Law of Independent Assortment A law of genetics, first described by Mendel, which states that genes on different chromosomes are distributed to gametes and occur in subsequent generations independently of one another.

Mendel's Law of Parental Equivalence A law of inheritance, first described by Mendel, which recognizes that the expression of an allele is not dependent upon which parent transmitted that allele.

Mendel's Law of Segregation The concept, first described by Mendel, that the two hereditary particles (alleles) controlling a phenotypic trait separate (segregate) into different gametes during meiosis; each allele is transmitted unchanged to an offspring.

messenger RNA (mRNA) A type of RNA, complementary to the protein-encoding segments of the sense strand of DNA, which carries the genetic information to be translated into a polypeptide chain.

metacentric chromosome A chromosome with the centromere in the middle.

metaphase The second phase of mitotic division, when the duplicated chromosomes are arranged along the metaphase plate of the spindle, towards the middle of the cell.

metaphase I The second phase of meiosis I; paired duplicated homologous chromosomes (bivalents) line up at the metaphase plate.

metaphase II The second phase of meiosis II; unpaired duplicated chromosomes line up at the metaphase plate.

metastasize With reference to all diseases but especially cancer, to spread from a point of origin to other locations in the body.

middle repetitive sequences DNA sequences that occur 10–100,000 times in a genome.

missense mutation A substitution mutation that produces a triplet encoding a different amino acid than was encoded prior to the mutation.

mitochondrion Membrane-bound organelle in the cytoplasm of eukaryotes; mitochondria produce energy for the cell and have their own DNA, separate from that of the nucleus. Mitochondria are inherited only from the female.

mitosis A type of cell division in which a diploid cell divides once to produce two diploid daughter cells, each identical to the parent; the type of cell division found in all non-germinal cells.

mitotic cell cycle. *See* **cell cycle (mitotic cell cycle)**

MN blood group A blood group of humans characterized by the presence or absence of M and N antigens on red blood cells.

monoculture The cultivation of a single genetic strain over large geographic areas.

monohybrid cross Mating between individuals homozygous for different alleles at one locus.

monomorphic Having only one form; having only one kind of allele at a locus.

monozygotic twins Twins derived from the splitting of a single zygote shortly after fertilization; identical twins.

mosaic An individual having two or more genetically different cell lines within the body.

mRNA. *See* **messenger RNA (mRNA)**

multiple sclerosis An autoimmune disease in which an individual produces antibodies to the myelin tissue of his or her nervous system.

muscular dystrophy (Duchenne muscular dystrophy) An X-linked recessive genetic disease causing severe muscular deterioration and early death; primarily affects young males.

mutagen Any substance or process that causes a genetic mutation.

mutation Any heritable change in the sequence of nucleotide bases in DNA; may affect only one or two bases (point mutation) or long segments (chromosomal mutation).

nail-patella syndrome An autosomal dominant trait, producing absent or malformed fingernails, toenails, and kneecaps.

natural resistance A set of innate, nonspecific defense mechanisms that protect an individual against a broad spectrum of toxic and pathogenic agents.

natural selection Non-random reproduction; differential survival and reproduction on the basis of genotype, often popularly designated "survival of the fittest."

negative eugenics The application of programs that reduce the number of deleterious or undesirable genes in a population, often by prohibiting individuals carrying such genes from breeding.

negative feedback A type of system control in which there is an inverse relationship between present and future levels of function of the system; in other words, present high levels of production result in decreased future levels of production, and vice versa.

neoplasm A cancerous tumor.

neurofibromatosis A deforming disease caused by an autosomal dominant allele, resulting in dozens or hundreds of nonmalignant tumors over the entire body.

newborn screening The testing of individuals at birth for genetic disorders.

nitrogenous base One of four molecules (adenine, cytosine, guanine, and thymine) that link together the two chains of a DNA molecule; in RNA, uracil instead of thymine comprises the fourth base.

nondisjunction The failure of homologous chromosomes to separate during meiotic division, producing gametes with abnormal numbers of chromosomes.

nonoverlapping A feature of the genetic code; the genetic code is read in consecutive groups of three nucleotide bases (a codon).

nonsense codon (terminator codon) A codon that does not specify the addition of any amino acid to a polypeptide chain; therefore, it terminates the reading of the genetic message and ends the production of that protein.

normal distribution A pattern of quantitative variation in which a set of phenotypic values shows a "bell-shaped" pattern of dispersion about a mean value; most individual values are close to the mean, while decreasingly small numbers show large variation from the mean; an example would be the distribution of heights of corn plants in a field.

norm of reaction The range of all possible phenotypic responses of a particular gene or set of genes over all environments.

nuclear transplantation The replacement of the nucleus of one cell with the nucleus from a second cell.

nucleic acid DNA and RNA; long molecules composed of a chain of nucleotides; the genetic material.

nucleolus "Little nucleus;" a dark-staining body within the nucleus.

nucleotide A fundamental building block of DNA or RNA, consisting of a five-carbon sugar, a phosphate, and a nitrogenous base.

nucleus The largest organelle within an eukaryotic cell; usually appears as a darkened, central area; contains chromosomes.

oncogenes Genes associated with the cancerous transformation of cells.

one gene-one polypeptide hypothesis The concept that each gene encodes information for the production of a single polypeptide chain; the single polypeptide chain may be either a complete protein itself or an entire chain that combines with other chains to form a protein.

oogenesis The development of ova.

oogonium A precursor cell within an ovarian follicle, which develops into a primary oocyte.

operator gene A regulatory gene in an operon that exerts control over transcription of adjacent structural genes.

operon A model of a structural and functional unit of the DNA of prokaryotes; includes a set of adjacent structural and regulatory genes involved in a particular metabolic pathway.

organelle "Little organ;" one of many structural and functional sub-units within a cell.

ovary Female gonad; produces mature sex cells (ova).

overlapping genes A phenomenon whereby a segment of DNA is transcribed twice, but with the reading frame in one transcription offset from that in the other; thus, one nitrogenous base participates in two codons, but occupies a different position (i.e., first, second, or third); the two codons often encode different amino acids.

oviduct The duct through which an ovum passes from the ovary to the uterus.

ovulate To release a secondary oocyte from the ovary.

ovum Egg.

P A designation for the original or first generation in a pedigree.

parthenogenesis A type of reproduction in which an egg undergoes complete development without fertilization.

partial dominance. *See* **incomplete dominance (partial dominance)**

passive immunity Immune protection that results from the introduction of antibodies from one person into another.

pattern baldness A phenotype showing sex-influenced inheritance, which produces early baldness; the individual retains a distinct rim of hair around the head.

pedigree A diagram representing a set of familial relationships; a genealogy.

penetrant Expressed; a gene showing *any* phenotypic manifestation is said to be penetrant; with respect to an individual, a gene is either penetrant or not; a qualitative measure of gene expression in an individual.

peptide bond A linkage between adjacent amino acids in a polypeptide.

pesticide treadmill The necessity, when using pesticides to control insect and other crop pests, to increase continually the dose of pesticides applied to maintain the same level of pest control.

phage. *See* **bacteriophage**

phenocopy A phenotype caused by an environmental factor that resembles a phenotype normally caused by a genetic factor; for example, sun-bleached hair (environmentally caused) resembles blond (genetically caused) hair.

phenotype Any expression of genes; phenotype may refer to any trait of an organism, such as its structure, physiology, biochemistry, behavior, or to all traits collectively; the phenotype is determined by the interaction of genes and environment.

phenylketonuria (PKU) A genetic defect characterized by inability to produce the enzyme phenylalanine hydroxylase; phenylketonurics are unable to metabolize phenylalanine properly and without a controlled diet become mentally retarded.

pheromone A type of hormone that is released into the environment as a mechanism of chemical communication between individuals.

Philadelphia (Ph1) chromosome A shortened chromosome 22 of humans resulting from translocation of the long arm, usually to the long arm of chromosome 9; usually found in persons with leukemia.

placenta A structure formed by the complex growth and intertwining of adjacent tissue from mother and fetus; provides a means for the passage of nutrients from mother to fetus and waste products from fetus to the mother; also secretes some hormones.

plasma The liquid portion of the blood.

plasma cell A type of cell formed by the differentiation of B cells; produces antibodies.

plasmid A small ring of DNA found in the cytoplasm of bacteria.

pleiotropic The multiple phenotypic effects of a single allele.

ploidy The number of sets of chromosomes possessed by a cell or individual.

point mutation. *See* **gene mutation (point mutation)**

polar body A daughter cell of either the first or second meiotic division in females that has a normal chromosomal complement, but almost no cytoplasm.

polygenic inheritance The control of a single phenotypic trait by many different genes.

polymorphic Having many forms; having two or more alleles at a locus.

polypeptide A sequence of three or more amino acids linked by peptide bonds; may be a complete protein.

polyploid A cell or individual possessing entire extra *sets* of chromosomes; designated with reference to the number of haploid sets present, such as triploid (3N), tetraploid (4N), etc.

polyribosome A structure formed by the attachment of many ribosomes to a single mRNA molecule during translation.

polytene chromosome A giant chromosome consisting of many identical DNA molecules, arranged side by side, resulting from chromosome replication without subsequent cell division.

position effect Where the expression of a gene depends upon its physical location within the genome; a change in the chromosome position may change the phenotypic effect.

positive eugenics The application of techniques that selectively increase or otherwise favor beneficial genes in a population.

prenatal screening Biochemical tests for genetic diseases conducted on tissue samples taken from a fetus or its surrounding amniotic fluid *in utero*.

primary nondisjunction Nondisjunction of homologous chromosomes during meiosis I.

primary oocyte A cell within an ovarian follicle, derived from an oogonium, that enters into the first meiotic division, giving rise to a secondary oocyte and the first polar body.

primary sex ratio The sex ratio at conception.

primary spermatocyte A cell within the seminiferous tubules derived from a spermatogonium; divides in meiosis I to produce two secondary spermatocytes.

primates The order of mammals including monkeys, Great Apes, and humans.

probability A measure of the uncertainty of occurrence of a specific event in the future; usually expressed as a number between zero (the event is impossible) and one (the event is certain to occur); likelihood.

progeria An autosomal recessive genetic disease characterized by extremely rapid aging and a very short life span.

progesterone A female sex hormone.

prokaryotic cell (prokaryote) One of two basic types of cells; smaller and simpler in appearance than eukaryotic cells; lacks a nucleus and has few distinguishable features under a light microscope; has only one circular chromosome.

promoter That region of the bacterial operon to which RNA polymerase attaches to initiate gene transcription.

prophase The first stage of mitosis, beginning when chromosomes become visible under the microscope.

prophase I The first of the four phases of meiosis I.

prophase II The first phase of meiosis II.

protein A major category of molecules required by living organisms; consists of one or more chains of amino acids, sometimes combined with other types of molecules.

protoplast fusion The production of hybrid cells in tissue culture, as a result of the union of cells from different species or different cell lines.

pseudohermaphrodite An individual possessing only one type of gonad (or none at all), but externally showing some secondary sex characteristics of both sexes.

pure breeding A genetic lineage in which all members are homozygous for a particular gene or set of genes and, hence, show little or no phenotypic variation for the trait or traits controlled by that gene or genes; such a lineage is said to "breed true."

quantitative variation A type of phenotypic variation in which individual phenotypes differ on a finely graded, or continuous scale.

r. *See* **coefficient of relationship (r)**

race In biological terms, almost any population or subgroup of a species that can be distinguished from other subgroups (a definition so broad as to be almost meaningless); in genetics, race often refers to a population differing in the frequency of certain alleles, chromosomes, or phenotypic features from other populations.

rad A measure of the amount of radiation energy absorbed by living tissues.

random mating In a genetic sense, mating without respect to genotype.

recessive An allele whose phenotypic effects are suppressed by its homologous partner in a heterozygote; may also refer to the phenotypic trait that is suppressed.

reciprocal cross A hybrid mating involving the same genetic traits as in a previous cross, but in which the sexes of the individuals of each type are reversed; for example, a mating of a short female pea plant and a tall male pea plant is the reciprocal cross of a tall female pea plant and a short male pea plant.

reciprocal translocation A type of chromosomal mutation in which acentric portions are exchanged by non-homologous chromosomes.

recombinant DNA techniques Genetic engineering techniques; artificial alteration of DNA sequences or individual genotypes; may include transfer of genetic material from one individual to another or insertion of artificially constructed DNA sequences; or combination in hybrid progeny of genetic material from two or more parents of different genetic types; or any other technique that produces new genetic combinations or altered expression of preexisting combinations.

recombination The exchange of alleles among homologous chromosomes as a result of crossing over in meiosis I; produces new allelic combinations on chromosomes; may also refer to or include the results of independent assortment, which have the similar consequence of producing gametes with allelic combinations found in neither parent.

reduction division The first meiotic division, which produces two daughter cells, each containing one duplicated member of each pair of chromosomes.

regulator (regulator gene) A gene in an operon that encodes a repressor protein and thus controls transcription.

regulatory gene A gene that controls the expression of structural or other regulatory genes.

rem Roentgen equivalent man; the amount of biological damage done by radiation; one rem = 10 millisieverts.

repressor In an operon, a protein produced by a regulator gene that, when bound to the operator, inhibits transcription.

restriction endonucleases Bacterial enzymes that each cut DNA at specific base sequences. These enzymes are used in many modern genetic engineering methodologies.

restriction fragment length polymorphism (RFLP) A large variety of patterns of the DNA after it has been incubated with a series of bacterial enzymes that cut the DNA at specific base sequences. Each individual has his or her own unique RFLP pattern, which can be used for both identification of genetic disease and for a variety of other medical and forensic purposes.

retrovirus One of a group of RNA viruses that causes tumors in vertebrates.

reverse transcriptase An enzyme produced by RNA viruses; used to construct a complementary DNA copy of the viral RNA for insertion into the host's genome.

RFLP. *See* **restriction fragment length polymorphism (RFLP)**

Rh blood group A blood group of humans characterized by the presence or absence of the Rh antigen on red blood cells.

Rh incompatibility Where mother and fetus have different Rh blood types; incompatibility between an Rh-negative mother and Rh-positive fetus often causes erythroblastosis fetalis.

ribonucleic acid (RNA) A molecule comprised of a chain of nucleotides containing the sugar ribose; RNA is the hereditary material in a few viruses, and is involved in protein synthesis in all organisms.

ribose A five-carbon sugar; one of the structural components of RNA.

ribosomal RNA (rRNA) A type of RNA that combines with proteins in the cytoplasm to form ribosomes.

ribosome A cytoplasmic organelle formed by the association of rRNA and proteins; the site of translation.

ring chromosome A circular chromosome in eukaryotes; formed after both ends of a chromosome are lost by terminal deletion, and the free ends then join.

RNA. *See* **ribonucleic acid (RNA)**

RNA polymerase An enzyme that catalyzes the formation of RNA (polymerization); facilitates the union of ribonucleotides in a complementary sequence to that of a DNA molecule.

rRNA. *See* **ribosomal RNA (rRNA)**

same sense (same sense mutation) A type of point mutation in which one nucleotide base replaces another, but because of the redundancy of the genetic code, the new codon encodes the same amino acid as the old codon; therefore, same sense mutations do not change the protein encoded by that gene.

sarcoma A type of cancer affecting bones and muscles.

satellite DNA A highly variable region of the human genome that can be isolated by a cDNA probe and then RFLP mapped for individual human "genetic fingerprinting."

secondary nondisjunction Nondisjunction of sister chromatids of a duplicated chromosome during meiosis II.

secondary oocyte A cell within an ovarian follicle resulting from the first meiotic division of a primary oocyte; if fertilized, divides to produce ovum and second polar body.

secondary sex ratio The sex ratio at birth.

secondary spermatocyte Either of two cells produced during meiosis I by the division of a primary spermatocyte; subsequently divides in meiosis II to produce two spermatids.

semi-conservative replication The type of replication found in DNA, in which each daughter double helix receives one complete parental strand of DNA, which in turn serves as a template for the construction of the complementary strand.

seminiferous tubule Extremely long and convoluted tubule within the testis; the site of spermatogenesis.

sense strand The strand of DNA that is transcribed to mRNA; the complementary strand is not transcribed.

sex chromosomes In eukaryotes, an X or Y chromosome, variation in which is related to sex determination; usually, in one sex both members of the pair are homologous, while in the other sex they differ in structure; in some species, one sex has two sex chromosomes, the other sex has only one.

sex-influenced inheritance A pattern of inheritance in which the expression of alleles is different in each sex; usually, an allele is dominant in one sex, recessive in the other; refers only to autosomal alleles.

sex-limited inheritance A pattern of inheritance in which autosomal genes are expressed only in one sex, but not the other.

sex-linked genes Genes located on a sex chromosome, usually the X.

sex ratio The proportion of males to females in a population; often expressed as the number of males per 100 females.

sexual reproduction Reproduction usually involving the fusion of gametes from two (or more) parents to produce offspring; the key aspect is reorganization in the progeny of genetic material derived from the parents.

sickle-cell anemia Hereditary disease caused by an autosomal recessive allele; causes hemoglobin to crystallize, deforming red blood cells and producing numerous phenotypic abnormalities.

sister chromatid exchange (SCE) The exchange of identical segments between sister chromatids during cell division; does not produce genetic recombination such as occurs due to crossover between non-sister homologues in meiosis. *See also* **chromatid**

SLE. *See* **Systemic lupus erythrematosus (SLE)**

somatic "Of the body;" sometimes refers to all bodily tissues, and sometimes used to distinguish between germinal and all other bodily (i.e., somatic) tissues.

speciation The process of evolutionary change in which one species evolves into one or more new species.

species A group of individuals of one type that can freely interbreed in nature and produce fertile offspring.

specific immunity Resistance of an individual to a particular microorganism or pathogen.

spermatid Either of two cells that are the result of division by a secondary spermatocyte in meiosis II; spermatids subsequently develop into spermatozoa.

spermatogenesis The development of spermatozoa.

spermatogonium A precursor cell within the seminiferous tubules that initiates spermatogenesis by dividing mitotically, becoming two primary spermatocytes.

spermatozoon (sperm) Male gamete; sex cell.

sperm banking The collection and preservation of sperm in a dormant but living condition for later use.

spindle apparatus A set of fibers, shaped like a spool, that become visible in the cytoplasm during cell division; in animal cells, the spindle extends between two centrioles; in plants, the spindle is similar to that in animals, but no centrioles are present.

spleen An amorphous abdominal organ, part of the lymphatic system, that serves as a reservoir for blood and functions in the immune response.

sterile male technique The release of large numbers of sterilized males of a specific insect pest species for the limitation or control of that species; the sterile males mate with wild females, who subsequently produce no offspring.

structural gene A gene encoding the amino acid sequence of a polypeptide.

submetacentric chromosome A chromosome with the centromere located slightly off-center.

substitution (substitution mutation) A type of point mutation in which one nucleotide base replaces another; affects only the codon in which substitution occurs.

superfecundation The fertilization of two ova by sperm from different males during the same menstrual cycle; twins resulting from superfecundation are only half siblings, in spite of sharing a uterus during gestation.

superovulation The ovulation of more eggs than normal; induced artificially by injecting a female with gonadotropic hormones.

surrogate mother A female who receives an implanted conceptus or embryo from another female (the genetic mother) and who then carries that embryo for the remainder of its gestational period.

symbiosis "Living together;" any biological relationship or interaction involving two or more species; symbionts may be benefited, harmed, or not affected by the relationship.

synapsis The close side-by-side pairing of duplicated homologous chromosomes in meiosis I; does not normally occur in mitosis.

syndrome A set of symptoms associated with one disease or pathological condition.

systemic lupus erythematosus (SLE) An autoimmune condition resulting from the production of antibodies to one's own DNA.

T cell A type of lymphocyte responsible for cellular immunity.

TDF. *See* **testis determining factor (TDF)**

telocentric chromosome A chromosome whose centromere is located at the very end.

telophase The fourth and last stage of mitosis; nuclear membranes enclose each set of daughter chromosomes, and the cytoplasm of the parent cell divides to form two complete daughter cells.

telophase I The fourth and last phase of meiosis I.

telophase II The fourth and last phase of meiosis II.

teratogen Any chemical, radiation, or other factor that causes physical deformity.

terminator codon. *See* **nonsense codon (terminator codon)**

tertiary sex ratio Sex ratio measured at any specified time after birth.

testcross The mating of a suspected heterozygote to a recessive homozygote.

testicular feminization A syndrome probably due to an X-linked recessive mutation, in which individuals with a normal male karyotype (46, XY) possess small internal testes, but no other internal reproductive structures; externally, they have an apparently normal female phenotype.

testis Male gonad; produces mature sex cells, spermatozoa; the primary sexual character of males; usually paired in vertebrates.

testis determining factor (TDF) A gene (or genes) that maps to the short arm of the human Y chromosome and is responsible for induction of testis formation.

testosterone The male sex hormone.

test tube baby An evocative but misleading phrase popularized by the public news media for *in vitro* fertilization; freshly ovulated eggs are collected from a female and these eggs are then fertilized *in vitro* (usually in a petri dish, *not* a test tube) and subsequently reintroduced into the mother's uterus, where normal gestation occurs; a child produced in this manner differs from another child only in that his or her conception occurs outside the mother's body.

thalassemia A group of diseases in which affected individuals produce a reduced amount of the alpha chain or the beta chain of hemoglobin.

thymine dimer The bonding to one another of adjacent thymine molecules on a single strand of DNA, rather than to the adenines on the complementary strand.

thymus gland An amorphous gland, part of the lymphatic system, located in the upper chest; the site of maturation of T cells.

tissue culture The growth of selected cells or tissues in defined culture medium in the laboratory.

totipotent A cell with the capacity to develop into a complete organism.

transcription The production of mRNA from a DNA template.

transduction Transfer of a segment of DNA from one cell to another by a virus; the DNA from the first cell can be incorporated into the genome of the second.

transfer RNA (tRNA) The smallest type of RNA, only about eighty nucleotides long, which transports an amino acid to a ribosome for incorporation into the growing protein.

transformation A cell incorporating free DNA from the surrounding medium into its own genome; also refers to the change of normal cells into cancerous ones.

translation The reading of mRNA by ribosomes, resulting in the production of a polypeptide chain; the sequence of nitrogenous bases in the mRNA specifies the sequence of amino acids in the polypeptide.

translocation A type of chromosomal mutation in which a segment of DNA is detached from its normal chromosome and attached to a non-homologous chromosome.

translocation Down's syndrome A type of Down's syndrome caused by the translocation of the long arm of chromosome 21 to one of the other autosomes, most frequently to chromosome 14.

transposon A gene that moves between locations on one or more chromosomes.

triplet A codon.

trisomy The presence in a cell of three homologues of a particular chromosome; produces a diploid number of 2N+1.

tRNA. *See* **transfer RNA (tRNA)**

Turner's syndrome A set of phenotypic anomalies associated with females having only one X chromosome; the genotype is designated 45, XO.

unconventional viruses A group of infectious agents, extremely small in size, but bearing few similarities to normal viruses; they produce diseases involving slow degeneration of nervous tissue and ultimately death.

unique sequences DNA sequences that occur only once in a genome.

vector or **cloning vehicle** Plasmids or phages used in molecular cloning to move genes and/or pieces of DNA from one cell to another.

vertebrates Animals with backbones; includes fish, sharks, amphibians, reptiles, birds, and mammals (including humans).

virus A structure with some but not all features normally attributed to living organisms; consists of nucleic acid, either RNA or DNA, enclosed in a protein coat; reproduces only when parasitizing living cells.

Wilms tumor Cancer of the urinary bladder; associated with a deletion on chromosome 11.

X chromosome The sex chromosome found in both sexes in species with two sex chromosomes and in which the male is the heterogametic sex; the larger of the two sex chromosomes in humans; females have two X chromosomes, males only one.

xeroderma pigmentosum A disease resulting in numerous skin cancers caused by a recessive autosomal allele.

X inactivation The "turning-off," or inactivation, of an X chromosome during development.

X-linked genes Any or all genes on the X chromosome; often used synonymously with *sex-linked.*

XX-XO system The chromosomal system of sex determination, found in some insects, in which the female has two sex chromosomes (XX) and the male has only one (XO).

XX-XY system A chromosomal system of sex determination, in which females are homogametic (XX) and males heterogametic (XY).

Y chromosome The sex chromosome found only in males in species with two sex chromosomes and in which the males are the heterogametic sex; the smaller of the sex chromosomes in humans; males have one Y chromosome, females none.

Y-linked genes Any or all genes on the Y chromosome.

zygote A fertilized egg; also refers to any developmental stage of an organism prior to hatching or birth; in humans, most commonly refers to period from conception to end of second week of gestation.

ZZ-ZW system The chromosomal system of sex determination found in birds and some insects, in which the male is the homogametic sex (ZZ) and the female heterogametic (ZW).

Index

Ocular albinism, 101
Oncogenes, and cancer, 240–41
One gene-one polypeptide
	hypothesis, 179
Oocyte, 118, 124–25
Oogenesis, 118, 124–25
Operator, gene regulation,
	187–88
Operon hypothesis, 187–89
Organellar DNA, 193
Organelles, 29
Organic molecules, 26
Organs, and cells, 27
Organ systems, 27
Osteogenesis imperfecta, 11
Ovaries, 36, 118, 119
Overcompensation, 77
Overlapping genes, 181
Oviduct, 120
Ovulation, 120, 125
Ovum, 36, 118

Paracentric inversion,
	chromosomes, 227
Parental Equivalence, Law of,
	49
Parkinson's disease, 274
Parthenogenesis, 98
Partial dominance, 56–57, 195
Partial sex linkage, 100
Particulate radiation, 250
Passive immunity, 270
Patau's syndrome, 220, 224
Paternity, testing, 341–42
Pattern baldness, 111
Pauling, Linus, 164
Pedigree analysis, 68–69
Penetrance, and phenotypic
	variation, 198
Penis, 121, 127
Peptide bond, 159, 178
Pericentric inversion,
	chromosomes, 227
Pesticide resistance, 304–5
Pesticide treadmill, 304
P generation, 48
Phagocytic cells, and body
	defenses, 263
Phenocopy, 76, 237
Phenotype, 44, 45, 204
Phenotypic variation, 204–14
Phenylalanine, 74, 76
Phenylketonuria (PKU), 10,
	73–74, 76, 150, 179,
	180, 196–98
Pheromones
	and eukaryote regulation,
		195
	and pest control, 195, 306,
		340
Philadelphia chromosome,
	230–31
Piebald trait, 81
Placenta, 120

Plant patenting laws, 309
Plants, classification, 354
Plasma, and blood, 264
Plasma cells, 267, 268
Plasmapheresis therapy, 279
Plasmids, 158, 281, 330–32
Platelets, blood, 262, 264
Pleiotropic gene, 79, 195
Ploidy, and sex determination,
	98
Point mutation, 230–32
Polar body, 118, 124, 125
Polio virus, 272, 279
Polydactyly, 68, 198–99
Polygenic inheritance, 56, 203
Polypeptides, 159
Polyploid, 35, 221–23
Polyribosome, and RNA, 176,
	178
Polytene chromosomes, 186, 192
Population genetics
	gene pool, 16, 296
	gene pool manipulation
		assortive mating, 320
		dysgenics, 321
		eugenics, 321–22
		euphenics, 321
	Hardy-Weinberg equilibrium
		equations, 297–99
	Hardy-Weinberg law, 297–99
	races
		characterized by skin
			color, 314–15
		and electrophoresis, 316
		genetic differences among,
			206
		genetic distance, 316–17
		genetic variability, 296,
			316–17
		racial groups, 296
	racism, 205–9, 317
Porphyria, 81
Porter, R. R., 269
Position effects, 227
Positive eugenics, 321
Prenatal diagnosis, 145, 148–50,
	341–42
Prenatal screening, 80, 145
Primary nondisjunction, 130,
	131, 132
Primary oocyte, 118, 124
Primary sex ratio, 112
Primary sexual characteristics,
	96
Primary spermatocyte, 123
Primary structure, amino acids,
	159
Primate, 14
Probability, 359–61
Proband, 69
Progenotes, 357
Progeria, 68, 73, 236–37
Progesterone, 118
Project Head Start, 211–12

Prokaryotes, and gene regulation
	lac operon, 187
	lac system regulatory
		characteristics, 187–89
	operon and genetic
		engineering, 339–40
	operon hypothesis, 187–89
Prokaryotic cells, 26, 27
Promoter, gene regulation,
	187–88
Prophase, meiosis, 59–63
Prophase, mitosis, 38
Protan-type color blindness, 102
Protein
	colinearity of, 176
	as informational molecule,
		169
	structure of, 158–59
	synthesis, 176–78
	viral ghost, 163
Protists, 352–53
Proto-oncogenes, 240
Protoplast fusion, 303
Przewalski's horse, 313
Pseudohermaphrodites, 129
Ptosis, 68
Punnett square, 50, 51
Pure breeding, 48

Qualitative variation, 46
Quantitative inheritance, 56–57,
	201–3
Quantitative variation, 46
Queen Victoria, and hemophilic
	descendants, 105

Races, 296, 314–17
Racial groups differences,
	intelligence, 207–9
Racism, 205, 317, 320
Rad, 250, 251
Radiation, and mutation,
	233–34, 249–56
Radon, and radium, 249, 250
Random mating, 71, 297
Randomness, 50, 55, 110,
	359–61
Recessiveness, 48, 49, 72–80,
	195
Reciprocal crosses, 45
Reciprocal translocation,
	chromosomes, 228–29
Recombinant DNA technology,
	329–36, 339
Recombinant organisms, 346–47
Recombination, 59
Red blood cells, 262, 264
Red Data Books, 308
Red-green color blindness,
	102–3, 110
Reduction division, 59
Regulatory genes, function of,
	187–91